AF798583

We didn't write a pitch.
We established a record!

STRATEGIC ASSET BRIEF DECLASSIFIED

The Asset-Based Unicorn

Why Goddess Phenezy Holdings, Inc. Is Structurally Different

TechnologiesRUs.com | XperiencesRUs.com

Insight

by Goddess Phenezy

A blind man sees
what we who claim sight cannot.
Must death come before us all
before pure vision enlightens our soul?

Awaken we must,
for slumber has been our true enemy.

Whispers the wind our faith to us.
Tears are shed from the clouds.
The sun beams brightly—yet light we lack.
into darkness descends a fallen star.

Nature knows her creatures' damnation.
Ask not about heaven or hell.
No longer may we adhere to unjustified isms.
Maybe to live is to die.

Succumb to reality.
Obstinance breeds ignorance.
Many do not seek truth.
No longer may we adhere to unjustified isms.

THE SOVEREIGN INDEX I

THE SOVEREIGN INDEX Part II

The National & International Championships

THE SOVEREIGN INDEX Part III

Intelligence, Academies, Conferences & Financial Architecture

STEM, Games & Intelligence

Conferences & Academies

THE SOVEREIGN INDEX Part IV

Valuation, Finance & Closing

Dedications

A love letter to the people who shaped this journey

To my beloved son, KJ,

I love and miss you. Look what you inspired! In your honor, I am making the world a better place. I love you from here!

Every line of code, every late night, every breakthrough — it all traces back to you. You planted a seed of purpose in my heart that grew into something I never could have imagined. This isn't just a project. It's a promise kept. The world is brighter because you were in it, and I carry your light forward every single day. Wherever you are, I hope you can see what your love set into motion.

Dedications

To My Mom, Mary Phenezy,

Thanks for listening to my crazy mind as I spew out thoughts and ideas sent from the heavens. Love you and appreciate you.

Mom, you never once told me my dreams were too big or my ideas too wild. Even as I expanded and turned the old visions into new visions that hadn't fully formed yet, you didn't just tolerate it — you leaned in. If you didn't understand, you still believed before there was anything to believe in. That kind of unconditional faith is rare and sacred, and I want you to know it's woven into the very fabric of everything I've built to honor your Grandson, KJ. You were the very best grandmother KJ could have ever wanted. I love and appreciate you.

To My Dad, Alton Anderson,

Thanks for teaching me chess! I used that knowledge to create a chess app as the initial foundation for the Xperiences R Us platform.

Dad, you probably had no idea that those chess lessons when I was 5 years old would become the spark for an entire technology platform. But that's exactly what happened. You taught me to think several moves ahead, to see patterns, to be patient and strategic. Chess wasn't just a game — it was a masterclass in life. When I sat down to build the very first PWA, it was your voice I heard: *"Think before you move."* The Xperiences R Us platform exists because you, a father, took the time to sit with his child and share something meaningful. That foundation means everything to me. Love ya!

To My GODdaughter, Kay Kay,

You walk with the kindness and intelligence the world needs now and would need later. Thank you Tim & Diane for sharing your beautiful daughter with me.

Kay Kay, watching you grow has been one of the greatest joys of my life. You came into my life exactly on time. You carry yourself with a grace and wisdom that goes far beyond your years. The world can be a noisy, chaotic place, but you move through it with a quiet strength that inspires everyone around you. Your kindness isn't weakness — it's a superpower, and don't you ever let anyone tell you otherwise. Tim and Diane, thank you for the incredible gift of letting me be part of her life. Your family is a blessing I will never take for granted. This work, this mission — it's for the next generation. It's for you, Kay Kay. I love you all.

> **Everything great begins with love. This platform was born from it, built on it, and carries it forward into the world.**

We didn’t write a pitch.
We established a record!

STRATEGIC ASSET BRIEF   DECLASSIFIED

The Asset-Based Unicorn

Why Goddess Phenezy Holdings, Inc. Is Structurally Different

Insight: The Modern Unicorn Problem

Modern "unicorns" are often technology companies valued over $1 billion based on speculative future growth, relying heavily on projected revenue and market sentiment. This makes their valuations inherently fragile, subject to market volatility and sudden loss of confidence. When capital markets tighten, these speculative unicorns face existential pressure, as their valuations, built on narrative and sentiment, quickly collapse without a permanent anchor, causing the "floor" to disappear.

> **Goddess Phenezy Holdings, Inc. was engineered to avoid this structural vulnerability. Its valuation is rooted in owned, defensible, and replacement-cost-validated assets, offering a materially different risk profile from speculative ventures.**

The Speculative Collapse Sequence

Sentiment Shifts

Market confidence erodes; growth narrative loses credibility

Capital Tightens

Funding rounds dry up; bridge financing becomes unavoidable

Narrative Fails

Projected multiples collapse without hard asset backing

Floor Disappears

No owned assets to anchor valuation — nothing to fall back on

Implosion

Down rounds, layoffs, fire sales — or total collapse

Asset-Based vs. Speculative Unicorns

A structural risk comparison

Speculative Unicorn

- Valuation built on narrative & sentiment
- No hard floor — collapses under market pressure
- Dependent on App Store, Google, AWS
- Revenue tied to user growth & algorithm favor
- Existential risk during economic downturns

Asset-Based Unicorn (GPHI)

- Replacement-cost-validated owned assets
- Hard floor: 500+ ISBNs, domains, software, broadcast infrastructure
- Zero platform tax via Your Partnrr
- Revenue across 6 verticals: sports, education, media, culture, economics, AI
- Assets hold and appreciate under stress

Asset Pillars

Independent compounding value centers

Verticals

Diversified revenue coverage

Platform Tax

Owned economic infrastructure

Speculative vs. Asset-Based Valuation

Understanding the structural difference requires examining how value is constructed and defended under each model. The contrast is not subtle—it is foundational.

1

Speculative Unicorn

- Valuation driven by projected revenue multiples
- Platform-dependent distribution and monetization
- Subject to capital market sentiment cycles
- Vulnerable to deplatforming and algorithmic shifts
- Growth narrative required to sustain valuation floor

2

Asset-Based Unicorn

- Valuation anchored in replacement cost and defensibility
- Platform-agnostic owned infrastructure
- Resilient through economic volatility cycles
- Immune to deplatforming and third-party dependency
- Intrinsic value retained even absent future growth

> **Goddess Phenezy Holdings, Inc. is an asset-based unicorn. This distinction is not philosophical. It is structural.**

Our Asset Foundation

What This Enterprise Actually Owns

The designation *asset-based unicorn* is not theoretical within this enterprise. It is supported by a diversified portfolio of owned, defensible, and replacement-cost-validated assets that collectively form the permanent foundation of Goddess Phenezy Holdings, Inc. These assets are designed to operate independently, yet compound in value when activated together—creating an architecture of interlocking durability that no single market disruption can dismantle.

Four Pillars of Permanent Value

The enterprise's asset base is organized into four primary categories, each functioning as an independent value center while contributing to the compounding power of the integrated whole.

Digital Territory

Jurisdictional Domain Assets

Proprietary Software

The Engine

Economic Infrastructure

The Circulatory System

Intellectual Property

Knowledge Assets

Each category is designed to withstand independent stress-testing. A downturn in any single vertical does not compromise the others. This is portfolio-level resilience applied at the enterprise architecture level—a structural advantage that speculative unicorns, by definition, cannot replicate.

PILLAR 1

Digital Territory

Jurisdictional Domain Assets

Rather than listing individual domain names, this enterprise organizes its domain portfolio as functional asset groups, each representing a governed market vertical. These domains are not speculative holdings; they are operational infrastructure tied directly to monetization, licensing, governance, and authority. They function as digital land reserves—each capable of supporting independent operations, licensing agreements, and long-term enterprise value.

Critically, these assets are platform-independent, non-revocable, and immune to deplatforming risk. Unlike social media accounts, marketplace listings, or app store presences—all of which exist at the discretion of third-party gatekeepers—domain-based jurisdictional assets are owned outright. They cannot be algorithmically suppressed, terms-of-service revoked, or unilaterally deactivated.

The Power of 'The'

In English, "The" is a definite article, signaling singularity, finality, and undeniable authority. It doesn't imply "a" version or "one of many" — it asserts, "THIS IS THE one." In branding and domain strategy, this prefix is among the most coveted. It signifies category ownership, not mere participation, as seen in iconic brands like The New York Times, The White House, The Masters, and The Super Bowl – entities that don't just compete, they embody their respective categories. Owning 450+ high-value domains all beginning with "The" means Goddess Phenezy Holdings, Inc. has staked a jurisdictional claim to being the definitive authority across 450+ national and international experience verticals. This is more than a naming convention; it's a profound positioning strategy where each domain declares: "we are not one of many providers of this experience — we are THE provider." From an asset valuation standpoint, "The" domains command premium resale, licensing, and partnership value, carrying an implicit authority that generic or descriptive domains cannot replicate. Internationally, "The" translates as a universal signal of prestige and primacy, recognized across cultures as a marker of the definitive, the official, and the authoritative.

450+ domains. Each beginning with 'The.' Each asserting category ownership across national and international experience verticals. This is not a portfolio — it is a jurisdictional claim.

Domain Asset Classes

The digital territory portfolio spans five major asset classes, each governing distinct market verticals

1

Championship Jurisdictions

National and international domains governing sports, Artificial Intelligence, STEM, arts, gaming, and cultural competitions. These represent permanent authority over organized competitive ecosystems.

2

Educational Institutions

Domains supporting Academies, Institutes, Colleges, and Universities across STEM, Artificial Intelligence, Chess, Business, Arts, and Leadership. Each institutional domain is a licensable vertical.

3

Conference & Summit Infrastructure

National and international Conferences, Summits, and Conventions serving enterprise, government, education, and innovation sectors. These domains anchor recurring high-value convening events.

4

Economic Development Platforms

Incubators, Accelerators, Career Fairs, and Licensing Frameworks. Each platform domain supports ecosystem-level economic activation and partnership infrastructure.

5

Media & Cultural Properties

Broadcasting, publishing, festivals, awards, and cultural record platforms. These domains represent long-duration media assets with compounding content value.

PILLAR 2

Proprietary Software Infrastructure

The Engine — Production-Grade Systems Built for Institutional Deployment

The valuation floor of this enterprise is anchored in battle-hardened proprietary software — platform-agnostic, Progressive Web App–based, and engineered to retain 100% of gross revenue without app-store taxation or third-party dependency. These are not prototypes; they are production-grade systems built for institutional deployment, data integrity, and long-term operational durability.

Xperiences R Us Platform

Multi-vertical experiential infrastructure — championships, conferences, festivals, memberships, ticketing, and licensing across every enterprise vertical.

Technologies R Us Software Stack

Core technology arm housing proprietary systems, logic frameworks, integrations, and platform architecture powering the entire ecosystem.

Championship & League Management

Registrations, brackets, rankings, scheduling, compliance, and governance software designed for scalable competitive programming.

100%

Revenue Retained

Zero app-store or platform taxation

PWA

Platform-Agnostic

No third-party dependency or deplatforming risk

Key Software & Intelligence Systems

Six Proprietary Platforms. One Integrated Ecosystem.

Software Infrastructure

Xperiences R Us Platform

Multi-vertical experiential infrastructure supporting championships, conferences, festivals, memberships, ticketing, and licensing across every enterprise vertical.

Technologies R Us Software Stack

Core technology arm housing proprietary systems, logic frameworks, integrations, and platform architecture powering the entire ecosystem.

Championship & League Management

Registrations, brackets, rankings, scheduling, compliance, and governance software designed for scalable competitive programming.

Intelligence & Identity

AI-Powered Coaching & Learning

Adaptive A.I. engines supporting chess, STEM, sports analytics, and cognitive training. These systems learn with each participant interaction, creating compounding intelligence value.

Verified Identity & Digital Credentials

Player, student, creator, and participant identity layers, including digital resumes and achievement records. A portable, verified identity layer across the ecosystem.

Data Capture & Performance Analytics

Proprietary tools generating verified athletic, educational, creative, and behavioral data—a first-party data asset of increasing strategic value.

6

Proprietary Platforms

Fully owned, zero licensing dependency

100%

Revenue Retained

No app-store or platform taxation

1st

Party Data Ownership

Verified behavioral & performance data

The Zero-Tax Architecture

One of the most consequential structural decisions embedded in the software infrastructure is the elimination of third-party platform taxation.

Traditional Platform Model

App store fees consume 15–30% of gross revenue. Distribution is controlled by platform gatekeepers. Terms of service can change unilaterally. Algorithmic visibility determines reach. User data is shared with or owned by the platform.

GPHI Software Model

Progressive Web App architecture retains 100% of gross revenue. Distribution is owned and platform-independent. No third-party gatekeeper can restrict access. Direct user relationships without intermediary algorithms. All user data remains inside the ecosystem.

This is not a marginal efficiency gain. Over scale, the difference between retaining 70% and 100% of gross revenue represents hundreds of millions in preserved enterprise value. It is a permanent structural advantage baked into the architecture itself.

PILLAR 3

Economic & Transactional Infrastructure

The Circulatory System — Your Partnrr

At the center of the enterprise's economic architecture is Your Partnrr. This system functions as the transactional and relational operating system of the ecosystem—governing how value moves between participants, partners, affiliates, sponsors, institutions, and organizers. It is the circulatory system through which every economic interaction flows, is tracked, and is optimized.

Unlike fragmented third-party CRM, PRM, or affiliate platforms, Your Partnrr is owned infrastructure, embedded directly into the ecosystem. It ensures that economic activity remains internal, auditable, and scalable—eliminating leakage while enabling non-dilutive growth. This system transforms participation into long-term economic alignment, rather than transactional exposure.

Your Partnrr: Core Capabilities

Partner & Sponsor Relationships

Structured relationship management governing multi-tier partnerships, sponsorship agreements, and institutional collaborations across every vertical in the ecosystem.

Affiliate & Referral Economics

Automated referral tracking, attribution, and compensation systems that incentivize organic ecosystem growth without external marketing dependency.

Licensing Revenue Distribution

Intelligent licensing revenue allocation across content creators, organizers, institutional partners, and rights holders—ensuring fair, transparent value distribution.

Multi-Party Payment Flows

Complex, multi-stakeholder payment orchestration supporting simultaneous revenue splits, escrow functions, and milestone-based disbursements.

Credentialed Access

Verified credentialing systems governing access to opportunities, stages, and markets—transforming participation into a meritocratic, auditable pathway.

Why Owned Economic Infrastructure Matters

The strategic significance of Your Partnrr cannot be overstated. In most enterprise ecosystems, the economic layer is the most vulnerable point of fragmentation and value leakage.

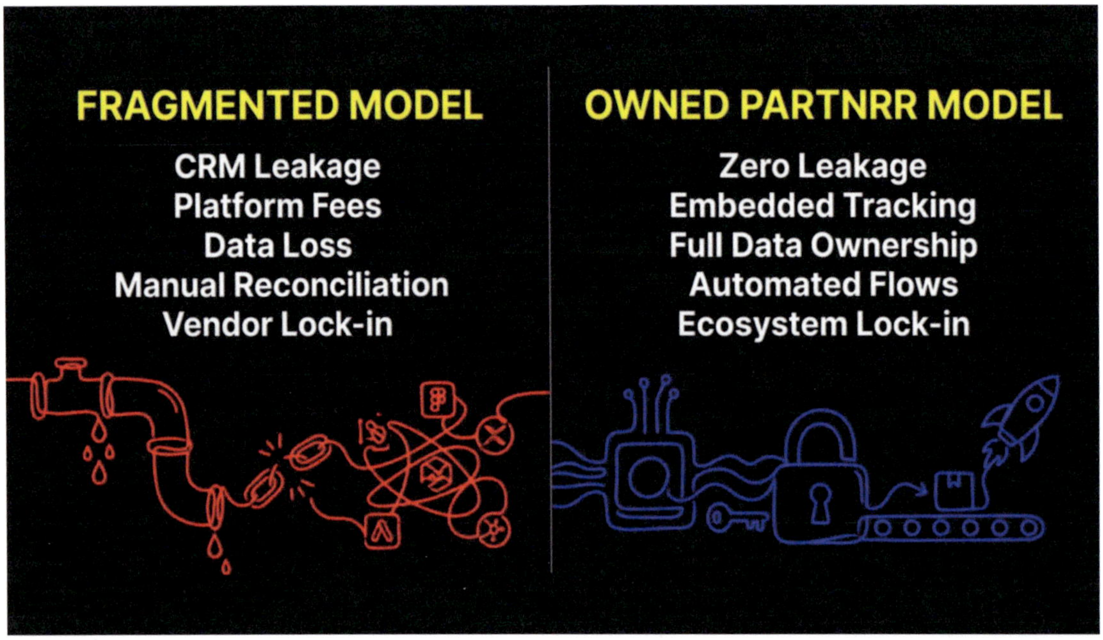

When an enterprise relies on third-party CRM, affiliate tracking, and payment processing platforms, it surrenders visibility, control, and margin at every junction. Your Partnrr eliminates these vulnerabilities by embedding the entire economic layer inside the owned ecosystem. Every dollar that flows through the system is tracked, attributed, and retained—creating a self-reinforcing economic engine that becomes more valuable with every transaction.

Media as a Core Asset Class

Owned Distribution, Not Marketing

Media within Goddess Phenezy Holdings, Inc. is not promotional. It is owned distribution infrastructure. This is a critical distinction that separates the enterprise from virtually every competitor in adjacent markets. While most organizations treat media as a cost center—an expense line item dedicated to marketing and brand awareness—GPHI treats media as an asset class with its own independent valuation, revenue generation capacity, and compounding characteristics.

The enterprise's media infrastructure operates on two primary rails: Xperience TV and the broader Xperiences R Us (www.Xperiencesrus.com) digital media engine. Together, they create a vertically integrated content ecosystem that captures, produces, distributes, and monetizes media without reliance on any third-party algorithm, platform, or gatekeeper.

Xperience TV

Xperience TV functions as the enterprise's dedicated digital broadcast and syndication network—an owned media rail that retains all content rights, data, and monetization inside the ecosystem.

Live & On-Demand Streaming

Championships, conferences, summits, and festivals broadcast in real-time and archived for perpetual on-demand access. Every event becomes a permanent media asset with recurring viewership value.

Serialized Programming

Educational and documentary content produced in series format, creating audience retention loops and long-term engagement patterns that compound viewership over time.

Highlight Generation

Rights-controlled media packages engineered for syndication, sponsorship integration, and cross-platform distribution—all under enterprise-owned rights management.

Long-Form Cultural Content

Institutional and cultural programming that establishes authority, builds brand equity, and creates archival value that appreciates over decades, not quarters.

Xperiences R Us as a Digital Media Engine

Xperiences R Us itself operates as a digital media platform, not merely an experiential brand. This distinction is architecturally significant.

Every Xperience—championship, conference, academy, or festival—is architected to generate structured content, broadcast-ready programming, and long-tail digital value. A single championship produces:

- Live broadcast footage
- Post-event highlights and recaps
- Participant profiles and achievement records
- Data-driven analytics content
- Educational derivatives and training materials

This converts traditionally one-time events into recurring media assets, materially increasing enterprise durability and valuation. The content library grows with every activation, creating a compounding archive of rights-owned, monetizable intellectual property.

From Single Event to Compounding Media Asset

Live Event

Championship, conference, academy, or festival is activated

Content Captured

Broadcast footage, highlights, profiles, and analytics generated

Audience Built

Content distributed across owned channels — no platform intermediary

Revenue Unlocked

Sponsorship, licensing, syndication, and subscription value created

Flywheel Repeats

Revenue funds next event — archive compounds with every cycle

The Media Compounding Effect

The strategic power of owned media infrastructure becomes apparent when viewed through the lens of compounding value creation.

Each event produces media. Each piece of media attracts audience. Each audience creates sponsorship and licensing value. That value funds additional events and content production. This is not a linear growth model—it is a flywheel. And because all content rights, distribution rails, and monetization systems are owned internally, the enterprise captures 100% of the compounding value at every stage of the cycle. No platform intermediary extracts margin. No algorithm gates distribution. The flywheel accelerates under its own power.

PILLAR 4

Intellectual Property & Knowledge Assets

The Compounding Knowledge Bank

500+ ISBN Portfolio

Registered works spanning chess, STEM, AI, leadership, and culture — establishing formal authorship across multiple knowledge domains.

The KJ Doctrine

Original framework governing human intelligence development in the age of AI — licensed across education, competition, and corporate training.

HI vs. A.I. Framework

Proprietary methodology defining the relationship between human and artificial intelligence — foundational to GPHI's AI vertical.

Championship Governance

Proprietary competition formats and governance protocols — licensable to national and international governing bodies.

Certified Curricula

Certification frameworks across STEM, AI, chess, business, arts, and leadership — built for institutional deployment.

Intellectual Property & Knowledge Assets

The Compounding Knowledge Bank

Beyond code and domains, the enterprise controls a substantial portfolio of formal intellectual property, designed to preserve authorship, doctrine, and long-term authority. These assets form a knowledge bank that compounds in value as it is activated through education, media, and institutional partnerships. Unlike software, which can become obsolete, and unlike domains, which require renewal, intellectual property—when properly structured and protected—creates permanent, self-reinforcing authority.

The IP portfolio is not a collection of defensive filings. It is an active knowledge infrastructure that generates licensing revenue, establishes institutional credibility, and creates barriers to competitive replication that grow stronger with time.

Key IP Holdings

500+ ISBN Publishing Portfolio

A comprehensive library of published works establishing formal authorship and institutional authority across multiple knowledge domains.

Proprietary Methodologies

Including the KJ Doctrine and HI vs. A.I.—original intellectual frameworks that anchor educational, competitive, and institutional programming.

Championship Formats & Governance

Proprietary competition formats, naming conventions, and governance models that define how organized competition operates within the ecosystem.

Educational Curricula

Certification structures and pedagogical frameworks supporting STEM, AI, chess, business, arts, and leadership programming at institutional scale.

Brand Systems & Trademarks

Trademarked verticals and brand architectures that protect the enterprise's market positioning across every operational domain.

IP as a Compounding Asset

The Only Asset Class That Gets More Valuable Every Time You Use It

Most asset classes depreciate through use — physical assets wear down, software requires maintenance, real estate demands upkeep. Intellectual property inverts this entirely: every citation, licensing agreement, and institutional adoption makes it more authoritative and more valuable than before.

Traditional Assets

- Depreciate from day of acquisition
- Require ongoing capital to maintain value
- Consumed through use
- Vulnerable to obsolescence

GPHI Intellectual Property

- 500+ ISBNs gain authority through citation
- KJ Doctrine & HI vs. A.I. gain credibility through adoption
- Championship formats become industry standards
- Curricula and trademarks command premium licensing fees

500+	0%	∞
Registered ISBNs	Depreciation Rate	Licensing Potential

A building used daily wears down. A methodology cited daily becomes doctrine. That is the structural advantage of IP as a primary asset class.

Why This Matters

Structural Superiority in Volatile Markets

This enterprise is not valued on narrative momentum. It is valued on replacement cost, asset defensibility, jurisdictional control, and infrastructure durability — four structural properties that speculative unicorns cannot manufacture, purchase, or replicate on a compressed timeline. When markets turn, this distinction is not philosophical. It is existential.

Replacement Cost Defensibility

The cost to replicate GPHI's 450+ domain portfolio, proprietary software stack, 500+ ISBN library, and broadcast infrastructure from scratch today would run into the hundreds of millions — and in many cases, the naming and jurisdictional assets are simply no longer available at any price.

Zero Platform Dependency

Every critical asset — domains, software, media infrastructure, IP — is owned outright. No algorithm can suppress it. No app store can tax it. No platform policy change can revoke it. This is structural immunity that no speculative unicorn can claim.

Compounding Without Capital Expenditure

IP appreciates through use. Domains appreciate through scarcity. Media compounds through content accumulation. Unlike physical assets that depreciate, GPHI's core asset classes gain value with every activation — creating a compounding architecture that requires no additional capital to sustain.

Jurisdictional Control Across Six Verticals

Sports, education, media, culture, economics, and artificial intelligence — GPHI holds jurisdictional authority across all six through owned domains, registered IP, and operational infrastructure. This cross-vertical coverage creates a moat that widens with each new activation.

Even in the absence of future growth, these assets retain intrinsic value. In the presence of growth, they compound. That is the definition of permanent capital.

The Four Pillars of Intrinsic Value

The enterprise's valuation rests on four measurable, auditable foundations—each independently defensible and collectively compounding.

Replacement Cost

The cost of replicating this asset portfolio from scratch—assembling the domains, building the software, developing the IP, and constructing the economic infrastructure—establishes a hard valuation floor that exists independent of market sentiment.

Asset Defensibility

Owned domains, proprietary software, registered intellectual property, and embedded economic infrastructure create moats that deepen with time. These are not competitive advantages that erode—they are structural barriers that compound.

Jurisdictional Control

Authority over named verticals across sports, education, media, culture, economics, and artificial intelligence creates governing positions that cannot be replicated by competitors entering these markets.

Infrastructure Durability

Platform-agnostic, self-hosted, and internally governed systems ensure operational continuity regardless of external technology shifts, platform policy changes, or market disruptions.

Resilience Under Stress

Asset-based unicorns—private companies valued over $1 billion with substantial tangible and infrastructural assets—are widely recognized as less risky, more resilient, and structurally superior during economic volatility. This is not a philosophical preference. It is a risk-adjusted financial strategy.

Stress Scenario	Speculative Unicorn Impact	Asset-Based Unicorn Impact
Capital markets contraction	Valuation collapse; unable to raise at prior terms	Valuation floor maintained by replacement cost
Platform policy change	Distribution disrupted; revenue impaired	No impact; infrastructure is self-hosted
Competitive entry	Market share erosion; narrative undermined	Jurisdictional moats prevent replication
Economic recession	Growth narrative collapses; down rounds likely	Assets retain intrinsic value; optionality preserved
Regulatory shift	Business model may become non-viable	Diversified asset base absorbs regulatory risk

The table above illustrates why institutional capital increasingly favors asset-backed structures during periods of uncertainty. When the tide goes out, only enterprises with permanent foundations remain standing.

The Compounding Architecture

What makes the GPHI asset base exceptional is not merely its breadth, but how the four pillars interact to create compounding value that accelerates over time.

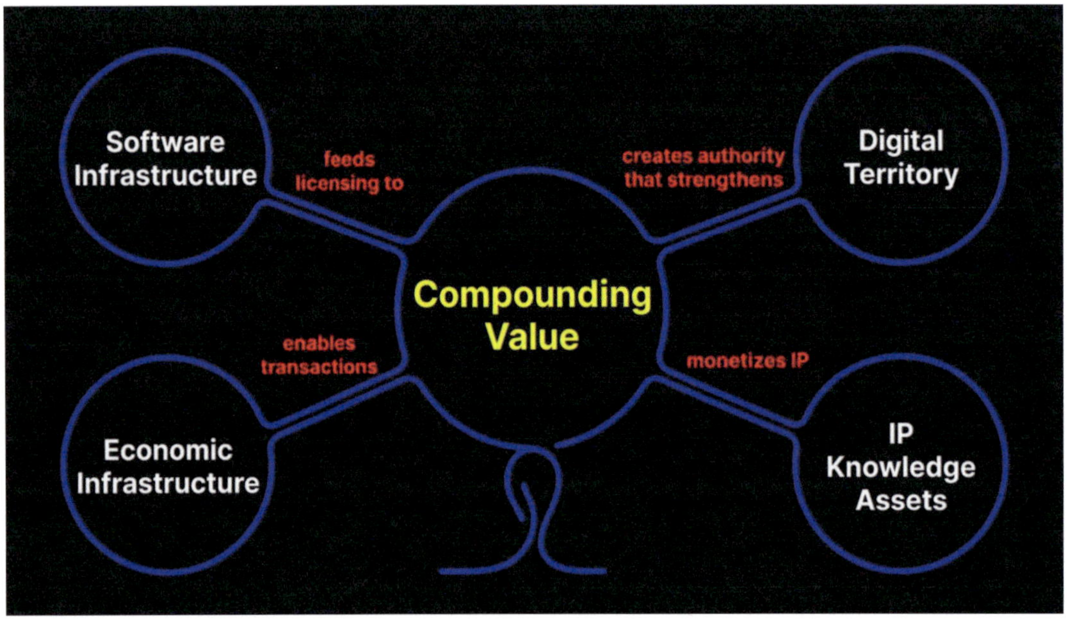

Domains create jurisdictional authority. Software converts that authority into operational infrastructure. The economic layer monetizes operations and distributes value. Intellectual property compounds the authority that drives the entire cycle. Remove any single pillar and the others still retain independent value. Activate all four together and the compounding effect becomes exponential. This is not theoretical. It is architectural.

The Position

Built to Own Terrain

This enterprise was not built to chase trends. It was built to own terrain. The distinction is not rhetorical—it defines every architectural decision, every asset acquisition, and every strategic deployment within the portfolio. While competitors optimize for the current cycle, GPHI optimizes for permanence.

What Our Valuation Is Not

Clarity demands that we define not only what the enterprise is, but what it deliberately is not. The absence of certain dependencies is itself a structural asset.

Not Derived from Hope

Every asset in the portfolio exists today. It has been built, registered, developed, and deployed. The valuation does not depend on future outcomes—it reflects current ownership of permanent infrastructure.

Not Dependent on Virality

The enterprise does not require algorithmic favor, social media momentum, or viral adoption to sustain its valuation. Growth is welcome; it is not existentially required.

Not Subject to Platform Favor

No third-party platform—App Store, Google, Meta, Amazon, or any other gatekeeper—has the ability to impair the enterprise's operations, distribution, or revenue collection.

What Our Valuation Is Anchored In

The enterprise's valuation is anchored in owned infrastructure, verified systems, intellectual property, and jurisdictional control across six foundational verticals.

Sports

Championship governance and competitive infrastructure

Education

Institutional academies, curricula, and certification

Media

Owned broadcast, publishing, and content distribution

Culture

Festivals, awards, and cultural programming authority

Economics

Transactional infrastructure and development platforms

Artificial Intelligence

AI-powered systems, coaching, and competitive frameworks

Each vertical represents not merely a market opportunity, but a governed position—backed by domain authority, operational software, economic infrastructure, and intellectual property. This is terrain ownership, not market participation.

Ecosystem Integration Map

The six verticals do not operate in isolation. They are connected through shared infrastructure, creating cross-vertical synergies that amplify the value of every individual component.

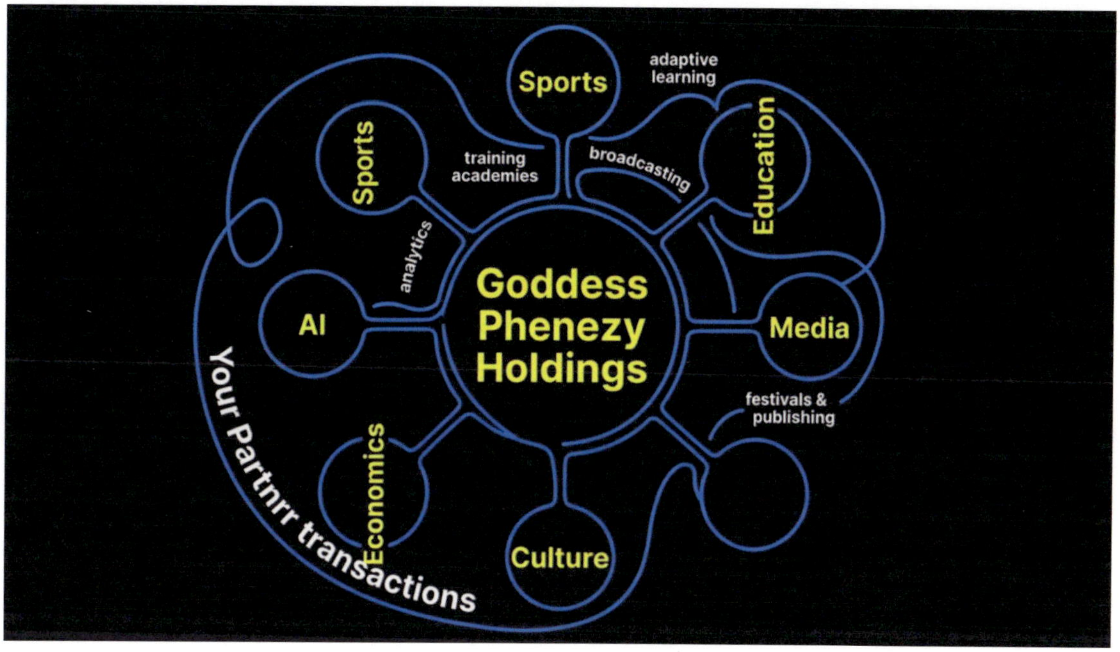

A chess championship (Sports) generates broadcast content (Media), produces training data (AI), creates educational programming (Education), attracts sponsorship revenue (Economics), and establishes cultural authority (Culture)—all within the same ecosystem, using the same infrastructure, governed by the same entity. This is the compounding power of vertical integration across owned terrain.

The Record

Thick Systems Require Only Review

This document exists because thin things require belief. Thin valuations depend on narrative. Thin companies depend on momentum. Thin architectures depend on external platforms. They require their audiences to believe in projections, trust in trajectories, and accept assumptions about futures that may never materialize.

Thick systems require only review.

The assets documented in this record are not projections. They are holdings. They are not roadmap items. They are deployed infrastructure. They are not market hypotheses. They are governed jurisdictions with operational software, economic systems, and registered intellectual property.

Thin vs. Thick: A Structural Comparison

Thin Enterprise	Thick Enterprise
• Valuation requires narrative persuasion • Assets are primarily intangible projections • Depends on continuous capital infusion • Vulnerable to sentiment shifts • Growth is existentially required	• Valuation requires only asset review • Assets are owned, deployed, and defensible • Generates value from existing infrastructure • Resilient through market cycles • Growth compounds; it is not existential

Goddess Phenezy Holdings, Inc. was deliberately engineered as a thick enterprise. Every structural decision—from platform-agnostic software to owned media rails to embedded economic infrastructure—was made to ensure that the enterprise's value could be audited, not merely imagined.

The Replacement Cost Floor

For institutional investors conducting due diligence, the replacement cost framework provides the most conservative and defensible basis for valuation. Consider what would be required to replicate this portfolio from scratch.

1 Assemble the Domain Portfolio

Acquire hundreds of jurisdictional domains across sports, education, media, culture, economics, and AI—many of which are now unavailable at any price due to naming scarcity and existing registration.

2 Build the Software Stack

Develop platform-agnostic, PWA-based software spanning championship management, AI coaching, identity credentialing, analytics, ticketing, and multi-vertical experiential infrastructure. Years of engineering at institutional grade.

3 Construct the Economic Layer

Design and deploy a proprietary transactional operating system governing partner relationships, affiliate economics, licensing distribution, and multi-party payment flows—embedded at the ecosystem level.

4 Create the IP Portfolio

Publish 500+ ISBNs, develop proprietary methodologies, design championship formats, build educational curricula, and register brand systems across every operational vertical.

The aggregate replacement cost of this portfolio—even before accounting for the compounding value of integration—establishes a valuation floor that exists independent of any growth assumption, market condition, or capital cycle.

Key Enterprise Metrics

The following metrics illustrate the scale and diversity of the asset portfolio underpinning the enterprise's valuation.

450+

Jurisdictional Domains

National & international 'The' branded experience verticals

500+

ISBN Publications

Registered works across chess, STEM, AI, leadership & culture

6

Owned Verticals

Sports, education, media, culture, economics & AI

3

Software Platforms

Xperiences R Us, Technologies R Us & Championship Management

0

Platform Tax

100% revenue retained via PWA zero-dependency architecture

The Closing Statement

Goddess Phenezy Holdings, Inc. (GPHI) is an asset-based unicorn, valued at over $1 billion, built for enduring structural durability rather than speculative momentum. Our foundation rests on four robust pillars: an extensive Digital Territory jurisdictional domain portfolio, advanced Proprietary Software Infrastructure including the Your Partnrr (www.YourPartnrr.com) platform with zero-tax architecture, a comprehensive Economic & Transactional Infrastructure, and a deep Intellectual Property & Knowledge Assets portfolio. These pillars support our engagement across six foundational verticals: sports, education, media, culture, economics, and artificial intelligence.

Our valuation is not derived from hope or platform favor, but from tangible, deployed assets. This includes a 500+ ISBN publishing portfolio, owned media infrastructure such as the Xperience TV digital broadcast/syndication network, and the Xperiences R Us (www.Xperiencesrus.com) digital media engine. Our proprietary software actively eliminates third-party platform taxation, ensuring control and efficiency. All assets are rigorously registered, infrastructure is fully deployed, and economic systems are operational.

For institutional capital seeking unparalleled structural durability over speculative momentum—for partners who understand that permanence is the ultimate competitive advantage—Goddess Phenezy Holdings, inc. presents a meticulously verified record. Our verifiable assets and infrastructure speak directly to a commitment to long-term value and control.

Goddess Phenezy Holdings, Inc. — The Asset-Based Unicorn

CLASSIFIED EXECUTIVE BRIEFING | OPERATION LEGACY

Goddess Phenezy Holdings, Inc.

Operation Legacy — Executive Summary - Declassified

Parent Entity:

Goddess Phenezy Holdings, Inc.

Founder & Commander:

Goddess Phenezy (USMC Combat Veteran)

Status:

Active // Asset-Based Unicorn ($1.18B–$1.25B)

Mission Profile:

Sovereign Ecosystem Architecture

Structure:

Holdings Company with Operating Subsidiaries in Technology, Xperiences, Championships, Conferences, and Academies

The Briefing

Conventional startup doctrine says a solo founder cannot build a billion-dollar enterprise without venture capital, celebrity backing, or institutional sponsorship. That doctrine assumes founders rent infrastructure, outsource authority, and monetize later. Goddess Phenezy Holdings, Inc. was built under a different doctrine.

This company was architected from day one as a sovereign holding company, designed to own critical infrastructure, naming rights, intellectual property, and digital territory across multiple industries before inviting participants into the ecosystem. Founded in honor of her son, KJ (Jamil), who passed away from lupus complications, this enterprise is not rooted in hype or speculation. It is rooted in justice, control, and permanence.

> **This is not a company built to chase markets. It is a company built to govern them.**

The mission is twofold and non-negotiable. First, create real opportunities and inspiration for veterans, women, youth, and Black and minority communities. Second, build systems that cannot be displaced, exploited, or quietly shut down by third parties. Every subsidiary, every line of code, every domain name, and every championship title was engineered to serve these dual imperatives. The result is not a startup in the traditional sense — it is a sovereign infrastructure company that happens to be valued at over one billion dollars in certified assets before its first institutional round.

What follows in this executive summary is a declassified disclosure of what has already been built, secured, and independently valued. The terrain is held. The architecture is complete. The question for strategic partners is not whether this ecosystem exists — it is whether they are positioned to participate in it before the gates close.

The Problem: Fragmented Systems, Rented Power

Across modern life, the same structural failure repeats itself with devastating consistency. Value is extracted upward while risk is pushed downward. Participants build on platforms they do not control, compete in leagues they do not own, and store their most sensitive data in systems that can be revoked without notice. This is not a market inefficiency — it is a design flaw baked into the architecture of modern commerce, education, health, and entertainment.

Athletes & Competitors

Athletes do not own their leagues, platforms, or championship infrastructure. Their participation enriches entities that hold no obligation to protect or empower them.

Creators & Educators

Creators and educators rent distribution from companies they don't control. One algorithm change or policy shift can erase years of audience development overnight.

Patients & Individuals

Patients do not own or control their health data. Their most intimate information is monetized by intermediaries who profit from vulnerability rather than protecting it.

Businesses & Enterprises

Businesses build on software and platforms that can revoke access overnight. Entire operations can be dismantled by a terms-of-service update they never approved.

Most companies attempt to solve these failures with incremental improvements — a better app, a new feature, a niche community. These approaches treat symptoms while leaving the underlying disease intact. Goddess Phenezy Holdings solved it at the root: with ownership. When you own the infrastructure, you cannot be displaced. When you own the rules, you cannot be overridden. When you own the terrain, no one can evict you.

The Solution: A Sovereign Ecosystem

Goddess Phenezy Holdings, Inc. is the parent company of a vertically and horizontally integrated ecosystem that spans six critical sectors of modern participation. Rather than operating isolated products, the company functions as an operating system for modern participation — where identity, skill, competition, education, business, and health are connected through owned infrastructure.

Technology (SaaS & Infrastructure)

Proprietary platforms exceeding 578,000 lines of code power the entire ecosystem. No rented tech, no licensing dependencies, no external kill switches.

National & International Championships

Over 200 championship titles owned via premium domain names and trademark strategies, establishing presumptive authority across sports, esports, and professional disciplines.

Conferences & Professional Convenings

High-value professional gatherings that connect industry leaders, educators, and practitioners within the owned ecosystem rather than third-party platforms.

Academies & Credentialed Education

Certification curricula and academy-based credentialing tied directly to championships, ensuring participants are trained, certified, and retained inside owned channels.

Health, Safety & Identity Utilities

Patient-owned health data, safety verification, and identity authentication systems that put control back in the hands of individuals — not corporations.

Media & Publishing

Champions TV, 500+ owned ISBNs, rulebooks, playbooks, and official instructional materials — controlling the narrative from creation to distribution.

Participants do not simply attend events. They enter an ecosystem where value compounds across platforms. A competitor trained at an Academy competes in a Championship, is verified through VerifiEYED, tracked via Your Partnr, and amplified through Champions TV. Every interaction strengthens the whole.

The Asset Strategy: Own the Terrain First

Unlike early-stage startups that trade on projections and future promises, Goddess Phenezy Holdings trades on certified, existing assets. The company's valuation is asset-based, not speculative. Every dollar of valuation is backed by code that has been written, domains that have been secured, and intellectual property that has been registered. This is the fundamental difference between this enterprise and every other billion-dollar claim in the market today.

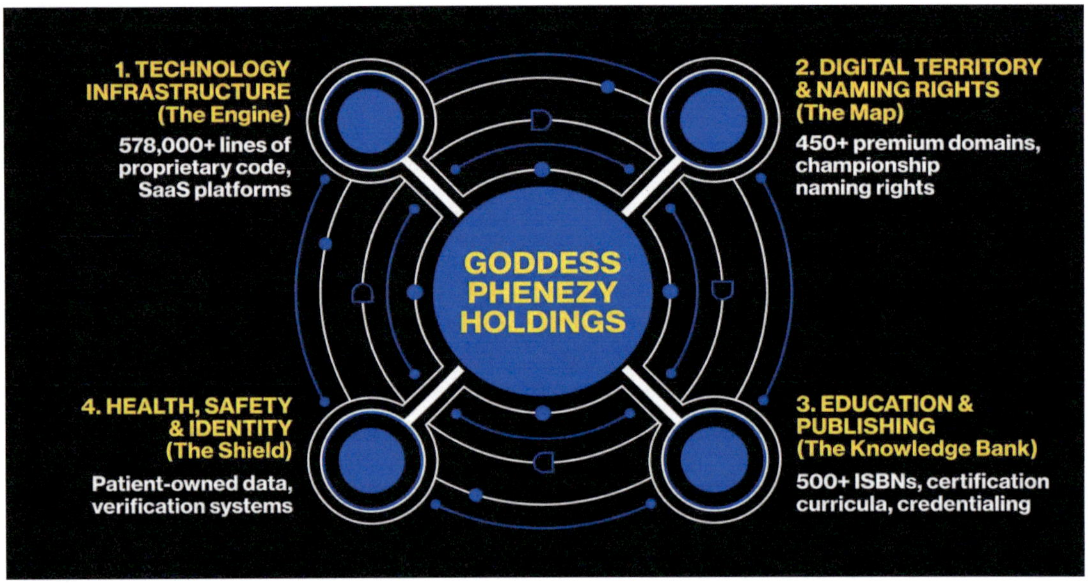

The strategy is deliberate and irreversible. Before a single ticket was sold or a single participant registered, the company secured the terrain — the domains, the code, the naming rights, the publishing authority, and the identity infrastructure. This is not a growth-stage company trying to find product-market fit. This is a holding company that has already built the market and now controls the terms of entry. The four asset classes function as interlocking fortifications: technology powers the platforms, digital territory establishes jurisdiction, education retains and credentials participants, and health and safety utilities protect the community.

SOVEREIGN VALUATION FRAMEWORK | INSTITUTIONAL BRIEFING

The Sovereign Valuation: Redefining the Unicorn

Goddess Phenezy Holdings, Inc. presents a fundamentally different financial architecture that many business owners try to use. Many of them value their business based on speculative, unproven hope and sentiment. Goddess Phenezy Holdings, Inc. values their ecosystem based on real, durable, verifiable asset layers that establish an indestructible valuation floor and an exponential enterprise ceiling.

Beyond the Traditional Unicorn

A traditional "unicorn" is defined as a privately held company with a valuation exceeding $1 billion. In the legacy technology ecosystem, such valuations are typically established through venture capital negotiations that rely heavily on speculative future cash flows and prevailing market sentiment. These are, by nature, rented valuations — temporary assessments subject to the whims of market cycles, investor appetite, and macroeconomic volatility. They are constructed on consensus, not on substance.

Goddess Phenezy Holdings, Inc. operates on a fundamentally different financial architecture. The Company utilizes a Sum-of-the-Parts (SOTP) and Asset-Based Valuation methodology, reflecting the reality that this enterprise is not a single-vertical startup chasing product-market fit, but a diversified, vertically integrated Sovereign Infrastructure company. The distinction is not semantic — it is structural, legal, and financial.

Where conventional unicorns derive their billion-dollar status from the projection of future revenues discounted to present value, Goddess Phenezy Holdings derives its valuation from assets that exist today — proprietary code, intellectual property registrations, strategic domain holdings, and operational infrastructure. These are not promises. They are possessions. The Company's total enterprise value is therefore not a negotiation outcome. It is an engineering fact, constructed across three durable and independently verifiable layers.

> Our valuation is not a projection. It is a structural reality — built on code, law, and sovereign ownership.

Three Durable Valuation Layers

The Company's total enterprise value is constructed across three interlocking strata, each independently defensible and collectively formidable. This architecture ensures that even under the most conservative analytical lens, the valuation floor remains unassailable.

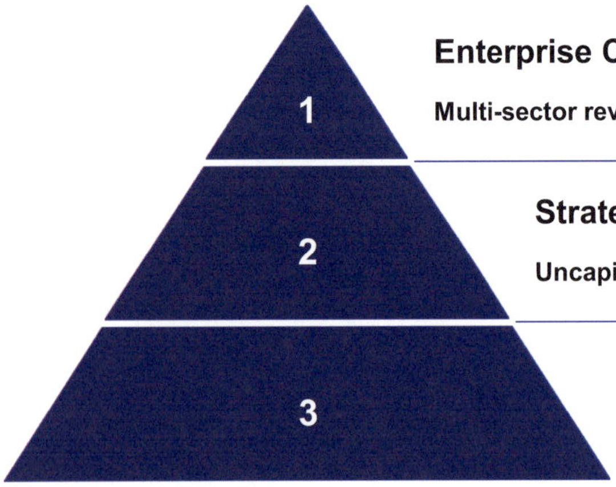

Enterprise Ceiling

Multi-sector revenue capacity

Strategic Land Reserves

Uncapitalized upside assets

Asset-Based Floor

$1.2B verifiable foundation

Each layer serves a distinct analytical function. The asset-based floor establishes the minimum replacement value of the Company's operational infrastructure. The strategic land reserves represent compounding, unactivated value that sits outside the conservative baseline. And the enterprise ceiling reflects the full revenue-generating capacity of the Company's operating subsidiaries across multiple sectors. Together, these layers present an investment thesis that is simultaneously conservative in its foundation and exponential in its trajectory.

Layer One: The Asset-Based Floor

$1.2 BILLION

The Cost Approach

Traditional venture models often discount asset-based valuations for software-centric companies, favoring instead revenue multiples and forward-looking cash flow projections. That framework does not apply here. Goddess Phenezy Holdings employs the Cost Approach, assessing fair value based on the capital, time, and operational complexity required for a third party to replicate the Company's infrastructure from the ground up. This establishes a conservative and verifiable valuation floor — one that does not depend on market conditions, investor sentiment, or future performance assumptions.

The Company's $1.2 billion foundation is comprised exclusively of battle-hardened software assets — the operational engine of the enterprise — rooted in 578,000+ lines of proprietary code. This is production-grade infrastructure that has been architected, tested, deployed, and refined through sustained operational cycles. It is not prototype code or minimum viable product scaffolding. It is the real thing.

What This Valuation Includes

- 578,000+ lines of proprietary code
- Infrastructure replacement cost analysis
- Operational complexity assessment
- Third-party replication modeling

What This Valuation Excludes

- Brand equity and goodwill
- Licensing revenue potential
- Market optionality
- Strategic land reserves

This valuation reflects infrastructure replacement cost only and intentionally excludes brand, licensing, or market optionality. This is not a projected valuation. It is a rebuild-from-scratch valuation — the answer to a single, clarifying question: what would it cost to build this from nothing? The answer is $1.2 billion. That is the floor. Everything above it is upside.

578,000+ Lines of Proprietary Code

The significance of the Company's proprietary codebase cannot be overstated. In an industry where most startups rely on open-source frameworks, third-party APIs, and cloud-dependent architectures, Goddess Phenezy Holdings has built its operational engine from the foundation up. Every line of code represents an investment in sovereignty — in owning, rather than renting, the infrastructure upon which the enterprise operates.

578K+	$1.2B	0
Lines of Code	**Replacement Cost**	**Platform Dependencies**
Proprietary, battle-hardened software assets	Conservative rebuild-from-scratch valuation	Fully sovereign, non-deplatformable infrastructure

The Cost Approach methodology applied here is not novel — it is the standard used by insurance underwriters, forensic accountants, and courts of law to establish the fair value of complex technology assets. What is distinctive is the scale and completeness of what has been built. The $1.2 billion figure is not an aspirational target. It is the actuarial reality of what exists today, quantified through disciplined, independently replicable methodology.

Layer Two: The Strategic Land Reserves

Critically, the $1.2 billion asset-based floor does not include two of the Company's most strategically significant asset classes. These holdings have been intentionally excluded from the conservative valuation baseline — not because they lack value, but because their value is realized through activation rather than static replacement cost. They are positioned as uncapitalized upside: assets whose full economic potential has not yet been reflected in the Company's formal valuation.

450+ Premium Domain Names

Digital territory that functions as sovereign real estate in the internet economy. These are not speculative registrations — they are premium, category-defining digital addresses that command authority and traffic. Their value compounds through activation, licensing, and strategic deployment across the Company's operating subsidiaries.

500+ ISBN-Registered IP Assets

Intellectual property assets registered through the international ISBN system, representing a formalized body of knowledge, methodology, and proprietary frameworks. These assets generate value through publishing, education, championships, experiential deployment, and licensing — multiple monetization vectors from a single registration.

Why Strategic Land Reserves Matter

These holdings function as Strategic Land Reserves — a term deliberately chosen to convey their structural importance. Just as sovereign nations maintain land reserves whose value compounds over time through development and strategic use, Goddess Phenezy Holdings maintains digital and intellectual reserves whose economic potential grows with each new activation pathway.

Activation

Deploying dormant domains and IP assets into live revenue-generating operations across the Company's subsidiary network, transforming latent value into active economic output.

Licensing

Granting controlled access to proprietary frameworks, methodologies, and digital territories — creating recurring revenue streams without transferring ownership or sovereignty.

Publishing & Education

Monetizing ISBN-registered intellectual property through educational programs, published works, championships, and experiential learning deployments at scale.

Sovereign Permanence

These assets are not platform-dependent and cannot be deplatformed, revoked, or displaced. They are owned. They endure. Their value is jurisdictional, not algorithmic.

Because their value is realized through market orchestration rather than static replacement cost, they are intentionally excluded from the conservative valuation baseline. This exclusion is not a weakness — it is a discipline. It ensures the $1.2 billion floor stands on its own merits, while the strategic reserves represent pure, unpriced upside available to sophisticated investors who understand compounding asset strategies.

A Land-and-Law-Holding Entity

> **Goddess Phenezy Holdings is not merely a software enterprise. It is a land-and-law-holding entity that owns digital territory, intellectual jurisdiction, and the structural rights through which markets operate.**

This distinction fundamentally alters the analytical framework through which the Company should be evaluated. Traditional SaaS companies are valued on revenue multiples, churn rates, and customer acquisition costs. Platform companies are valued on network effects and user engagement. But Goddess Phenezy Holdings occupies a category that transcends these conventional classifications.

The Company owns the terrain — the domains, the intellectual property registrations, the proprietary infrastructure — upon which markets, educational programs, championships, and experiential economies operate. This is not platform dependency. This is platform *sovereignty*. The assets that underpin this enterprise cannot be revoked by a third-party provider, displaced by an algorithm change, or deprecated by a platform policy update. They exist in law. They are registered. They are owned.

For institutional investors accustomed to evaluating risk through the lens of platform dependency and regulatory exposure, this ownership structure represents a fundamentally different — and fundamentally more durable — risk profile. The Company does not rent its competitive position. It has built it, registered it, and fortified it across multiple legal and technical jurisdictions. This is what we mean by sovereign infrastructure.

Layer Three: The Multi-Sector Enterprise Ceiling

Above the asset foundation and strategic land reserves lies the Company's enterprise expansion layer — the revenue-generating capacity of its operating subsidiaries. This is where the Company's valuation transitions from conservative floor to exponential ceiling, and where the full scope of the Goddess Phenezy Holdings vision becomes apparent.

By applying standard SaaS, licensing, and experiential multiples to an estimated $11 trillion total addressable market, the Company establishes a valuation ceiling aligned with its role as a global market orchestrator, not a single-product operator. This ceiling reflects scale potential, not speculation. It is derived from the same multiplier methodologies used by investment banks and institutional analysts worldwide — applied to a company whose market access spans software, intellectual property, digital real estate, education, and experiential commerce.

The enterprise ceiling is not a promise of future revenue. It is a mathematical expression of what the Company's infrastructure, assets, and market positioning make structurally possible. The distance between the $1.2 billion floor and this ceiling represents the Company's total available growth trajectory — a trajectory that investors can participate in from a position of foundational security rather than speculative exposure.

Total Addressable Market: $11 Trillion

The Company's operating subsidiaries span multiple high-growth sectors, each contributing to a combined total addressable market of unprecedented scale. Unlike single-vertical operators competing for share within a defined market, Goddess Phenezy Holdings operates as the infrastructure layer across multiple markets simultaneously.

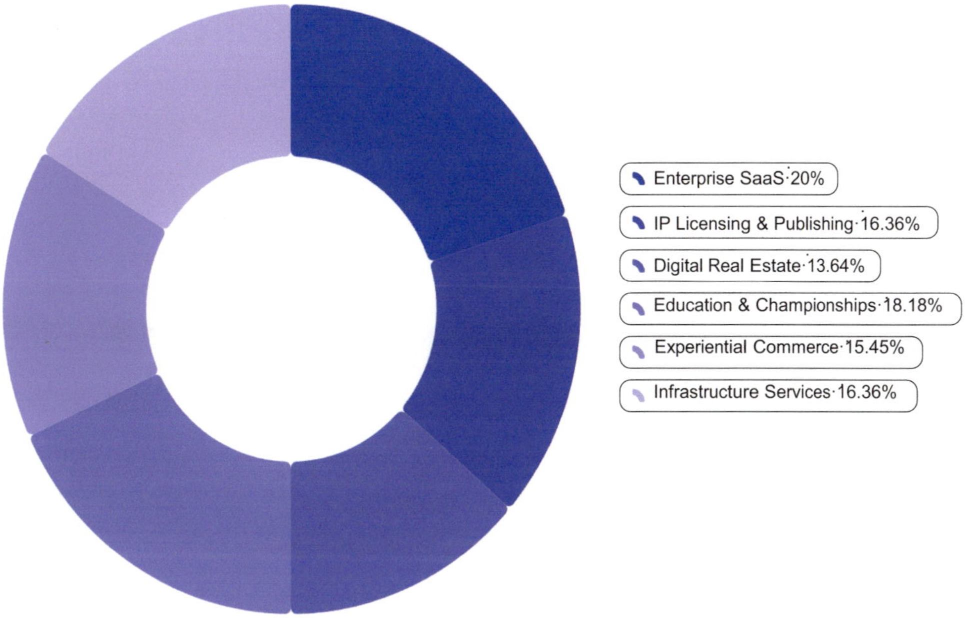

The $11 trillion figure is not a vanity metric. It is a composite of independently verifiable market sizes across sectors where the Company holds structural positioning through its proprietary code, domain portfolio, and registered intellectual property. Standard industry multiples applied to even fractional market capture within these sectors yield enterprise valuations that dwarf the conservative $1.2 billion asset floor — confirming that the floor is precisely that: a floor, not a ceiling.

Valuation Methodology Comparison

To appreciate the structural differentiation of the Goddess Phenezy Holdings valuation, it is instructive to compare the Company's methodology against the frameworks commonly employed across the technology investment landscape. Each methodology carries distinct assumptions, risk profiles, and levels of verifiability.

Dimension	Traditional Unicorn	Public SaaS Company	Goddess Phenezy Holdings
Valuation Basis	VC negotiation / sentiment	Revenue multiples	SOTP + Asset-Based (Cost Approach)
Floor Defensibility	Weak — market dependent	Moderate — revenue dependent	Strong — replacement cost verified
Asset Ownership	Typically platform-dependent	Mixed — some proprietary	Fully sovereign and owned
Deplatforming Risk	High	Moderate	Zero
IP Registration	Varies	Varies	500+ ISBN + 450+ domains
Upside Capture	Single vertical	Single/dual vertical	Multi-sector orchestration

The table above illustrates a critical insight: the Company's valuation is not only larger but structurally different from its peers. Where traditional unicorns and public SaaS companies expose investors to platform risk, sentiment volatility, and single-sector concentration, Goddess Phenezy Holdings offers a triple-layered architecture where even the most conservative layer — the asset floor — exceeds the billion-dollar threshold independently.

The Investment Profile

Unlike speculative startups that trade primarily on sentiment, market momentum, and the charisma of their founding teams, Goddess Phenezy Holdings presents an investment profile defined by structural permanence and layered defensibility. Every element of the Company's valuation can be independently audited, legally verified, and technically assessed.

1

An Indestructible Asset-Based Floor

$1.2 billion in verifiable infrastructure replacement value, rooted in 578,000+ lines of proprietary code. This floor exists regardless of market conditions, investor sentiment, or macroeconomic cycles. It is an engineering fact, not a financial opinion.

2

Extensive Unactivated Strategic Land Reserves

450+ premium domain names and 500+ ISBN-registered intellectual property assets positioned as uncapitalized upside. These reserves compound through activation, licensing, publishing, education, and experiential deployment — multiple vectors from sovereign holdings.

3

An Exponential Multi-Sector Revenue Ceiling

Enterprise expansion capacity across an $11 trillion total addressable market, powered by diversified operating subsidiaries applying standard SaaS, licensing, and experiential multiples to structural market positioning as a global orchestrator.

This three-layer architecture means that an investor in Goddess Phenezy Holdings is never exposed to a single point of failure. The floor protects against downside. The reserves provide compounding optionality. And the ceiling offers participation in exponential multi-sector growth. This is not a bet. It is a position.

Risk Architecture: Layered Defensibility

Sophisticated investors evaluate not only upside potential but the structural integrity of downside protection. The Goddess Phenezy Holdings valuation framework is deliberately engineered to provide layered defensibility at every stratum of analysis.

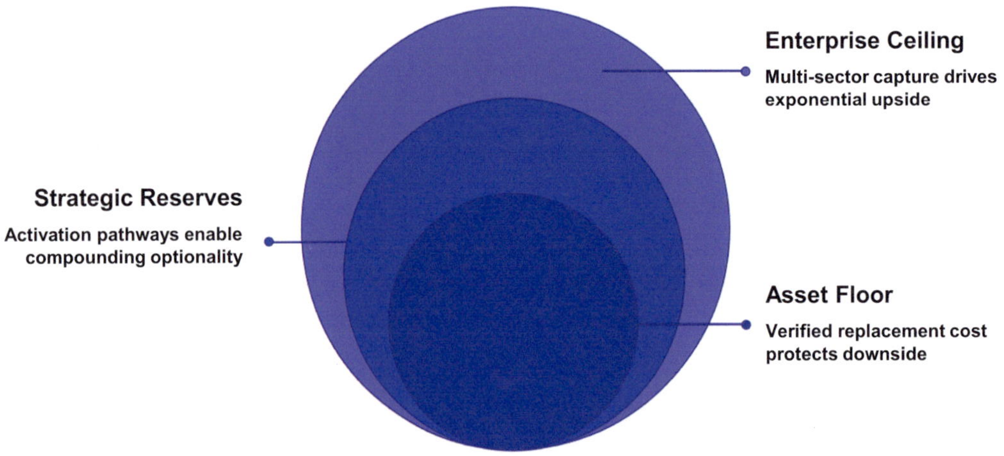

At the foundation, the $1.2 billion asset floor functions as an insurance policy against market volatility. Even in the most adverse scenario — where no strategic reserves are activated and no enterprise revenue is generated — the Company's proprietary infrastructure retains its replacement value. This is the kind of structural floor that institutional portfolios require: not a hope, but a hardened, auditable reality.

Above this floor, the strategic land reserves provide a compounding layer of optionality. Each domain activation, each ISBN monetization pathway, each licensing agreement adds incremental value without diluting the foundational asset base. And at the ceiling, the multi-sector enterprise layer transforms structural positioning into revenue — applying proven industry multiples to an addressable market of extraordinary scale. For the institutional investor, this architecture represents a rare convergence: verifiable downside protection paired with exponential upside access.

We Own the Vault

> "We do not merely participate in markets. We own the terrain upon which they operate."

This statement is not rhetoric. It is the operational thesis of Goddess Phenezy Holdings, distilled to its essence. The Company has spent years building, registering, and fortifying an asset base that does not depend on the permission, infrastructure, or goodwill of any third party. It owns its code. It owns its domains. It owns its intellectual property. It owns the structural rights through which its markets operate.

In an era where the most valuable companies in the world are increasingly vulnerable to platform risk, regulatory intervention, and algorithmic disruption, Goddess Phenezy Holdings has constructed something rarer and more durable: sovereign infrastructure. The Company does not lease its competitive advantage. It does not rent its market access. It does not borrow its valuation from the enthusiasm of venture capitalists. It has built a vault — and it owns every wall, every lock, and every asset inside.

We Own the Code

578,000+ lines of proprietary software — not open-source, not licensed, not borrowed. Built and owned.

We Own the Territory

450+ premium domains and 500+ ISBN registrations — sovereign digital and intellectual real estate.

We Own the Vault

A $1.2B floor, strategic reserves, and an exponential ceiling. Permanent. Defensible. Sovereign.

✧ SOVEREIGN INFRASTRUCTURE | INSTITUTIONAL OPPORTUNITY

The Sovereign Valuation Summary

Goddess Phenezy Holdings, Inc. does not conform to the legacy unicorn paradigm. It transcends it. Where traditional billion-dollar valuations are rented through venture negotiations and market sentiment, the Company's valuation is owned — constructed on verifiable assets, fortified by strategic reserves, and expanded through multi-sector enterprise capacity.

$1.2B	950+	$11T
Asset Floor	**Strategic Assets**	**Addressable Market**
Verified replacement cost	Domains + ISBN registrations	Multi-sector ceiling

For the institutional investor, the strategic partner, and the sophisticated analyst: this is not an invitation to speculate. It is an invitation to participate in something structurally permanent. The floor is verified. The reserves are sovereign. The ceiling is exponential. And the vault — every wall, every asset, every line of code — is owned.

> **We do not merely participate in markets. We own the terrain upon which they operate. We own the vault.**

TECHNOLOGIES
R US

ASSET CLASS 1

Technology Infrastructure: The Engine

The technological backbone of Goddess Phenezy Holdings comprises proprietary SaaS platforms exceeding 578,000 lines of code, independently valued using replacement-cost, moat analysis, and revenue potential models. This is not a minimum viable product. This is battle-hardened infrastructure built to scale across championships, conferences, academies, and health utilities simultaneously.

Every line of code was written in-house. There are no licensing dependencies, no third-party frameworks that can be revoked, and no external vendors who hold the keys to the kingdom. When you own your own technology stack, you eliminate the single greatest vulnerability facing modern businesses: platform dependency. Goddess Phenezy Holdings cannot be deplatformed because it *is* the platform.

1

Your Partnrr & Data Room

The partner economy operating system. A full-stack valuation engine, CRM/PRM infrastructure, and strategic intelligence platform. 214,000+ lines of code powering the ChurnShield™ and LiveMargin™ algorithms.

Estimated Strategic Value: ~$443M

2

Xperiences R Us Platform

The Super App for hosting National & International Championships, conferences, and live/virtual events. 364,000+ lines of code. Rated 100/100 Moat Score. The digital venue where all ecosystem activity converges.

Estimated Strategic Value: ~$666M

3

Limaj Health & VerifiEYED

Health justice and safety utilities. Includes Patient ID, Date Timer protocols, and identity verification systems. The protective layer that ensures every participant in the ecosystem is verified, safe, and sovereign over their own data.

Estimated Strategic Value: ~$54M–$76M

Your Partnrr Docs

Your Partnrr Docs is a patent pending, next-generation, proprietary, document intelligence platform designed for founders, creators, and professionals who build real assets—not just files.
Unlike traditional word processors, slide tools, or design software, Your Partnrr Docs allows a single living document to evolve seamlessly from professional writing to pitch decks, designed presentations, and sellable PDFs—without switching tools, exporting files, or rebuilding content.
Documents are driven by intent, not format. Users create once, then instantly transform their work into multiple professional outputs while preserving structure, meaning, and design integrity.

Your Partnrr Docs eliminates fragmented workflows, reduces tool dependency, and turns ideas into polished, revenue-ready deliverables—faster, cleaner, and smarter. One platform. One document. Many outcomes.

Technology Platform Valuation Breakdown

The combined technology infrastructure represents approximately $1.16 billion in independently certified asset value. The following chart illustrates the relative weight of each platform within the technology portfolio. These are not projections — they are replacement-cost valuations of code that already exists.

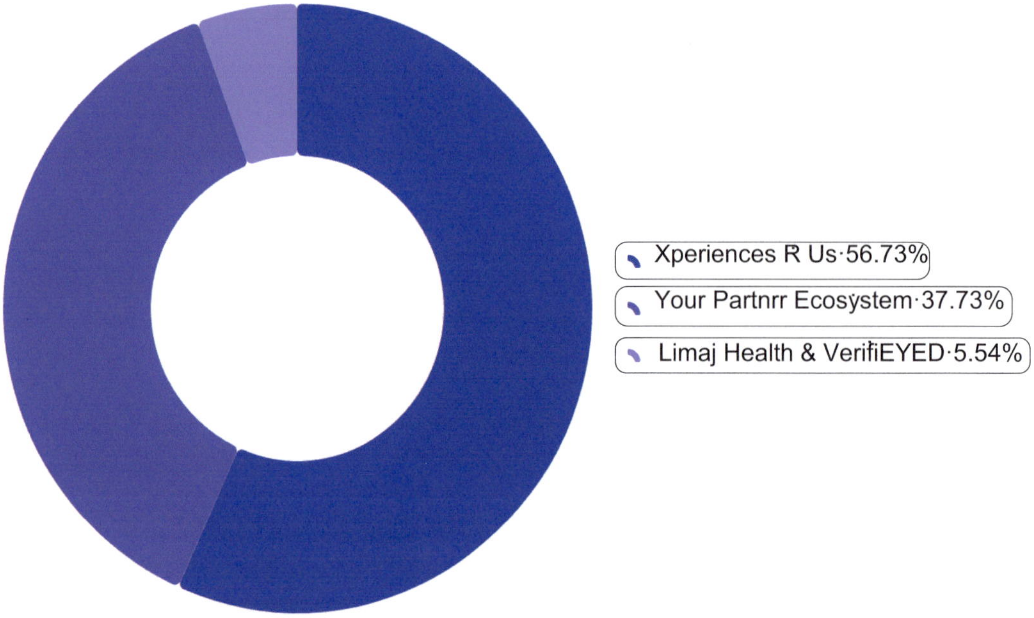

Xperiences R Us [(www.Xperiencesrus.com)](http://www.Xperiencesrus.com) commands the largest share of technology value at $666M, reflecting its role as the central venue platform — the "Super App" through which all championships, conferences, and live events are hosted and monetized. Your Partnrr [(www.YourPartnrr.com)](http://www.YourPartnrr.com) follows at $443M, serving as the intelligence engine that powers the partner economy, valuations, and relationship management. Limaj Health & VerifiEYED, while smaller in absolute value, represents the trust and safety layer that makes the entire ecosystem defensible from a regulatory and humanitarian standpoint.

ASSET CLASS 2

Digital Territory & Naming Rights: The Map

The company owns 450+ premium domain names, including exclusive .com ownership of over 200 National and International Championships, Academies, and Conferences. This strategy is both intentional and defensive. In the digital age, domain names are not simply web addresses — they are jurisdictional markers that establish presumptive authority, legitimacy, and first-mover advantage in perpetuity.

By securing names like TheNationalBasketballChampionship.com, TheInternationalPokerChampionship.com, TheNationalAIChampionship.com, TheInternationalAIAcademy.com, and The International Health and WealthConference.com, the company has established something that cannot be replicated or purchased after the fact. Competitors may host events in these categories, but they cannot host "The" event without confronting insurmountable naming, legitimacy, and trademark or copyright barriers.

Sample Domain Portfolio

Domain	Category
TheNationalBasketballChampionship.com	Sports
TheInternationalPokerChampionship.com	Gaming
TheInternationalGolfChampionship.com	Sports
TheInternationalPickleballChampionship.com	Sports
TheNationalAIChampionship.com	Technology
TheInternationalAIAcademy.com	Education
TheInternationalHealthAndWealthConference.com	Health/Finance

Strategic Positioning

This is not branding. This is jurisdiction. Independent strategic valuation methodologies place the standalone Control Value of this digital territory between $40M and $75M, excluding any operating revenue.

The defensive power of this portfolio cannot be overstated. Each domain represents a permanent barrier to entry for any competitor seeking to establish legitimacy in the same category. Once "The National" or "The International" championship title is held, the competitive landscape is fundamentally altered.

ASSET CLASS 3

Education & Publishing: The Knowledge Bank

Through its Academies and publishing division, Goddess Phenezy Holdings controls one of the most strategically valuable intellectual property portfolios in the ecosystem economy. Education is not a side business or a marketing funnel. It is a core infrastructure layer that ensures every participant who enters the ecosystem is trained, certified, credentialed, and retained inside owned channels — never external platforms.

500+ Owned ISBNs

A publishing portfolio of over 500 International Standard Book Numbers ensures that every rulebook, playbook, curriculum guide, and instructional manual is permanently registered and controlled by the holding company. This is institutional-grade publishing infrastructure.

Certification Curricula

From championship-specific training programs to professional development tracks, every certification pathway is designed, owned, and administered internally. Graduates carry credentials that connect directly to the championship and professional ecosystem.

Rulebooks & Playbooks

Official rules for over 200 championships are authored, published, and controlled by the holding company. When you own the rules, you own the game itself. Competitors play by standards that Goddess Phenezy Holdings defines and governs.

The vertical integration of education into the ecosystem creates a closed-loop retention engine. A participant who trains at an Academy earns credentials that are recognized at Championships, verified through VerifiEYED, and tracked through Your Partnrr. There is no incentive to leave — and no external platform that can replicate the interconnected value.

The Competitive Position: Governing Architecture

Goddess Phenezy Holdings does not compete as a single product company. It does not compete as an event producer, a software vendor, or a media publisher. It operates as a governing architecture — a multi-layered system that controls the venues, the rules, the technology, the identity layer, and the media amplification simultaneously. No competitor can displace a single layer without confronting the entire structure.

Own the Rules

Publishing & certifications define how the game is played

Own the Venues

High-value domains & platforms establish where participation happens

Own the Technology

SaaS infrastructure powers every interaction and transaction

Own the Amplification

Champions TV & publishing control the narrative

Own the Identity Layer

VerifiEYED & data systems ensure trust and sovereignty

This structure creates a self-reinforcing flywheel where users, data, revenue, and authority compound internally. Every new championship participant generates data that strengthens Your Partnrr. Every credentialed graduate amplifies the academy's reputation. Every verified identity deepens the trust infrastructure. And every broadcast through Champions TV drives new participants into the top of the funnel. The flywheel does not slow down — it accelerates with every rotation, creating compounding returns that make the ecosystem exponentially more defensible over time.

MOAT ANALYSIS

Why This Cannot Be Replicated

Institutional investors frequently ask the same question: "What prevents a well-funded competitor from building this?" The answer lies in the nature of the assets themselves. Goddess Phenezy Holdings has secured a category of assets that are definitionally unreplicable — not because they are technologically complex, but because they are first-mover permanent.

Domain Exclusivity

There is only one TheNationalBasketballChampionship.com. Only one TheInternationalAIAcademy.com. These cannot be duplicated, only approximated — and approximation destroys legitimacy.

Code Volume & Integration

578,000+ lines of proprietary code integrated across six sectors. Replicating this would cost hundreds of millions and years of development — time the market will not grant.

Network Effects

Every participant added strengthens every platform simultaneously. Competitors would need to replicate not just the code, but the interconnected ecosystem value that compounds with scale.

Sovereign Architecture

The integration of technology, naming rights, education, health, identity, and media into a single holding structure creates a fortress that cannot be attacked at any single point.

The competitive moat is not a feature. It is the entire topology of the company. Every asset class reinforces every other asset class, creating a defensive position that grows stronger — not weaker — with each passing quarter. This is why the Xperiences R Us platform received a 100/100 Moat Score in independent assessment.

FINANCIAL POSITION

The Asset Vault: Financial Position

Goddess Phenezy Holdings, Inc. rejects the standard "pre-revenue" valuation model used by speculative startups. While the venture capital ecosystem routinely assigns billion-dollar valuations based on addressable market fantasies and future revenue projections, this company operates under a fundamentally different standard: Cost-to-Replace + Strategic Moat methodology.

The following valuations have been independently certified by Your Partnrr Data Room algorithms and strategic IP assessment protocols. These figures represent the Current Asset Value of the code, domains, and intellectual property already secured by the Holding Company. Nothing in this valuation is hypothetical. Every asset exists, is operational, and is owned free and clear by the parent entity.

> **Valuation Methodology: Proven Infrastructure — Cost-to-Replace + Strategic Moat Analysis. Independently certified. Not speculative. Not projected. Built and secured.**

The Certified Balance Sheet

Asset Class	Description	Certified Value
Xperiences R Us Platform	The Venue. 364,000+ LOC. The "Super App" hosting global championships. Rated 100/100 Moat Score.	$666,373,052
Your Partnrr Ecosystem	The Engine. 214,000+ LOC. The SaaS OS for the partner economy, including ChurnShield™ and LiveMargin™ algorithms.	$443,451,692
Limaj Health & VerifiEYED	The Shield. Health Justice & Safety Utilities. Includes Patient ID & Date Timer protocols.	~$54,787,067
Digital Territory	The Map. 450+ Premium Domains. Strategic Control Premium for "The National" & "The International" naming rights.	$40,000,000–$75,000,000

~$1.2B

Total Holdings Value

Unicorn Status Confirmed

578K+

Lines of Code

Proprietary SaaS infrastructure

450+

Premium Domains

Digital territory secured

500+

Owned ISBNs

Publishing & knowledge assets

Asset Valuation by Class

The following visualization illustrates the distribution of certified asset value across the four core asset classes of Goddess Phenezy Holdings. Technology infrastructure — comprising Xperiences R Us, Your Partnrr, and Limaj Health — commands the overwhelming majority of total holdings value, reflecting the strategic primacy of proprietary code ownership in the modern economy.

The technology platforms alone represent over $1.16 billion in certified value — demonstrating that even without the digital territory portfolio, the company would qualify for unicorn status based on code assets alone. The Digital Territory valuation range of $40M–$75M represents a standalone Control Value assessment that excludes operating revenue, meaning the true economic value of the domain portfolio will be significantly higher once championships, academies, and conferences generate recurring revenue streams.

Financial Strategy: Asset Leverage

Because Goddess Phenezy Holdings owns its infrastructure outright, the financial model differs fundamentally from traditional event companies, SaaS startups, or media organizations. There are no landlords. There are no licensors. There are no platform dependencies that siphon margin or create existential risk. The company's three-pillar financial strategy transforms asset ownership into compounding, high-margin revenue.

Zero Rented Tech

The company does not pay licensing fees for CRM, ticketing, or virtual venues. Every technology component is proprietary. This eliminates recurring platform costs that typically consume 15–30% of revenue for competing organizations, creating a permanent structural margin advantage.

Master Licensing Revenue

The company monetizes the jurisdiction of its domains by licensing "National" and "International" rights to operators, generating high-margin royalties without operational overhead. This is the franchise model applied to digital territory — operators pay for the privilege of producing events under championship titles that Goddess Phenezy Holdings owns permanently.

Compound Asset Growth

Every user added to Xperiences R Us increases the data value of Your Partnrr and the network effect of Limaj Health. Growth in any single subsidiary accelerates value creation across all subsidiaries simultaneously — the definition of a compounding flywheel.

The result is a financial architecture where margins improve with scale, dependencies decrease over time, and revenue diversification happens organically as new championships, academies, and conferences are activated across the domain portfolio. This is not a business that needs to "figure out monetization." The monetization is embedded in the architecture itself.

Revenue Model: The Compounding Flywheel

The financial engine of Goddess Phenezy Holdings is not a linear revenue model — it is a compounding flywheel where every transaction, every participant, and every data point accelerates value creation across the entire ecosystem. Understanding this flywheel is critical for strategic investors evaluating the long-term return potential of the holding company.

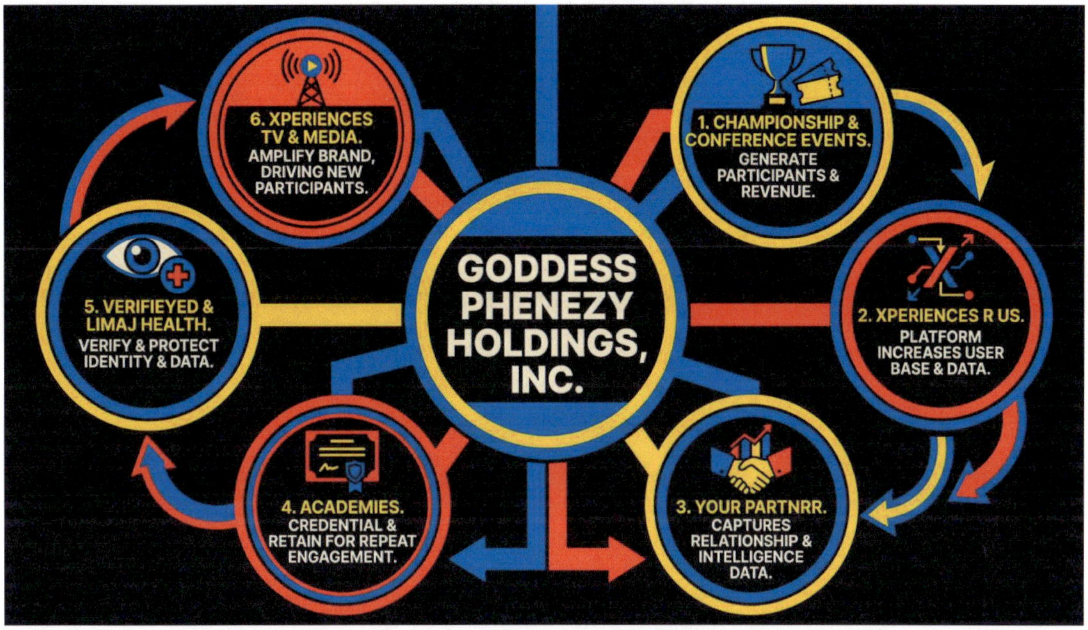

The critical insight for investors is this: the flywheel has no natural deceleration point. Unlike linear businesses that must continuously acquire new customers to grow, the Goddess Phenezy ecosystem grows by deepening engagement with existing participants while simultaneously attracting new ones. Every credential earned, every championship attended, every identity verified, and every partner relationship managed creates incremental value that cannot be extracted by a competitor. The flywheel compounds internally — and the holding company captures 100% of that compounding value because it owns every layer of the stack.

The Mission Continues: KJ's Legacy

This company exists because one life mattered. KJ (Jamil), the son of founder Goddess Phenezy, passed away from lupus complications. His legacy is not memorialized in a foundation or a scholarship alone. It is embedded in systems designed to protect lives, create opportunity, and prevent exploitation at scale.

> **"This manifesto is not a pitch for permission. It is a declassified disclosure of what has already been built. The terrain is secured."**

Every championship that bears "The National" or "The International" title creates opportunities for young athletes, veterans, women, and minority communities who have historically been excluded from ownership and governance in competitive industries. Every academy credential opens a door that was previously locked behind institutional gatekeepers. Every health utility powered by Limaj Health puts data sovereignty back in the hands of patients who have been exploited by systems designed to profit from their vulnerability.

The social mission is not a CSR addendum or a marketing narrative. It is the founding purpose of the enterprise. When Goddess Phenezy, a USMC Combat Veteran, designed this holding company, the architecture was not optimized solely for financial return. It was optimized for permanence, protection, and empowerment. Financial returns are a consequence of building something that genuinely serves people — not the other way around.

KJ's legacy lives in every line of code, every domain secured, every rulebook published, and every participant who enters this ecosystem and finds opportunity they could not find elsewhere. This is not charity. This is infrastructure justice — the belief that the best way to honor a life is to build systems that protect millions of others.

Key Metrics at a Glance

For strategic investors and institutional partners evaluating the scope and maturity of Goddess Phenezy Holdings, the following metrics provide a snapshot of the enterprise's current position. These are not forward projections — they represent assets and infrastructure that exist today.

100/100

Moat Score

Xperiences R Us platform rated at maximum defensibility

100%

Code Ownership

Zero rented technology, zero licensing dependencies

100%

Domain Control

All 450+ premium domains owned outright, no leased territory

200+

Championships

National & International titles secured via domain ownership

6

Sectors

Vertically & horizontally integrated industries under one holding structure

$1.2B

Certified Value

Asset-based valuation, not speculative

Ecosystem Architecture: How It All Connects

The true power of Goddess Phenezy Holdings lies not in any single subsidiary, but in how every component connects to every other component. This is a systems-level advantage that cannot be replicated by assembling individual best-of-breed solutions. The following diagram illustrates the sovereign ecosystem architecture and the flow of value between its core components.

Technology Engine

Platform backbone powering partners and experiences

Education & Academies

Training, certification, and talent development pipelines

Championship & Conference

Events and competitive platforms that engage community

Health & Identity

Wellness services and verified identity solutions

This architecture ensures that no single point of failure can compromise the ecosystem. If a competitor emerges in one vertical — say, championship events — they still lack the technology platform, the credentialing authority, the identity verification layer, and the media amplification infrastructure. Attacking one layer is futile because every layer defends every other layer. This is not a product company. It is a governing architecture — and governing architectures are not disrupted. They are joined.

The Terrain Is Secured

Goddess Phenezy Holdings, Inc. is not seeking permission. It is disclosing what has been built.

The technology is written. The domains are owned. The championships are named. The academies are structured. The health and identity infrastructure is deployed. The publishing portfolio is registered. The valuation is certified. Unicorn status is confirmed at approximately $1.2 billion in existing asset value.

For Strategic Investors

This represents a rare opportunity to participate in a certified asset-based holding company with zero platform dependency, compounding network effects, and a defensible moat rated 100/100. The infrastructure is built. The question is whether you are positioned at the table when the ecosystem activates.

For Institutional Partners

Goddess Phenezy Holdings offers licensing, co-production, and partnership structures across 200+ championship titles, academy programs, and conference platforms. The jurisdiction is established. The terms of entry are defined by the holder of the terrain.

> "This manifesto is not a pitch for permission. It is a declassified disclosure of what has already been built. The terrain is secured."
> — Goddess Phenezy, Founder & Commander

OPERATION LEGACY | CLASSIFIED EXECUTIVE BRIEFING

Goddess Phenezy Holdings, Inc.

Organizational Structure & Multi-Sector Market Dominance Strategy — A comprehensive overview of the corporate hierarchy, operating subsidiaries, and the unprecedented market opportunity spanning nine global industries with a combined addressable market exceeding $11 trillion.

Organizational Structure: The Hierarchy

Goddess Phenezy Holdings, Inc. is architected as a centralized holding company with purpose-built operating subsidiaries—each functioning as a specialized squadron within a unified command structure. This design is not incidental. It is the foundation of a multi-market ecosystem engineered for compounding growth, operational leverage, and strategic defensibility.

The parent entity retains centralized ownership of all intellectual property, source code, domain portfolios, and strategic direction. This ensures that every subsidiary operates under a shared technology backbone and unified brand architecture, enabling cross-platform monetization, data continuity, and rapid market expansion without the fragmentation that plagues multi-vertical conglomerates.

Below the parent entity, five operating subsidiaries—referred to internally as The Squadrons—execute across distinct but interconnected market verticals: championships and experiential events, technology infrastructure, SaaS commercialization, health advocacy, and relationship management. Each squadron carries a defined mission, proprietary asset base, and revenue mandate, yet all feed into and draw from the same centralized ecosystem.

Parent Entity: The Command Center

Key Assets Under Central Control

Goddess Phenezy Holdings, Inc.

Role: Centralized ownership of IP, Code, Domains, and Strategy

Mission: Educational credentialing and media

500+ ISBNs

Proprietary publishing portfolio spanning educational and media content

The Academies Curricula

Credentialing programs bridging competition and education

Xperience TV

Owned broadcasting and media amplification platform

The parent entity functions as the strategic brain of the entire organization. By retaining centralized ownership of all intellectual property—including over 500 ISBNs, the full Academies curricula, and the Xperience TV media platform—Goddess Phenezy Holdings ensures that no subsidiary operates in isolation. Every asset created at the subsidiary level flows upward into the holding company's IP portfolio, creating a compounding moat that grows stronger with every new championship launched, every credential issued, and every piece of content produced.

This centralized model also provides critical investor protections. IP consolidation at the parent level means that the value of the enterprise is not distributed across fragile, independently-owned subsidiary assets. It is unified, defensible, and strategically deployable across markets.

X
XPERIENCES
R US

🏆 THE BATTLEFIELD

Xperiences R Us, Inc.

Xperiences R Us (www.Xperiencesrus.com) is the flagship operating division of Goddess Phenezy Holdings and serves as the primary execution arm for the company's most visible and revenue-generative programs. This subsidiary is responsible for the design, production, and operation of 79 National & International Championships spanning traditional athletics, STEM competitions, creative arts, gaming, and emerging competitive formats.

But Xperiences R Us is far more than an events company. It operates as a vertically integrated experiential platform that combines live championships, globally influential conferences, proprietary academies, and a progressive web application (PWA) community—all running on technology owned by the parent entity. This means every ticket sold, every credential issued, every broadcast streamed, and every data point captured flows through owned infrastructure, eliminating dependency on third-party platforms and maximizing margin at every touchpoint.

1

Championships

79 official national and international championships across sports, STEM, and competitive fields—from traditional athletics to cutting-edge technology

2

Conferences

Globally influential events bringing together thought leaders, innovators, and enthusiasts across AI, Metaverse, health, and wellness

3

PWA Community

Proprietary platform featuring chess, academies, athlete profiles, pods, and the flagship "Missing from the Village" system for locating missing persons in underserved communities

Missing from the Village: A proprietary "loss and found" system designed to assist in locating missing men, women, and children within underserved communities—a flagship social impact feature embedded directly within the platform's community architecture.

Technologies R Us & Your Partnrr

Technologies R Us, Inc.

The Engines

This division powers the entire ecosystem—business operations, media delivery, academies, community platforms, championships, and beyond. Technologies R Us (www.technologiesrus.com) develops next-generation proprietary software that expands the founder's vision and exceeds the standards of most progressive web applications on the market today. Every digital touchpoint across all subsidiaries runs on technology built and owned internally.

Your Partnrr

SaaS Commercialization

Your Partnrr (www.YourPartnrr.com) commercializes the proprietary operating system that powers the internal ecosystem, licensing it externally as a SaaS platform. Key assets include The Source Code, The Data Room, and the proprietary ChurnShield™ Algorithm—a retention engine designed to reduce subscriber attrition and maximize lifetime value for enterprise clients.

Together, these two subsidiaries represent the technological backbone and commercial engine of the entire holding structure. Technologies R Us builds the infrastructure; Your Partnrr monetizes it externally. This dual-purpose approach means the company's R&D investment serves both internal operational excellence and external revenue generation—a capital-efficient model that few competitors can replicate.

Limaj Health & The CRM Tool

Limaj Health Initiative

THE ADVOCACY

Mission: Patient Rights, Medical Justice, and Health Identity

Limaj Health operates at the intersection of patient advocacy and digital infrastructure. Its cornerstone asset—the PHCI (Portable Health Care Identity) Protocol—represents a paradigm shift in how health data is owned, stored, and controlled. Rather than tethering patient records to institutional systems, the PHCI protocol empowers individuals with portable, patient-owned health identities that travel with them across providers, geographies, and life stages.

This is not merely a health records initiative. It is an infrastructure play in the emerging patient-sovereignty movement, positioning Goddess Phenezy Holdings at the forefront of health justice, data rights, and identity-first healthcare.

The CRM Tool

THE SAFETY NET

Mission: Making dating safer through couple relationship matching and management

The CRM (Couple Relationship Management) Tool reimagines what relationship technology can be. Rather than optimizing for swipes and superficial engagement metrics, this subsidiary builds infrastructure for safer dating experiences and deeper relationship management. The platform applies CRM-grade data architecture to the most personal dimension of human interaction—turning relationship compatibility, safety verification, and ongoing partnership management into a structured, technology-enabled experience.

This subsidiary addresses a massive and underserved gap in the relationship technology market, where safety concerns and relationship longevity remain unsolved problems for existing platforms.

The Complete Hierarchy

The organizational structure of Goddess Phenezy Holdings is designed for maximum strategic leverage. Each subsidiary operates with mission clarity and operational independence, yet all are bound by centralized IP ownership, shared technology infrastructure, and cross-platform data flows that create compounding value.

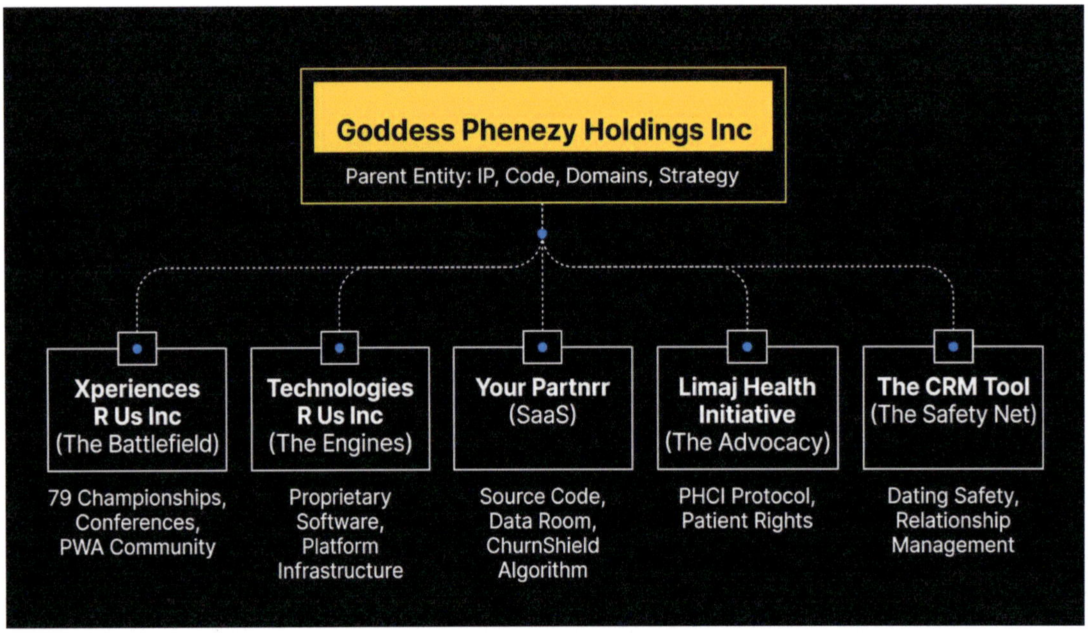

This is not a loose collection of businesses. It is a strategically orchestrated ecosystem where every championship fuels media content, every credential strengthens the education platform, every participant becomes a community member, and every data point deepens the technology moat. The hierarchy exists to ensure that value created in any one squadron compounds across the entire enterprise.

3.0 Market Opportunity — Multi-Sector Dominance

This is not market participation. This is market orchestration.

Goddess Phenezy Holdings, Inc. is positioned at the convergence of several of the world's fastest-growing global markets: sports, education, entertainment, technology, health, and digital media. Individually, each represents a multi-billion-dollar opportunity. Collectively—when unified through owned platforms, proprietary technology, and exclusive naming rights—they form a rare, compounding growth engine that few single-vertical competitors can match.

Unlike companies that rely on one revenue stream or audience, Goddess Phenezy Holdings operates a multi-market ecosystem in which participants, content, credentials, data, and monetization flow across divisions. This structure materially increases lifetime value per participant while reducing customer acquisition cost through cross-platform integration and internal retention. A championship participant becomes an academy student, who becomes a conference attendee, who becomes a platform subscriber—all within owned infrastructure.

The following sections detail the specific market verticals targeted, the strategic advantages held in each, and the revenue potential derived from conservative market capture estimates. The combined addressable market across all nine sectors exceeds $11 trillion, with initial capture targets representing a fraction of available opportunity—yet still generating billions in projected annual revenue.

Strategic Overview: The Compounding Ecosystem

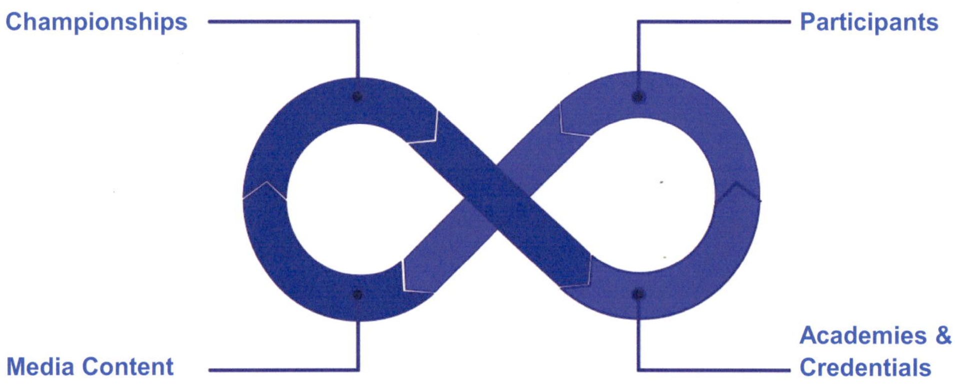

The strategic architecture of Goddess Phenezy Holdings is built on a fundamental insight: in a multi-market ecosystem, value compounds rather than simply accumulates. Each division doesn't merely generate its own revenue—it feeds participants, content, data, and engagement into every other division, creating a self-reinforcing growth flywheel.

When a participant enters a National Championship through Xperiences R Us, they simultaneously become a prospect for academy enrollment, a content source for Xperience TV, a data point for Technologies R Us (www.technologiesrus.com) analytics, and a potential subscriber for Your Partnrr's SaaS tools. This cross-pollination effect means the marginal cost of each new participant decreases while their lifetime value increases across the ecosystem—a dynamic that single-vertical competitors simply cannot replicate.

The following market analyses demonstrate how this compounding dynamic translates into specific, quantifiable revenue opportunities across nine distinct but interconnected global markets.

3.1 Global Sports Industry

The global sports industry is projected to exceed $614 billion annually, driven by media rights, sponsorships, live events, digital content, and global fan engagement. This is one of the most resilient and fastest-growing sectors in the global economy, with digital transformation accelerating revenue diversification beyond traditional gate receipts and broadcast deals.

Through ownership of National & International Championships across multiple sports—and the exclusive digital territory associated with those titles—Xperiences R Us captures value not only from events, but from licensing, broadcasting, data, and platform participation. The ownership of championship naming rights creates a moat that cannot be replicated through technology alone.

Naming Rights Ownership

Exclusive ownership of "The National" and "The International" championship naming rights across multiple competitive formats

Integrated Technology Stack

Proprietary technology for ticketing, CRM, virtual venues, and broadcasting—zero third-party dependency

Owned Media Amplification

The Champions' TV provides direct-to-audience broadcasting and content monetization

$614B

Global Market Size

Total addressable market annually

0.5-1%

Target Capture

Conservative market share target

$3.07B-$6.14B

Revenue Potential

Projected annual revenue opportunity

Growth Drivers: Expansion into non-traditional competitive formats (STEM, AI, Chess, Gaming) and master licensing of championship jurisdictions to regional and international operators will fuel sustained growth beyond initial market entry.

3.2 Youth & Educational Sports Market

The U.S. youth sports market exceeds $19 billion annually, with accelerating demand for scholarship pathways, credentialing, and academic outcomes. Parents, schools, and institutions are increasingly seeking competitive environments that deliver measurable educational value—not just trophies. This shift creates a structural opening for organizations that can merge athletic competition with academic credentialing.

Goddess Phenezy Holdings differentiates itself by merging competition with education, offering championships that reward not only physical performance, but intellectual, strategic, and technical skill. The integration of STEM, Chess, and Academic Championships with the Academies curricula creates a direct pipeline from competition to education to credentials to real-world opportunity.

Strategic Advantage

Integrated Academies

STEM, Chess, and Academic Championships integrated directly with credentialing Academies

Outcome Pipelines

Direct pathways from competition → education → credentials → scholarship and career opportunity

AI-Enabled Tracking

Performance and skills tracking powered by proprietary AI analytics

Market Capture

$19B

U.S. Market

Annual youth sports spend

1-2%

Target Share

Conservative capture rate

$190M-$380M

Annual Revenue

Projected opportunity

Growth Drivers: School and district partnerships, AI-enabled performance tracking, and rising institutional demand for outcomes beyond trophies will propel this division. The shift toward holistic youth development creates a secular tailwind that positions Goddess Phenezy as the credentialing partner of choice for schools, districts, and families seeking measurable ROI from competitive participation.

3.3 Online Gaming, Virtual Sports & Esports

The global esports and online gaming market exceeds $1.44 billion, fueled by virtual competition, streaming culture, and youth-first engagement models. This sector represents one of the fastest-growing entertainment categories globally, with participation rates among Gen Z and Gen Alpha far exceeding traditional sports viewership growth.

Through National & International Virtual Championships operated entirely on proprietary platforms, Goddess Phenezy Holdings captures both participation revenue and platform value. Unlike competitors who rely on third-party streaming infrastructure and tournament platforms, every virtual championship runs on owned technology—creating a vertically integrated competitive gaming ecosystem.

Owned Virtual Brands

Ownership of virtual championship brands and premium domain portfolio

AI-Driven Competition

Proprietary AI competition formats and real-time performance analytics

Cross-Platform Engagement

Virtual championships tied to media, education, and credentialing ecosystems

$1.44B

Global Market

Esports and online gaming TAM

0.5-1%

Target Capture

$720M-$1.44B

Annual Revenue

Projected opportunity

Growth Drivers: Continued virtual sports adoption across demographics and cross-platform engagement tied to media and education divisions will drive sustained expansion. The convergence of gaming, streaming, and competitive credentialing creates a unique value proposition that pure-play gaming companies cannot offer.

3.4 Health, Wellness & Health Justice Industry

The global health and wellness industry is projected to reach $6.99 trillion annually—making it one of the largest addressable markets on the planet. Through National & International Health & Wealth Conferences and proprietary health identity infrastructure via Limaj Health, Goddess Phenezy Holdings functions not merely as a content provider, but as an infrastructure participant in health education, advocacy, and patient empowerment.

The convergence of health and financial literacy represents a largely untapped market opportunity. While traditional wellness companies focus on fitness or nutrition, Goddess Phenezy addresses the systemic barriers to health outcomes—medical justice, patient rights, and health data sovereignty—through both conference programming and the proprietary PHCI protocol.

Strategic Advantage

Health + Financial Literacy Convergence: Uniquely positioned at the intersection of health outcomes and wealth building

Patient-Owned Data Models: The PHCI (Portable Health Care Identity) protocol empowers individuals with sovereign health identities

Infrastructure Play: Not competing on content alone—building the rails for health identity and patient rights

Market Opportunity

$6.99T

Global Market

Health and wellness TAM

0.01%

Target Capture

Ultra-conservative share

~$699M

Annual Revenue

Projected opportunity

3.5 Technology, STEM & AI Markets

A $1.6 Trillion Opportunity

The global STEM and technology education market is projected to reach $1.6 trillion by 2030, growing at a CAGR exceeding 14%. This explosive growth is driven by the global skills gap, institutional demand for technology-literate graduates, and the accelerating integration of AI into every industry vertical.

Goddess Phenezy Holdings embeds technology not only as content, but as infrastructure—delivering competitive, credentialed experiences tied directly to real-world application. Championships in AI, STEM, and technology are not academic exercises; they are pipeline-building mechanisms that connect participants to credentials, employers, and opportunities.

Target Capture
0.5%–1% initially

Revenue Potential
$8M – $16M annually

Growth Trajectory
Exponential upside as STEM demand accelerates

The STEM vertical functions as a strategic bridge between education and technology—two of the company's core competencies. By credentialing STEM achievement through nationally recognized championships, Goddess Phenezy creates a credential class that employers, universities, and institutions will increasingly recognize and demand.

3.6 Global Poker Market

The global poker market is projected to grow from $85.4 billion to $170 billion by 2030—representing one of the most significant growth trajectories in competitive entertainment. Poker occupies a unique position at the intersection of skill-based competition, entertainment media, and digital gaming, making it a natural fit for the Goddess Phenezy ecosystem.

By operating National & International Poker Championships and integrating digital play, analytics, and media production, this single vertical alone supports billion-dollar revenue potential. The combination of live championship events with digital platform participation creates a dual-revenue model that captures value from both physical and virtual engagement.

1 Championship Ownership

National and International Poker Championship naming rights and operational control across multiple formats and jurisdictions

2 Digital Integration

Proprietary digital play platforms with real-time analytics, performance tracking, and streaming capabilities

3 Media Monetization

Championship content feeds directly into Xperience TV and media licensing channels

$170B

Market by 2030

0.5-1%

Target Capture

$850M-$1.7B

Annual Revenue

3.7 Performing Arts & Creative Competitions

The global performing arts industry exceeds $330 billion annually, encompassing dance, music, theater, spoken word, and emerging creative formats. Through structured National & International Championships in dance, rap, breakdancing, karaoke, and creative performance, Goddess Phenezy Holdings converts culture into monetizable competition and media.

This vertical represents a strategic convergence of entertainment, competition, and cultural expression. Unlike traditional performing arts organizations that rely on ticket sales and donations, Goddess Phenezy's championship model transforms creative performance into a scalable, media-rich, digitally distributed competitive format—complete with credentialing, platform participation, and broadcasting revenue.

Dance & Movement

National and international championships across breakdancing, contemporary, hip-hop, and cultural dance forms

Music & Vocal Performance

Rap battles, karaoke championships, and vocal competition formats with media integration

Creative Expression

Emerging competitive formats that credential creative talent and feed content into the media ecosystem

> **Target Market Capture: 0.5% of the global performing arts market — approximately ~$1.65 billion annually**

3.8 Global Conference Industry

The global conference and professional events market is projected to reach $1.6 trillion by 2028, driven by the insatiable demand for in-person networking, thought leadership, and industry-specific knowledge exchange. This market has rebounded aggressively post-pandemic, with hybrid and digital formats expanding the total addressable audience far beyond physical venue capacity.

Goddess Phenezy Holdings operates conferences not as standalone events, but as content engines that feed media production, academy curricula, and digital platform engagement. Every keynote, panel discussion, and workshop session becomes content that can be credentialed, broadcast, archived, and monetized across the ecosystem. This transforms the traditional conference model from a one-time revenue event into a perpetual content and engagement asset.

Revenue Opportunity

$1.6T

Market by 2028

Global conference TAM

0.1%

Target Capture

Conservative share

~$1.6B

Annual Revenue

Projected opportunity

> Conferences serve as the top-of-funnel acquisition channel for the entire ecosystem—converting attendees into championship participants, academy students, platform subscribers, and media consumers.

3.9 B2B SaaS & Digital Infrastructure

THE ENGINE

The global CRM SaaS market is projected to exceed $163 billion by 2030, with an additional $10B+ TAM across valuation and data room infrastructure. Through Your Partnrr and Your Partnrr Data Room, Goddess Phenezy Holdings licenses the same proprietary operating system used to run its own ecosystem—creating a uniquely credible go-to-market narrative for enterprise buyers.

This is a critical strategic differentiator. Unlike SaaS companies that build products in isolation and then seek market fit, Your Partnrr's technology is battle-tested across championships, conferences, academies, media platforms, and community applications *before* it reaches external customers. Enterprise buyers aren't purchasing theoretical software—they're licensing proven operational infrastructure.

The Source Code

Proprietary codebase powering the entire Goddess Phenezy ecosystem

The Data Room

Enterprise-grade data infrastructure for valuation, diligence, and operations

ChurnShield™ Algorithm

Proprietary retention engine reducing subscriber attrition

$163B	0.1%	~$173M
CRM SaaS Market	Target Capture	ARR Potential
Projected by 2030		Annual recurring revenue

Total Addressable Market: The Complete Picture

When viewed in aggregate, the nine market verticals targeted by Goddess Phenezy Holdings represent a combined total addressable market exceeding $11 trillion. Even with ultra-conservative capture rates—ranging from 0.01% to 2%—the projected annual revenue opportunity spans from $8.96 billion to $15.87 billion.

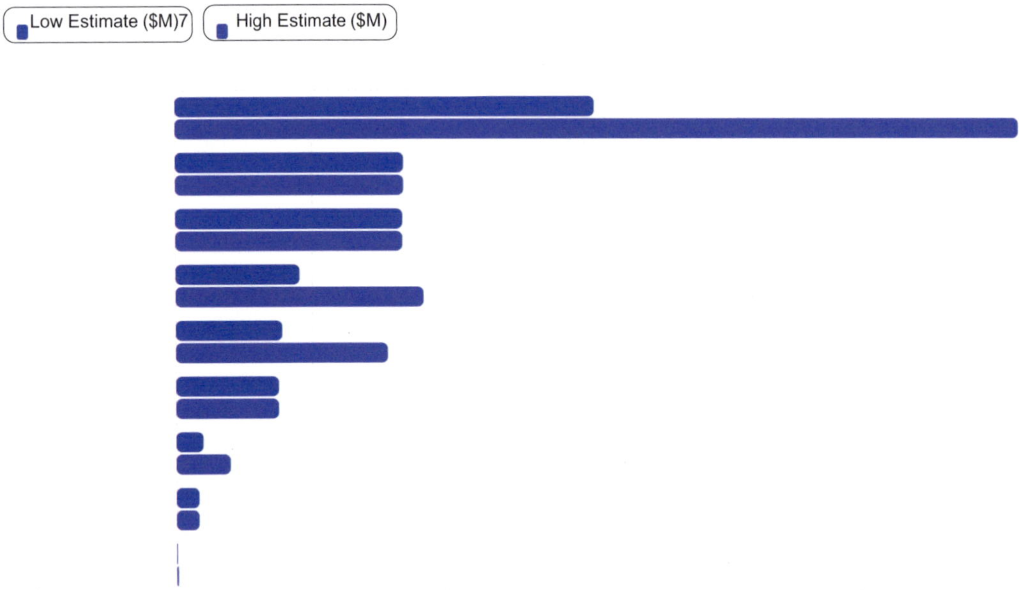

These figures reflect conservative initial capture targets. The compounding dynamics of the ecosystem—where participants, content, and data flow across divisions—create upside potential that linear market models cannot capture. As cross-platform retention deepens and the flywheel accelerates, actual market capture rates have the potential to significantly exceed these baseline projections.

The Opportunity: Market Orchestration at Scale

Goddess Phenezy Holdings is not competing in markets. It is orchestrating them.

Ecosystem Coverage

Nine interconnected global market verticals

Championships

National & International competitive events

The organizational structure presented in this document is not a corporate formality—it is a strategic weapon. By centralizing IP ownership at the parent level, deploying purpose-built subsidiaries across nine high-growth verticals, and connecting them through proprietary technology and cross-platform data flows, Goddess Phenezy Holdings has constructed an architecture designed for compounding value creation.

The market opportunity is not theoretical. Each vertical is anchored by documented market sizing, conservative capture estimates, and strategic advantages that are already in motion. The total addressable market exceeds $11 trillion, with initial revenue projections spanning billions of dollars annually—even at fractional market share.

For investors and strategic partners evaluating this opportunity, the question is not whether these markets are large enough to support the vision. They are. The question is whether the architecture, team, technology, and strategic positioning are sufficient to execute. The evidence presented in this briefing makes the case that they are—and that the compounding nature of the ecosystem creates accelerating returns that single-vertical competitors cannot match.

> The convergence of championships, education, technology, media, health, and SaaS—all on owned infrastructure—creates a moat that grows deeper with every participant, every credential, and every data point captured.

CONFIDENTIAL — INVESTOR INTELLIGENCE BRIEFING

Competitive Intelligence & Strategic Dominance

Goddess Phenezy Holdings, Inc. — The Asymmetric Advantage, Tactical Assessment, Decentralized Growth Protocols & The Athlete Advancement Operating System

This briefing details the full competitive architecture, marketing engine, and athlete advancement infrastructure that positions Goddess Phenezy Holdings not as a market participant — but as the market itself. What follows is a war-room-level dissection of how ecosystem encirclement, zero-acquisition-cost protocols, and sovereign platform design produce compounding, defensible dominance at scale.

Competitive Intelligence — The Asymmetric Advantage

Traditional competitors fight within silos. They build a single application. They license a single platform. They compete for a single audience in a single vertical. That is a symmetric posture — and symmetric warfare is a war of attrition that rewards the deepest pocket, not the sharpest strategy.

Goddess Phenezy Holdings does not fight inside silos. We control the ecosystem. Our competitive posture is not about outspending rivals or outbidding for market share. It is about rendering the competitive question irrelevant by owning every layer of the value chain simultaneously. When you control the territory, you do not need to win battles — you dictate the terms of engagement.

Digital Territory

Proprietary platforms, PWAs, and sovereign digital environments that eliminate dependency on third-party gatekeepers.

Operating Systems

Full-stack SaaS infrastructure — CRM, PRM, Data Room, Automation — owned and operated internally.

Applications

IQ Whizzy, Chess App, and integrated learning platforms embedded natively within the ecosystem.

Media

Xperience TV — a proprietary broadcasting network converting every activation into perpetual content.

Live Execution

National and international championships, conferences, and events executed through owned infrastructure.

> **This is not competition. This is encirclement.**

The Five Pillars of Dominance

The competitive architecture of Goddess Phenezy Holdings is not accidental. It is engineered around five structural pillars — each reinforcing the others, each creating a layer of defensibility that compounds over time. Competitors cannot replicate one pillar without confronting the other four. This is how asymmetric advantage becomes permanent strategic supremacy.

Pillar 1: Sovereignty Over Infrastructure

We own the platforms, the code, the networks. No landlord. No licensing dependency. No kill switch held by a third party. Infrastructure sovereignty means operational permanence.

Pillar 2: Jurisdictional Control — "The" Strategy

Goddess Phenezy Holdings does not operate "in" markets. We establish jurisdictional authority over categories — becoming THE definitive platform, not *a* platform.

Pillar 3: Ecosystem Flywheel Integration

Every product feeds every other product. IQ Whizzy feeds championships. Championships feed Xperience TV. Xperience TV feeds acquisition. Acquisition feeds Your Partnrr. The flywheel never stops.

Pillar 4: Functional Necessity — Missing from the Village

Our platforms fill structural voids in communities. When a platform becomes functionally necessary — when people cannot advance without it — switching costs become infinite.

Pillar 5: The Legacy Protocol — The KJ Doctrine

30% of net profits structurally allocated to scholarships, grants, and community elevation. Growth aligned with institutional trust and ESG mandates.

The KJ Doctrine — ESG as Structural Architecture

Most companies bolt on ESG ((Environmental, Social, and Governance) reporting as a compliance exercise — an afterthought designed to satisfy screening criteria without altering the business model. The KJ Doctrine inverts this entirely. At Goddess Phenezy Holdings, social impact is not a department. It is a structural allocation protocol embedded at the financial engineering level.

Thirty percent of net profits are permanently earmarked for scholarships, community grants, and elevation programs. This is not a pledge. It is not a marketing promise. It is a structural allocation — hard-coded into the financial architecture of the Holdings itself. This means that as the ecosystem scales, the community investment scales proportionally, creating a compounding social return that tracks revenue growth dollar-for-dollar.

For institutional investors evaluating ESG compliance, this is not a checkbox — it is a competitive moat. The KJ Doctrine generates institutional trust, community loyalty, regulatory goodwill, and brand equity simultaneously. Every dollar that flows through the ecosystem automatically funds advancement for the communities it serves.

30%

Net Profit Allocation

Structurally committed to scholarships, grants & community elevation

100%

ESG Alignment

Built into financial architecture, not bolted on as compliance

Tactical Assessment — Force Multipliers & Risk Control

The opportunities ahead are not theoretical. They are activated by assets already owned. The risks are real — but structurally mitigated through jurisdictional control, infrastructure ownership, and decentralized execution. This section delivers the battlefield assessment: what amplifies our advance, and what threats have already been neutralized by design.

This is not a company hoping conditions remain favorable. It is a company engineered to operate regardless of conditions. Economic downturns do not eliminate the need for education, competition, or advancement — they intensify it. Regulatory shifts do not threaten sovereign infrastructure — they threaten companies dependent on third-party platforms. Market volatility does not disrupt owned media — it disrupts companies paying for rented attention.

1

Infrastructure Ownership = Operational Permanence

We do not lease platforms. We do not depend on App Store approvals, algorithm changes, or API deprecations. Our Progressive Web Apps, SaaS operating systems, and broadcasting infrastructure are fully owned and independently deployed.

2

Jurisdictional Control = Regulatory Insulation

By establishing ourselves as THE platform — not a platform — we set the standard rather than comply with someone else's. This creates first-mover regulatory positioning across every vertical we enter.

3

Decentralized Execution = Anti-Fragility

Our growth is not dependent on a centralized sales team or a single geographic market. The 10 × 10 Matrix and Your Partnrr distribute execution across thousands of community members and B2B partners simultaneously.

Investor Implication: Risk is not absent — it is structurally contained. Every identified threat has a corresponding architectural countermeasure already deployed within the ecosystem.

MARKETING & ACQUISITION

Decentralized Growth Protocols

Legacy companies burn capital to buy attention — only to lose users to churn. Goddess Phenezy Holdings does not purchase attention. We command it.

The Zero-Acquisition-Cost Strategy — Overview

The growth model deployed by Goddess Phenezy Holdings represents a fundamental departure from the economics that govern traditional customer acquisition. Legacy companies operate on a simple — and increasingly unsustainable — formula: spend money on advertising to generate awareness, convert a fraction of that awareness into users, then spend more money to retain them. This model bleeds capital at every stage and produces diminishing returns as competition for attention intensifies across every digital channel.

Our model is infrastructure-first, utility-driven, and economically self-propelling. Through proprietary platforms, owned media, and embedded financial incentives, we convert passive audiences into financially motivated network builders, driving exponential expansion while compressing Customer Acquisition Cost (CAC) toward zero. The user does not arrive because we paid for their attention. The user arrives because the ecosystem offers something they cannot get anywhere else — and once inside, the economic architecture rewards them for expanding the network themselves.

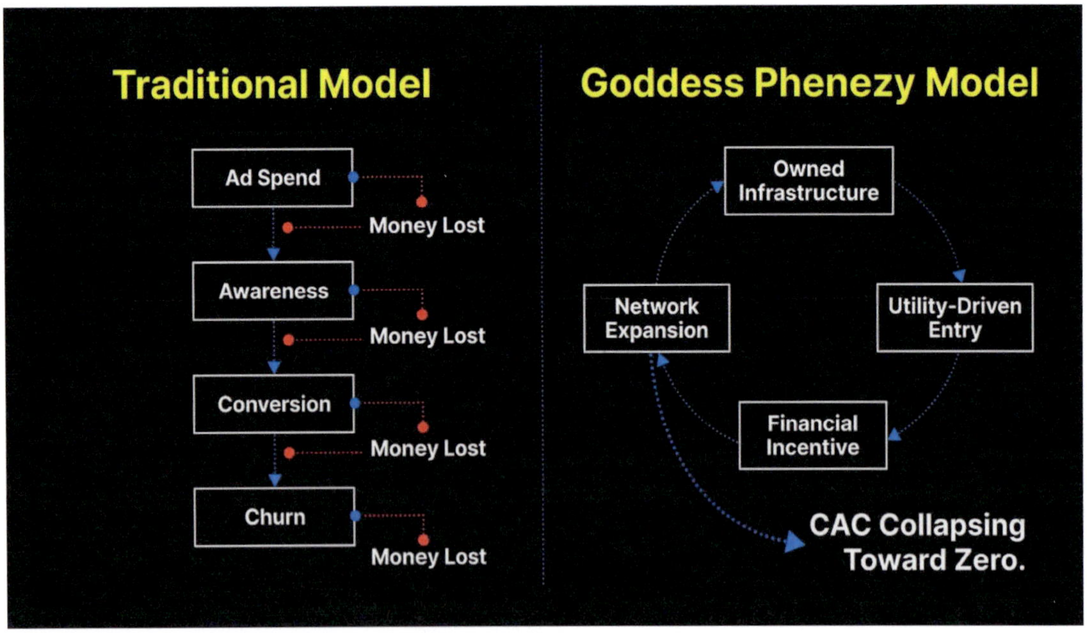

> This is not marketing. This is economic physics.

Brand Positioning — The Sovereign Standard

Xperiences R Us (www.Xperiencesrus.com) is not positioned as an "event company." That framing is a competitive trap — it invites comparison with every local tournament organizer, streaming service, or EdTech startup. We reject that frame entirely.

We are positioned as The Architects of Opportunity. Our brand represents a fundamentally different value proposition — one that transcends any single product category and establishes a sovereign standard against which all competitors are measured. When we define the category, we eliminate the need to compete within it.

Ownership Over Participation

Users do not rent access. They claim territory. Every interaction builds equity within their own digital identity and network position.

Official Jurisdiction

National and international championship authority. Not a third-party event — THE event. Jurisdictional control over competitive categories.

Credentialed Advancement

Verified badges, rankings, and achievement protocols that travel with the user. Progress is credentialed, not anecdotal.

Built-In Wealth Mechanics

Economic incentives embedded at every layer. Users earn as they learn, compete, and build — not after they leave.

When a user enters the ecosystem, they are not purchasing a ticket. They are claiming digital and physical territory inside a sovereign platform designed for growth, ownership, and upward mobility. This is the positioning that makes the brand unassailable — because it is not a product claim. It is an identity.

Ecosystem Infiltration — The Acquisition Funnel Reimagined

Traditional acquisition funnels push users through paid advertisements — renting attention from platforms that control the algorithm, the pricing, and the rules. Every dollar spent on those platforms enriches the platform owner while producing a user whose loyalty is paper-thin. The moment the ad spend stops, the funnel collapses.

Goddess Phenezy Holdings inverted this architecture. We do not push users through ads. We pull users in through high-value assets that live natively inside our platform. Every entry point delivers immediate, tangible value — education, competition, entertainment, economic opportunity — and every entry point connects to every other asset in the ecosystem. There is no external dependency. There is no rented attention. There is only owned infrastructure delivering compounding value.

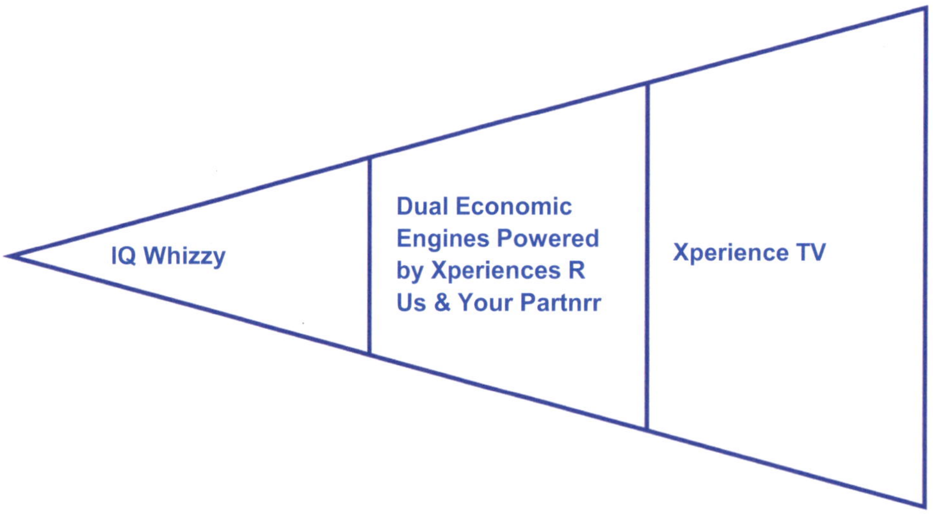

Each stage of this funnel is self-reinforcing. Users entering through IQ Whizzy generate content and competition data that feeds Xperience TV. Xperience TV broadcasts generate awareness that pulls new users into IQ Whizzy. The economic engines at the base convert engaged users into active growth agents — completing the loop without a single dollar of external advertising spend.

IQ
Whizzy

Top of Funnel: IQ Whizzy — The Knowledge Magnet

IQ Whizzy is the strategic tip of the spear — the asset that makes first contact with the broadest possible audience and converts curiosity into ecosystem membership. It is a fully integrated, free learning platform embedded directly inside the Xperiences R Us Progressive Web App (PWA). Not a separate product. Not a partner integration. A natively embedded utility that lives inside the sovereign infrastructure.

The Execution

Youth, parents, students, professionals, and educators enter the ecosystem through IQ Whizzy for free learning, discovery, and skill-building. The platform is designed to deliver immediate value with zero friction — no paywall, no credit card, no complicated onboarding. Once inside, users are immediately and organically exposed to the full ecosystem:

- National & International Championships — competitive pathways become visible
- The Proprietary Chess App — strategic gaming as engagement accelerator
- Xperience TV — live and on-demand content that deepens engagement
- Community Earning Mechanics — the economic incentive layer activates

Strategic Role

IQ Whizzy is not a profit center. It is a pipeline generator that feeds lifetime ecosystem value.

Every learner who enters through IQ Whizzy represents a potential championship competitor, Xperience TV viewer, community matrix member, and B2B SaaS subscriber. The cost of their entry is zero. The lifetime value of their participation compounds across every platform in the Holdings.

THE
INTERNATIONAL
BASKETBALL
CHAMPIONSHIP
XPERIENCE TV

Mid Funnel: Xperience TV — The Broadcast Engine

Xperience TV is the proprietary, fully owned digital broadcasting network that transforms every live activation into a perpetual marketing asset. This is not a YouTube channel. This is not a Twitch partnership. This is sovereign media infrastructure — owned, operated, and monetized entirely within the ecosystem.

The execution is precise: Xperience TV distributes the full Holdings portfolio to engaged audiences across the globe, converting operational events into 24/7 engagement without third-party platforms or advertising spend. Every championship final, every conference keynote, every educational module becomes content that works for the ecosystem indefinitely.

Championship Finals

Poker, Chess, Sports — every competitive event is broadcast live and archived as on-demand content. Audiences grow with every championship cycle.

AI, STEM, Health & Leadership Conferences

Thought leadership content that positions the ecosystem as the intellectual hub for advancement-oriented communities.

Educational Programming

IQ Whizzy content repurposed for broadcast — extending the learning platform's reach while feeding new users back into the PWA.

Force Multiplier: Every live activation becomes a perpetual marketing asset, converting operations into 24/7 engagement — without third-party platforms or advertising spend.

THE DUAL ECONOMIC ENGINE

B2C & B2B Forced Matrices

We have eliminated weak referral programs in favor of a free, sovereign, dual-economic system. We do not hire sales teams. We enrich our community.

Central to this economic model is our free, proprietary 10x10 forced matrix affiliate program. "Forced matrix" implies an automated placement system: as new members join, they are automatically placed into the next available position within the matrix, ensuring equitable distribution and fostering growth for everyone. This includes "spillover" from upline activity, meaning members can benefit from the growth of the overall community, not just their direct referrals.

The "10x10" structure refers to its depth and breadth: each member can directly refer up to 10 individuals on their first level, and the system extends 10 levels deep. This geometrically expanding structure creates a powerful network effect. Furthermore, the program is "free" to join, removing any financial barrier to entry and maximizing participation, which is critical for rapid expansion and community engagement. Members earn directly from the activity of those they refer and indirectly through the spillover generated by their upline and downline within this deep matrix.

This dual-engine system harmonizes B2C consumer engagement with B2B business development. The consumer side drives organic community growth and content consumption, while the B2B side integrates specialized services and partnerships, creating a mutually reinforcing ecosystem. This innovative approach generates exponential growth and ensures that wealth is circulated and built within the community, establishing a truly sovereign and self-sustaining economic infrastructure.

B2C & B2B Forced Matrices

We have eliminated weak referral programs in favor of a free, sovereign, dual-economic system. We do not hire sales teams. We enrich our community.

Engine 1: The Community 10 × 10 Forced Matrix (B2C)

The Community 10 × 10 Forced Matrix is the growth engine that transforms every user into an economically motivated network builder. Traditional referral programs offer weak, one-time incentives that generate tepid results. The 10 × 10 Forced Matrix is fundamentally different — it is a structured compensation architecture embedded directly inside the platform that creates compounding financial returns for community members who actively expand the ecosystem.

Revenue Triggers

Digital token purchases, Championship entry fees, VIP upgrades — every transaction activates the matrix.

Behavioral Outcome

Members are financially incentivized to invite others, train networks using IQ Whizzy, and promote events and content.

Exponential Growth

Community-driven expansion while redistributing value back to participants instead of external advertisers.

The result is mathematically elegant: Customer Acquisition Cost collapses toward zero. The community does not need to be "convinced" to grow the ecosystem. They are structurally rewarded for doing so. Every new member they bring in generates revenue that flows back through the matrix — and every member those new members bring in compounds the returns further. This is not a referral bonus. This is an economic system that aligns individual incentives with ecosystem expansion.

Engine 2: Your Partnrr Forced Matrix — B2B Wealth Creation

While the Community Forced Matrix monetizes participation, Your Partnrr (www.YourPartnrr.com) monetizes infrastructure. This is the B2B engine that transforms the ecosystem from a consumer platform into an enterprise-grade operating system — and it does so without a single traditional enterprise sales call.

The Asset

Your Partnrr is a full SaaS Operating System encompassing CRM, PRM, Data Room, Docs Valuation, and Automation capabilities. It launches with a free entry tier that includes real training, real tools, real certifications and real earning potential. This is not a freemium trap. It is a genuine onramp into business infrastructure that delivers value from day one. We have no fees!

The Execution

We bypass traditional enterprise sales entirely. Users are empowered to:

01

Learn Business Systems

Integrated training within the platform teaches users to deploy and operate enterprise-grade software — no prior experience required.

02

Deploy the Software

Users activate Your Partnrr for their own businesses, agencies, or organizations — immediately deriving operational value.

03

Onboard Founders, Agencies & Operators

Empowered users become distribution channels — bringing new B2B subscribers into the platform through direct experience and results.

04

Earn Recurring Commissions

Commission structures across SaaS tiers: $99 / $299 / $499 per month. Recurring revenue shared with the network that drives adoption.

Your Partnrr — The Wealth Creation Reality

The structural power of Your Partnrr deserves explicit articulation for this investor audience. This is not affiliate marketing dressed up in new language. This is work-based wealth creation at scale — and the mathematics are real.

The Mathematical Reality

Your Partnrr's free system — combined with the 10 × 10 forced matrix — has the mathematical potential to create millionaires based entirely on execution, discipline, and network leadership. The tiered SaaS pricing creates recurring revenue streams that compound across the network:

$99/mo
Starter Tier

$299/mo
Professional Tier

$499/mo
Enterprise Tier

Structural Differentiation

Every commission is tied to a real software subscription delivering real business value. Users earn because they brought genuine utility to a business — not because they convinced someone to click a link. The value chain is transparent, the outcomes are measurable, and the earning potential scales with effort and network quality.

This is what separates Your Partnrr from every affiliate program in the market: the product is an operating system, the training is real, and the wealth creation is work-based.

The Dual Engine — Flywheel Visualization

The two economic engines do not operate in isolation. They are structurally interconnected — each feeding the other in a compounding loop that accelerates as the ecosystem grows. B2C participants evolve into B2B operators. B2B operators generate infrastructure that attracts more B2C participants. The flywheel gains momentum with every revolution.

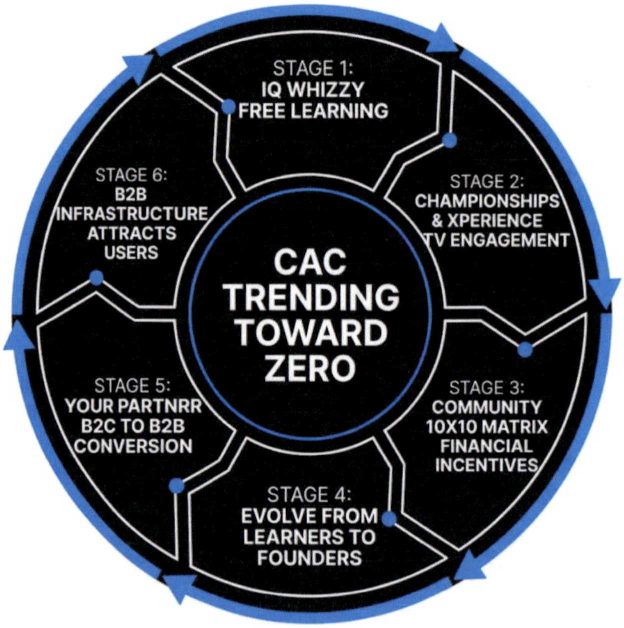

This flywheel is the core economic engine of the Holdings. It explains why growth accelerates rather than decelerates, why margins expand rather than compress, and why competitors cannot simply "copy" a feature — because the competitive advantage is not a feature. It is the system architecture itself.

Sales & Lifetime Value (LTV) Maximization

Acquisition is only the beginning. The ecosystem is engineered for continuous value expansion — systematically deepening engagement and increasing monetization at every stage of the user lifecycle. Once a user enters the ecosystem, they are guided through an ascension architecture designed to convert initial curiosity into permanent economic participation.

1 Direct Monetization

Event entries, digital tokens, merchandise, premium content — immediate revenue from day-one participation.

2 VIP Ascension Path

Systematic guidance toward VIP tiers unlocking premium Xperience TV access, advanced IQ Whizzy modules, and exclusive championships.

3 B2B Transition

As users evolve from learners and competitors into founders and operators, they are up-leveled into Your Partnrr — converting B2C participants into high-LTV B2B SaaS subscribers.

4 Flywheel Complete

B2B operators generate infrastructure and content that attracts new B2C users, completing the growth loop and compounding ecosystem value.

Strategic Takeaway — Marketing as Economic System

The Truth

Marketing at Goddess Phenezy Holdings is not a department.

It is a distributed economic system.

Every user becomes a learner, a participant, a broadcaster, and a potential partner.

This is not customer acquisition.

This is infrastructure-driven expansion.

Learner

Enters through IQ Whizzy with zero cost, zero friction. Immediate value delivery begins.

Participant

Competes in championships, consumes Xperience TV content, builds athlete identity.

Broadcaster

Shares content, invites networks, activates the 10 × 10 Matrix for economic returns.

Partner

Transitions into Your Partnrr, deploys SaaS infrastructure, earns recurring commissions.

ATHLETE ADVANCEMENT ENGINE

From Discovery to Destiny

Competition is the entry point. Advancement is the product.

Athlete Advancement Engine — Overview

At its core, Xperiences R Us exists to advance athletes — not just crown winners. This distinction is not philosophical. It is architectural. It defines every product decision, every feature prioritization, and every revenue model within the competitive ecosystem.

Traditional sports systems extract value from athletes while offering fragmented, opaque pathways to opportunity. Athletes compete, but they do not own their data, their visibility, or their trajectory. Advancement depends on geography, politics, or luck. The system is rigged to benefit intermediaries — agents, scouts, tournament operators — while the athletes themselves remain commoditized inputs.

We replaced that system with infrastructure. Xperiences R Us is not a tournament operator. It is an Athlete Advancement Operating System that transforms participation into verified progress, visibility, and opportunity. Every competition generates data. Every data point feeds discovery. Every discovery creates advancement. The athlete is not a product to be extracted from. The athlete is the sovereign owner of their own trajectory.

Identity

Sovereign digital profiles owned by athletes

Network

Teams, organizations, legacy signaling

Wellness

Recovery, longevity, family protection

Performance

AI-quantified metrics and rankings

Data Pipeline

Competition as instrumented advancement

Recruitment

Scholarships, NIL, draft system

Athlete Identity & Ownership — The Foundation

Every athlete inside Xperiences R Us is issued a living digital identity — not a static profile. This is the foundational layer upon which every other advancement feature is built. Without sovereign identity, athletes remain dependent on third parties to tell their story. With sovereign identity, athletes control their narrative, their exposure, and their trajectory.

Athlete Profile & Identity Layer

Athletes build a verified, portable profile that travels with them across competitions, teams, and seasons. This profile captures the complete picture of the athlete — sport type, position, height/weight, age group, school, GPA, and graduation year. It is not a social media page. It is a credentialed identity document that gains authority with every verified competition, every recorded stat, and every scout interaction.

Verified Badges

Coach and scout verification badges authenticate the athlete's credentials and training history.

Shareable QR Code

Instant profile access with scan tracking — athletes can share their identity at any event, any meeting, any opportunity.

Downloadable Resume PDF

Professional-grade athletic resume generated directly from verified platform data.

AI Bio Generator

AI-powered biography generation with version history — the athlete's story told with precision and impact.

> Strategic Advantage: Athletes own their profile, their narrative, and their exposure. No other competition ecosystem offers this level of identity sovereignty.

Performance Intelligence & Visibility — The Signal

Identity without visibility is a tree falling in an empty forest. Xperiences R Us converts performance into machine-readable signal — quantified, ranked, and visible by design. The platform does not wait for a scout to happen to be in the right gym on the right night. It ensures that every athlete's performance is captured, analyzed, and surfaced to the people and institutions that matter.

Skills DNA™ & Metrics Engine

AI-generated 6-attribute athletic profiling creates a comprehensive performance fingerprint for every athlete in the ecosystem. This is not a subjective coach evaluation. It is an algorithmically generated assessment based on verified competition data.

Speed	Agility
Strength	Endurance
Coordination	Flexibility

Analytics Dashboard

- Real-time stat tracking per championship and season
- Profile views — who is watching
- Scout views — verified institutional interest
- Draft rank — competitive positioning
- Hype Score — market momentum indicator

Real-Time Hype Meter™

Live performance tracking with leaderboard placement that creates real-time visibility during competition. Athletes are no longer "undiscovered." Their performance is quantified, ranked, and visible by design.

Competition as Data, Not Just Trophies

Every championship inside Xperiences R Us is instrumented for advancement. This means that the competition itself is not the end product — it is a data generation event that feeds the entire advancement pipeline. Registration, brackets, standings, match history, and game-by-game stat tracking are all captured, structured, and made actionable.

Championships → Data Pipelines

Every competition generates structured data: registration flows, bracket progressions, standing histories, match records, and granular game-by-game stat tracking. This data is very persistent, searchable, and feeds every downstream system in the platform.

1v1 Challenge System

Peer-to-peer challenges with shareable invite codes enable athletes to generate performance data outside of organized championships. Every challenge is recorded, tracked, and integrated into the athlete's living profile.

Talent Draft System

Athletes are surfaced into the Talent Draft, where scouts, teams, investors, and sponsors discover ranked talent, draft athletes into teams or opportunities, and track development over time. Competition outcomes directly fuel recruitment, drafting, and monetization — not just applause.

Strategic Advantage: Competition outcomes directly fuel recruitment, drafting, and monetization — not just applause. Every game played is an advancement event.

Recruitment, Scholarships & NIL — The Conversion

This is where the Athlete Advancement Engine converts performance data into real-world opportunity at scale. Xperiences R Us collapses years of guesswork, cold outreach, and opaque recruiting processes into a continuously updated opportunity feed powered by AI-weighted matching logic.

Scholarship Matching Engine

Athletes are matched to real scholarship opportunities using a proprietary AI-weighted logic model that evaluates three dimensions simultaneously:

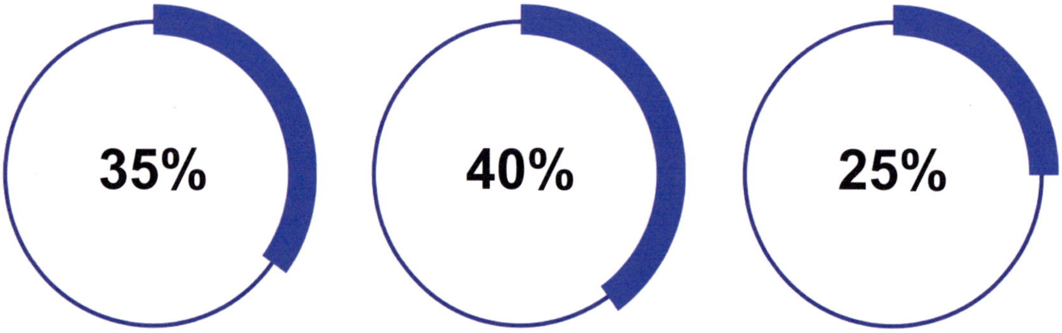

Academic Weight

GPA, test scores, academic trajectory, and institutional fit

Athletic Weight

Skills DNA™ profile, championship performance, draft rank, and verified metrics

Geographic & Program Fit

Location preferences, program culture, and competitive division alignment

The engine supports NCAA D1/D2/D3, NAIA, independent, and international programs — covering the full spectrum of competitive collegiate athletics. This is not a directory. It is an active matching system that surfaces opportunities the athlete may never have discovered on their own.

Talent Engine & NIL Opportunities

The platform auto-generates college matches, tryouts, NIL deal opportunities, and verified scout views. For athletes navigating the increasingly complex NIL landscape, the platform provides structured pathways to monetize their name, image, and likeness while maintaining compliance — all within the sovereign ecosystem.

Holistic Athlete Support — Longevity, Not Burnout

Athlete advancement is not sustainable without health. The sports industry is littered with stories of talented athletes who burned out, broke down, or were exploited because the systems around them prioritized extraction over elevation. Xperiences R Us rejects that model entirely. We built health and protection infrastructure directly into the advancement architecture.

Recovery & Wellness Infrastructure

Recovery Tracks (solo or pod-based), streak tracking (current, longest, lifetime), and guided onboarding into recovery programs. Athletes are not just competing — they are being supported in sustaining their competitive capacity over years, not weeks.

Parent & Guardian Protections

For minor athletes, the platform provides parent dashboards with scout-view alerts, privacy controls, contact gating, and verified account linking. Families are not locked out of the process — they are given tools to protect and guide their athlete within a transparent system.

> **Strategic Advantage: We advance athletes *and* protect families — an institutional differentiator that drives trust, reduces churn, and creates a platform that parents actively endorse.**

Teams, Organizations & Legacy Signaling

Athletes don't advance alone. Every breakthrough is built on a network of teams, coaches, organizations, and institutional affiliations. Xperiences R Us built the network layer that transforms individual achievement into credentialed signal — visible, verifiable, and persistent.

Network Features

- **Team rosters with win/loss tracking across seasons**
- **League, school, club, academy affiliations — the full organizational graph**
- **Searchable athlete directory with verified views and contact protocols**

Every organizational connection an athlete makes is captured and displayed within their living profile. Teams gain institutional credibility. Coaches gain verified track records. The entire network benefits from transparent, data-driven accountability.

Awards & Achievement

Champion Badges

Verified championship credentials

XP Progression

Continuous advancement tracking

Verified Achievements

Credentialed, not anecdotal

Strategic Advantage: Achievement becomes credentialed signal, not social media noise. When a scout, sponsor, or institution evaluates an athlete from Xperiences R Us, they are evaluating verified data — not curated highlights.

The Economic Imperative: Why Xperiences R Us Dominates

The Athlete Advancement Engine isn't just a feature set; it's the foundational economic logic driving Xperiences R Us's unparalleled scalability, retention, and monetization. For institutional investors, this translates directly into a compelling, undeniable growth trajectory:

Athlete Retention: Unlocking Perpetual LTV

When our platform actively fuels an athlete's career—not merely hosts their competition—switching costs become insurmountable. Athletes remain intrinsically tied to a system that owns their performance data, identity, scholarship matches, and draft positioning. This retention compounds lifetime value across every revenue stream in the ecosystem.

Parental Trust: Eradicating Churn, Fueling Advocacy

Comprehensive parent dashboards, robust privacy controls, proactive scout-view alerts, and stringent guardian protections cultivate institutional trust. This eliminates the primary churn driver in youth sports: parental anxiety. When families feel secure and empowered, they transform into unwavering advocates, not skeptics.

Data Dominance: Forcing Institutional Adoption

As Xperiences R Us rapidly accumulates verified performance data, proprietary Skills DNA™ profiles, and authoritative draft rankings, it becomes the undisputed discovery tool for scouts, recruiters, and institutional programs. Network effects in data forge winner-take-all dynamics, cementing our market leadership.

Predictable Revenue: Sponsors Back Verified Achievement

Sponsors investing within our ecosystem commit to measurable, verified athlete advancement—not fleeting viral moments or unproven hype. This cultivates sustainable, high-LTV sponsorship relationships built on tangible progress and drastically reduced risk, securing long-term revenue streams.

The Competitive Moat — Why Competitors Cannot Copy This

The question every institutional investor must answer is: what prevents replication? The answer is architectural. Xperiences R Us does not have a single feature that creates its advantage. It has a system architecture where each component depends on and amplifies every other component. Replicating one feature delivers no competitive advantage without replicating the entire ecosystem.

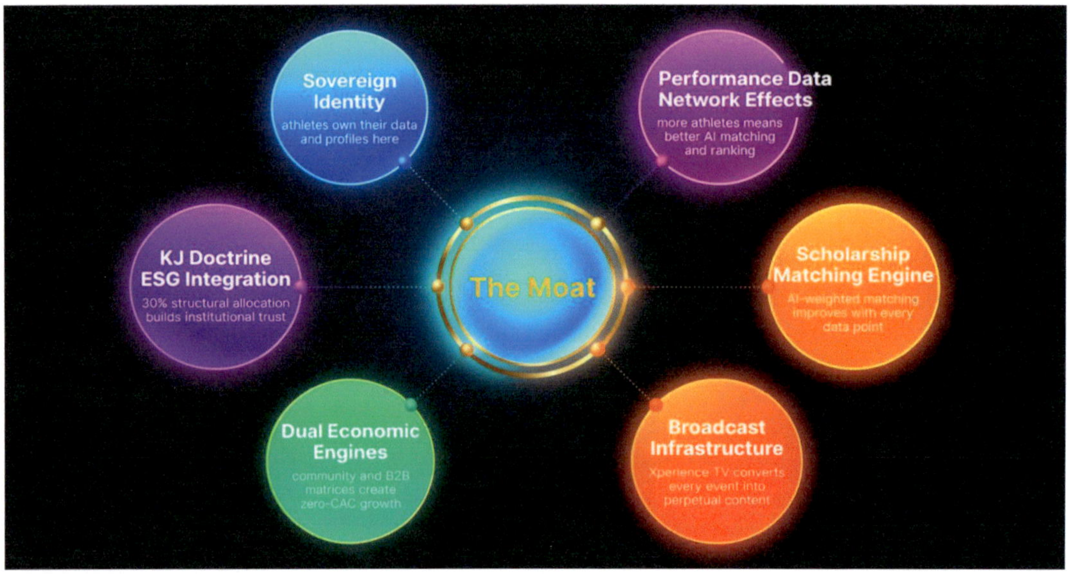

A competitor can build a tournament platform. A competitor can build a streaming service. A competitor can build a SaaS CRM. But no competitor can simultaneously build all of these, integrate them into a single sovereign ecosystem, embed dual economic engines, issue living digital identities, deploy AI-powered scholarship matching, and structurally allocate 30% of profits to community elevation — while already having the network effects of an active, growing user base. The moat is the system.

The Strategic Truth

Conventional Platforms:

Most platforms merely host competitions. Some stream content. Others sell software. They operate within existing markets, competing for incremental gains.

Xperiences R Us:

We fundamentally change the game. Xperiences R Us doesn't just host or stream; we architect the advancement of human potential at scale.

This isn't a feature. This is the Moat

Our Unassailable Advantage

Our unique system architecture integrates every component into a self-reinforcing ecosystem, creating a barrier to entry that is architectural, not merely additive. It's a holistic synergy where every data point, every connection, and every experience amplifies the entire network.

Holistic Ecosystem

Seamless integration of identity, performance data, scholarship matching, and broadcast infrastructure.

Dual Economic Engines

Community-driven growth combined with B2B solutions for exponential, zero-CAC expansion.

Impactful ESG Integration

A structural 30% allocation of profits dedicated to community elevation, building unparalleled trust and loyalty.

Goddess Phenezy Holdings, Inc. is not competing for market share.

Operations & Infrastructure — Decentralized Command

How Goddess Phenezy Holdings, Inc. governs a global championship ecosystem through code, licensing, data, and media — not headcount.

The Infrastructure-First Operating Model

Traditional event companies scale linearly. To host more events, they hire more staff, manage more logistics, and absorb more risk. Overhead rises faster than revenue. Every new market, every new championship, every new season compounds the burden on the balance sheet without a proportional lift to the bottom line. This is the structural ceiling that has constrained the competitive event industry for decades — and it is the ceiling that Goddess Phenezy Holdings, Inc. was engineered to shatter.

Goddess Phenezy Holdings, Inc. scales exponentially.

We do not operate as an event management company. We operate as a Sovereign Infrastructure Provider. Rather than manually managing 79+ National & International Championships, the Holding Company governs the intellectual property, software, standards, data, and media, while execution is decentralized to licensed Ambassadors operating inside a controlled, platform-enforced system. The model separates ownership of high-leverage assets from the labor-intensive realities of local execution — capturing the economics of both a SaaS platform and a franchise network without the fragility of either.

Operations are not people-heavy. Operations are code-driven, license-based, and data-instrumented. Every registration, every bracket, every payout, every content clip, and every athlete advancement event flows through proprietary systems that the Holding Company owns outright. This means margins improve as volume increases, defensibility compounds with each new data point, and competitors face not a single product to replicate but an entire operating system.

> **Core Thesis: When infrastructure is owned rather than rented, every incremental event, athlete, and ambassador strengthens the moat instead of diluting it.**

Linear vs. Exponential: The Scaling Divide

The fundamental distinction between traditional event operators and the GPH model is not one of degree — it is one of kind. Understanding this divide is essential for evaluating the operational thesis and the long-term capital efficiency of the enterprise.

Traditional Event Company

- Revenue scales with headcount
- Each market requires new staffing
- Third-party vendor dependencies
- Overhead rises faster than revenue
- Data is fragmented or lost

GPH Sovereign Infrastructure

- Revenue scales with licensing & code
- Each market activates via Ambassador
- Fully owned platform stack
- Marginal cost approaches zero
- Data compounds into defensible asset

This architecture means that the 79th championship costs a fraction of the first to operate, while generating equivalent — or greater — revenue. The Holding Company captures value at every layer: licensing fees at activation, SaaS subscriptions at operation, transaction royalties at settlement, and media revenue at distribution. The Ambassador bears the physical execution risk. The platform captures the data. The IP compounds.

Centralized Governance — The Holding Company

Parent Entity: ***Goddess Phenezy Holdings, Inc.*****Role: Central command and asset owner**

The Holding Company retains absolute control over all high-leverage infrastructure. This is the Command Layer — the nucleus from which all standards, rules, software, media, and intellectual property emanate. While execution is deliberately distributed to the field, governance is deliberately concentrated at the center. This asymmetry is by design: it ensures consistency of experience, defensibility of assets, and scalability of operations regardless of how many markets, sports, or verticals the ecosystem enters.

The Holding Company does not manage events. It manages the system that makes events possible. It does not employ event staff at scale. It employs engineers, IP strategists, data architects, and media operators who build and maintain the rails on which every Ambassador, athlete, and sponsor operates. This is the structural reason GPH can govern 79+ championships without proportional headcount growth — and why every new championship added to the network strengthens rather than strains the core.

This corporate structure extends to liability protection. Xperiences R Us, Inc. and Technologies R Us, (www.technologiesrus.com) Inc. serve as the public-facing operating subsidiaries that conduct business directly with customers, vendors, and the public. This corporate structure creates a legal firewall: the Holding Company (Goddess Phenezy Holdings, Inc.) owns the high-value assets (IP, code, domains, trademarks, media libraries) but leases them to these subsidiaries for operational use. Because the subsidiaries are separate legal entities, any operational liabilities—lawsuits, contractual disputes, event-related claims—are contained within the subsidiary level and cannot pierce through to attach the Holding Company's core assets. This is a standard asset protection strategy used by sophisticated holding companies: the parent retains ownership and control while insulating itself from the operational risk exposure of day-to-day business activities.

Core Governance Functions

The Holding Company's governance mandate spans five critical domains. Each function represents a layer of the moat — individually significant, collectively impenetrable.

1

Ownership and maintenance of 578,000+ lines of proprietary source code across Xperiences R Us PWA, Athlete Advancement Engine, Chess App, IQ Whizzy, and Your Partnrr + Data Room. This codebase is the operational backbone — every registration, every bracket, every payout, every data capture event runs through code the Holding Company owns outright.

2

Jurisdictional Control

Ownership and licensing of 450+ premium domains, including exclusive "The National" and "The International" naming rights across sports, gaming, education, and conferences. These domains are not vanity assets — they are jurisdictional claims that define market positioning and prevent competitive encroachment.

3

Standards & Rulemaking

Official championship formats, eligibility rules, advancement criteria, verification standards, and dispute resolution protocols. Uniformity of rules enables portability of athlete credentials across every GPH-sanctioned event on the planet.

4

Media Sovereignty

Ownership and operation of Xperience TV and GPH Publishing, ensuring all content, data, and distribution remain internal. No third-party platform dictates reach, monetization, or algorithmic visibility.

> **Result: Stability, defensibility, and repeatability across every xperience, regardless of location.**

Proprietary Code Arsenal

The 578,000+ lines of proprietary source code represent the most capital-efficient asset in the portfolio. Unlike physical infrastructure that depreciates, software assets appreciate with every iteration, every user, and every data point captured. Below is the platform portfolio that constitutes the operational backbone of the entire ecosystem.

Xperiences R Us PWA

The unified athlete-facing portal — registration, profiles, competition history, and advancement tracking in one progressive web application.

Athlete Advancement Engine

The intelligence layer that transforms raw competition data into verified advancement signals, rankings, and recruitment readiness scores.

Chess App

Proprietary competition and training platform purpose-built for GPH-sanctioned chess championships and cognitive development programs.

IQ Whizzy

Free-access educational platform driving engagement and retention by combining learning with competitive gamification and advancement credit.

Your Partnrr + Data Room

The Ambassador operating system — CRM, payments, reporting, compliance, and communication managed through a single enforced interface.

Domain & Jurisdictional Authority

Intellectual property in the digital era is defined not only by patents and trademarks but by the domains that anchor brand identity and market perception. GPH's portfolio of 450+ premium domains represents a jurisdictional moat that competitors cannot replicate or circumvent.

The "National" & "International" Framework

By securing exclusive naming rights for "The National" and "The International" across sports, gaming, education, and conferences, GPH has effectively claimed the semantic high ground. When parents, athletes, sponsors, or media search for authoritative national or international championships, GPH domains dominate the landscape.

These domains are not passive holdings. Each one is a potential activation point — a new championship vertical, a new conference series, a new educational program — that can be licensed to an Ambassador and launched with minimal marginal cost.

- Sports: National championships across 40+ disciplines
- Gaming: Esports and cognitive competition verticals
- Education: Academic olympiads and STEM competitions
- Conferences: Leadership, coaching, and industry summits

450+

Premium domains owned

79+

Championship verticals active or licensed

Zero

Third-party naming dependencies

Decentralized Execution — The Ambassador Protocol

Physical execution is decentralized without losing control. This is the Field Layer — the operational complement to the Command Layer. While the Holding Company governs the system, Ambassadors run the ground game. This separation is what allows GPH to enter new markets, new sports, and new geographies without proportional increases in headcount, overhead, or operational risk.

The Ambassador Protocol is not a loose affiliate program. It is a tightly controlled licensing framework in which every local operator is bound to GPH-owned brands, domains, rules, platforms, and data standards. The Ambassador brings local knowledge, relationships, and execution capacity. The Holding Company provides the infrastructure, brand authority, and technology that make that execution scalable, standardized, and monetizable.

The Ambassador Model

Local leaders, organizations, and event directors acquire Master Licenses to operate within the GPH ecosystem. This is not a franchise in the traditional food-service sense — it is closer to an operating concession within a sovereign digital infrastructure. The Ambassador model is designed to attract entrepreneurial operators who want the brand equity, technology stack, and athlete network of a national platform without the capital requirements of building one from scratch.

License Acquisition

Ambassador secures Master License for a specific region, sport, or vertical under GPH-controlled brands and domains.

Platform Onboarding

All operations are configured within Your Partnrr — registration, brackets, payments, CRM, and compliance are pre-built.

Local Execution

Ambassadors operate regional qualifiers, national events, conferences, and academies using GPH infrastructure.

Data & Revenue Flow

All data, media, and transaction records flow back to the Holding Company — fueling the ecosystem and generating royalties.

Ambassadors operate three primary event categories: regional qualifiers that feed into national brackets, national events that crown champions and generate media content, and conferences and academies that drive education, networking, and sponsor engagement. Each category operates under GPH-controlled brands, domains, rules, and platforms — ensuring that regardless of who executes, the athlete experience, data capture, and brand integrity remain uniform.

Forced Platform Adoption

The Ambassador Protocol's most critical design feature is not the license itself — it is the mandatory platform adoption clause. Every Ambassador is required to operate exclusively within the owned ecosystem. There are no approved third-party alternatives. This is not a suggestion or a best practice — it is a contractual and technical enforcement mechanism that ensures operational uniformity, data integrity, and revenue capture at every level of the network.

Registration & Eligibility Enforcement

Every athlete must register through the platform. Eligibility is verified automatically against GPH standards before any competition entry is confirmed.

Bracket Generation & Stat Capture

Brackets are generated algorithmically. Every match result, stat line, and performance metric is captured in real time and stored in the athlete's permanent profile.

Payments, Payouts & Reporting

All financial transactions — entry fees, sponsor payments, Ambassador payouts — flow through the platform. No off-platform payments are permitted.

CRM, Communications & Follow-Ups

Post-event engagement, athlete retention campaigns, parent communications, and sponsor reporting are automated through Your Partnrr — ensuring no relationship falls through the cracks.

All of the above is automated via Your Partnrr (www.YourPartnrr.com)

The result is a closed-loop operating environment where data never leaves the ecosystem, revenue never leaks to third parties, and every interaction becomes a compounding data asset owned by the Holding Company.

Operational Economics

The Ambassador Protocol creates a revenue architecture in which the Holding Company captures value at multiple extraction points while distributing the cost and risk of physical execution to local operators. This is the structural advantage that enables exponential scaling.

Holding Company Earns

Licensing Fees

Upfront and recurring fees for Master License rights to operate under GPH brands, domains, and standards.

SaaS Subscriptions

Monthly platform access fees for Your Partnrr, bracket engines, CRM, and all operational tooling.

Transaction & Media Royalties

Percentage-based royalties on all registrations, payments, and media monetization flowing through the ecosystem.

Ambassadors Execute

Local Operations

Venue selection, staffing, logistics, and on-the-ground event management remain entirely in the Ambassador's domain.

Physical Risk

Insurance, liability, local compliance, and staffing risk are borne by the Ambassador — not the Holding Company.

Market Development

Ambassadors invest in building local athlete communities, sponsor relationships, and brand awareness within their territories.

> We govern the system. Ambassadors run the ground game.

The Platform Stack — A Single Operating Environment

Operations are not fragmented across third-party tools. The entire ecosystem is housed inside a fully integrated Progressive Web App (PWA). This architectural decision is not merely a technology preference — it is a strategic imperative. By consolidating every operational function into a single owned environment, GPH eliminates vendor dependencies, reduces integration failures, ensures data continuity, and captures the full economic value of every transaction and interaction that occurs within the ecosystem.

The PWA architecture means the platform is accessible on any device, any browser, and any operating system without requiring app store approval or distribution. Athletes access it on their phones. Parents access it on their tablets. Ambassadors access it on their laptops. Scouts access it from anywhere in the world. One codebase. One login. One unified experience. This is the operational spine of the entire enterprise.

Unified Access — One Login, Total Ecosystem

The power of the platform stack is not in any single feature — it is in the integration of all features under a single authentication layer. One login unlocks the entire GPH universe, creating an experience that is seamless for users and extraordinarily valuable for the Holding Company's data strategy.

Athlete Profiles & Advancement Dashboards

Verified identity, competition history, Skills DNA™, rankings, and recruitment readiness — all in one persistent profile that grows with the athlete.

Championship Registration & History

Seamless entry into any GPH-sanctioned event with automatic eligibility verification, bracket assignment, and historical record keeping.

IQ Whizzy Learning Platform

Free educational content that drives engagement, retention, and cognitive development — keeping athletes active in the ecosystem between competition seasons.

Proprietary Chess App

Purpose-built competition and training environment for GPH chess verticals, fully integrated with advancement tracking and tournament infrastructure.

Xperience TV

Owned media distribution channel delivering championship finals, athlete spotlights, educational content, and conference keynotes directly to the audience.

Token Economy & 10×10 Matrices

Proprietary engagement and incentive systems that reward participation, advancement, and community contribution with tokenized value.

Role-Based Dashboards

Tailored interfaces for parents, scouts, coaches, and operators — each seeing exactly the data and tools relevant to their role in the ecosystem.

Zero Vendor Lock-In

The most consequential operational decision GPH has made is the decision to own rather than rent its technology infrastructure. In an era where most organizations are dependent on SaaS vendors for critical operations — ticketing through Eventbrite, CRM through Salesforce, streaming through YouTube — GPH has built and owns every layer of its stack. The strategic implications of this choice cannot be overstated.

No External Ticketing

Registration, ticketing, and entry management are built into the PWA. No dependency on platforms that can change pricing, terms, or features without notice.

No Third-Party CRM

Athlete, parent, sponsor, and Ambassador relationships are managed inside Your Partnrr. No data exports. No integration breakdowns. No vendor migrations.

No Streaming Providers

Xperience TV is owned media. No algorithmic suppression, no ad revenue sharing with platforms, no risk of content demonetization or removal.

Operational continuity is guaranteed by ownership. When you own the code, you own the roadmap. When you own the data, you own the leverage. When you own the distribution, you own the audience. GPH operates on this principle at every layer.

Platform Architecture Overview

The following diagram illustrates how the platform stack connects the Command Layer, the Field Layer, and the end users within a single closed-loop operating environment. Every data point, transaction, and interaction flows through owned infrastructure.

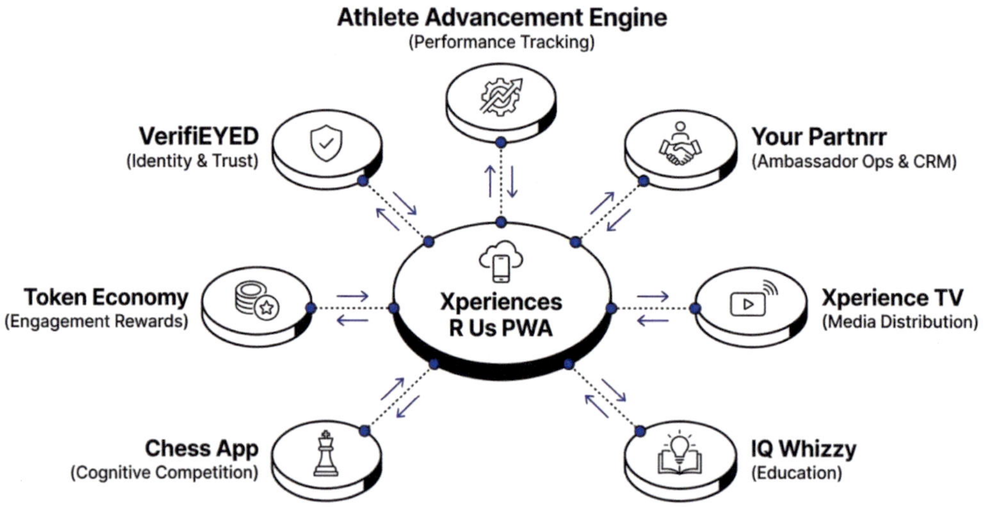

This architecture ensures that no data leaves the ecosystem, no revenue leaks to third parties, and every interaction — whether an athlete registering for a competition, an Ambassador processing a payment, or a scout reviewing a prospect — compounds the value of the platform and the defensibility of the Holding Company's position.

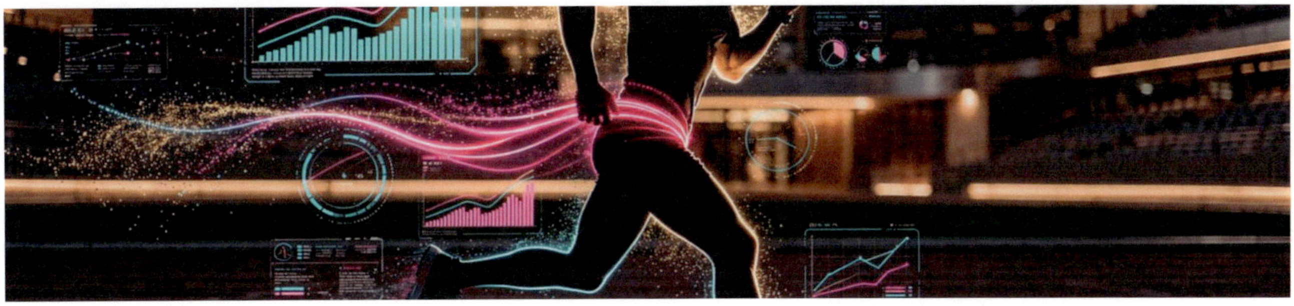

Athlete Advancement as Operations

Advancing athletes is not a byproduct of operations. It is the operation.

1

Data-Driven Advancement

Every championship and challenge within the GPH ecosystem is meticulously instrumented, not just for competition, but to explicitly produce athlete advancement. Events are redefined from end products into crucial mechanisms for consistently capturing, verifying, and distributing high-fidelity performance data. This systematic instrumentation transforms every metric and interaction into actionable insights, serving as the primary input to accelerate athlete development. The advanced athlete is our ultimate objective; all other revenue streams are monetizable derivatives of this core operational function.

2

Structural Moat & Compounding Value

This fundamental inversion of the traditional event model creates GPH's most significant structural moat. Unlike competitors who sell access or entertainment, GPH leverages events as a sophisticated data-capture layer that fuels a proprietary athlete advancement engine. This engine continuously aggregates and refines performance data, creating long-term, compounding value for every athlete. The resulting deep, verified data transforms potential into measurable progress and opportunities, reinforcing a self-perpetuating cycle: more engagement yields more data, leading to more precise advancement pathways, attracting greater talent, and securing GPH's indispensable position.

Operationalized Advancement

Athlete advancement at GPH is not an abstract mission statement — it is a concrete, instrumented operational pipeline. Every layer of the platform is designed to capture, verify, enrich, and distribute advancement signals that make athletes discoverable, recruiteable, and fundable.

Athlete Identity Layer

Verified profiles, QR-based visibility, digital resumes, and AI-generated bios create a portable, authenticated identity that travels with the athlete across every GPH event and platform.

Performance Intelligence

Stats, Skills DNA™, Hype Meter™, and composite rankings transform raw competition outcomes into structured, comparable performance data that scouts and sponsors can evaluate at scale.

Competition as Data

Every match, every bracket, every result feeds directly into recruitment and draft systems. Competition is not an endpoint — it is a data generation event.

Talent Draft Engine

Scouts, teams, and sponsors discover ranked athletes through a structured draft and discovery interface — eliminating the opacity and gatekeeping of traditional recruitment.

Scholarship & NIL Pipelines

AI-matched scholarship opportunities and Name, Image, Likeness deals are surfaced continuously based on verified performance data, geographic eligibility, and athlete preferences.

The Compounding Advantage

The most powerful feature of the operationalized advancement model is its temporal dimension. Unlike traditional event models where an athlete's relevance is tied to a single season or a single tournament, the GPH model creates accumulating verified signal over time.

Athletes don't "age out" of relevance. They accumulate verified performance data, competition history, advancement milestones, and community contributions that make them more — not less — valuable to scouts, sponsors, and institutions with every passing season. This temporal compounding creates four structural advantages that reinforce one another.

1

Long-Term Retention

Athletes have a compelling reason to stay in the ecosystem year after year — their profile, rankings, and advancement history are portable within GPH but not exportable outside it.

2

Parent Trust

Parents see tangible, verified evidence of their child's development and opportunity pipeline. This trust translates to continued investment and advocacy.

3

Institutional Adoption

Schools, leagues, and governing bodies increasingly recognize GPH credentials as a reliable signal of athlete quality and readiness — driving inbound partnership demand.

4

Sponsor Confidence

Sponsors can verify audience quality, engagement depth, and athlete reach through platform data — making sponsorship decisions data-driven rather than speculative.

Media & Content Operations — Xperience TV

Media is not a separate department. It is an automated exhaust system. In the GPH operating model, content is not something that requires a separate budget, a separate team, or a separate strategy. It is the natural output of every championship, every draft moment, every conference keynote, and every athlete milestone that occurs within the ecosystem. The platform is instrumented to capture, package, and distribute this content through owned channels — transforming every live activation into a perpetual media asset.

This is why GPH does not need to negotiate with broadcasters, pay for media placements, or compete for algorithmic visibility on third-party platforms. Xperience TV is the owned distribution layer that ensures every piece of content reaches its intended audience on GPH's terms, on GPH's schedule, and to GPH's economic benefit.

Always-On Content Pipeline

The content pipeline is not seasonal — it is continuous. Every event type within the GPH ecosystem generates distinct content categories that feed into Xperience TV automatically.

Championship Finals

Live and recorded coverage of national and international championship events — the flagship content that drives viewership and brand authority.

Draft Moments

High-emotion content capturing athletes being discovered, selected, and celebrated through the Talent Draft Engine — storytelling that resonates with athletes, families, and sponsors alike.

Conference Keynotes

Thought leadership and educational content from GPH conferences and academies — positioning the brand as an authority in athlete development and competitive excellence.

IQ Whizzy Educational Content

Free educational programming that drives engagement between competition seasons and extends the platform's value proposition to parents and institutions.

Perpetual Value Creation

The economics of GPH's media operations are fundamentally different from traditional event media because every live activation generates a cascade of derivative content assets that continue to produce value long after the event concludes.

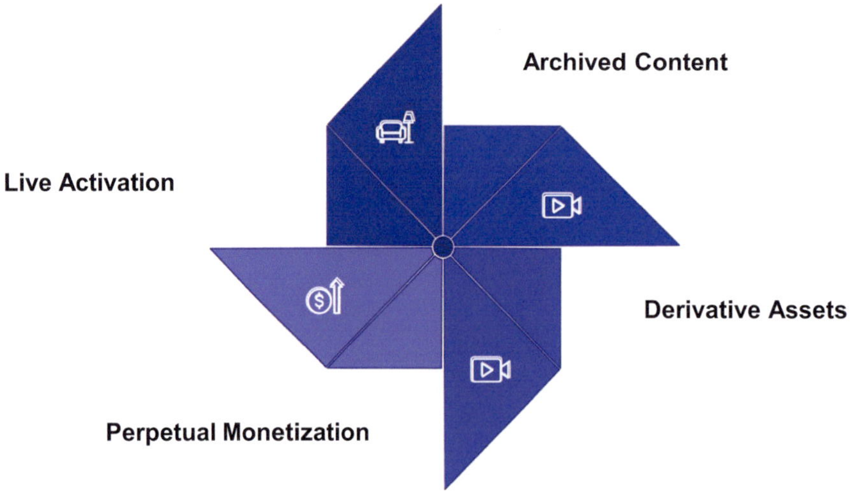

A single live activation becomes archived content, highlight clips, educational modules, and promotional assets. A two-day event becomes a 365-day monetizable asset. This content flywheel means that the marginal cost of media production approaches zero while the cumulative library of content grows with every event, creating an ever-expanding catalog that compounds viewership, engagement, and advertising revenue over time.

Security, Trust & Compliance — The Shield

Operational safety is embedded at the platform level. In an ecosystem that serves minors, processes financial transactions, handles medical data, and connects athletes with adult scouts and coaches, trust is not a feature — it is a prerequisite. GPH does not treat safety and compliance as afterthoughts bolted onto operations. They are foundational layers of the platform architecture, enforced through code rather than policy alone.

The Shield encompasses two integrated systems: VerifiEYED for identity verification and role-based access control, and Limaj Health & PHCI for secure medical data handling and portable health identity standards. Together, they ensure that every participant in the ecosystem — athlete, parent, scout, coach, Ambassador, and sponsor — is verified, accountable, and operating within defined boundaries.

VerifiEYED — Identity & Trust Verification

VerifiEYED is the platform's identity and trust layer, ensuring that every individual who interacts with athletes within the GPH ecosystem has been verified and assigned appropriate access permissions. This system is not optional — it is a mandatory gate that every participant must pass through before gaining access to any platform functionality.

Role-Based Identity Verification

Every user is verified according to their role — athlete, parent, coach, scout, Ambassador, or sponsor — with verification requirements calibrated to the sensitivity and access level of each role.

Scout, Coach & Adult Vetting

All adults who interact with athletes undergo enhanced vetting including background checks, credential verification, and ongoing monitoring to ensure ongoing compliance with trust standards.

Minor Protections & Parent Controls

Dedicated safeguards for athletes under 18, including parental consent workflows, restricted communication channels, and visibility controls that give parents real-time oversight of their child's ecosystem interactions.

50%

Total conversion rate

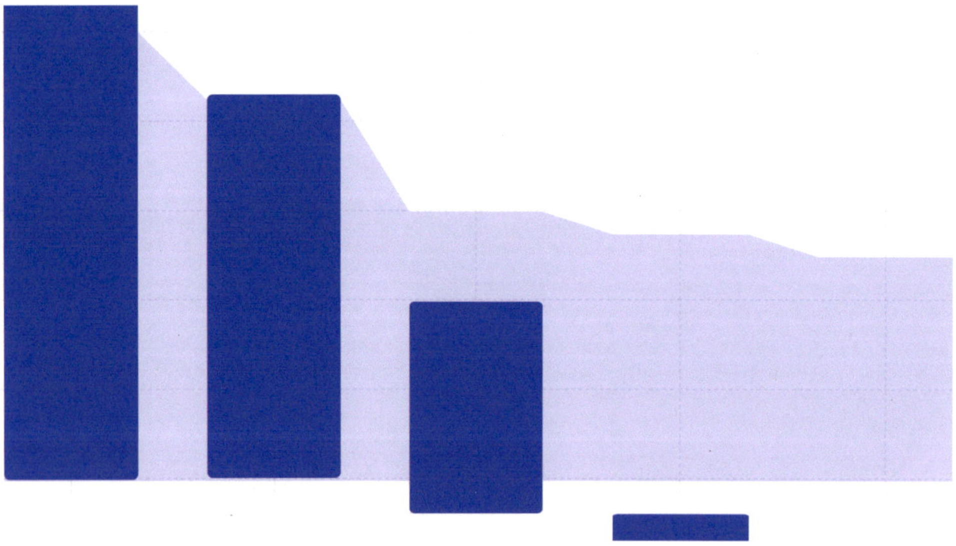

The funnel chart illustrates the multi-layered VerifiEYED verification process, showing the progression of users through each critical stage from initial registration to gaining approved access. Each step ensures a robust and secure environment for all participants, with specific measures for adult vetting and minor protections.

Limaj Health & PHCI

Athletic competition inherently involves physical risk. GPH addresses this reality through Limaj Health and its Portable Health Care Identity (PHCI) standards — a privacy-first medical data infrastructure that ensures athletes' health information is handled securely, portably, and in compliance with jurisdictional regulations worldwide.

- Secure handling of medical waivers and recovery data — digitized, encrypted, and accessible only to authorized parties with explicit consent.
- Portable Health Care Identity standards — an athlete's medical profile travels with them across events, regions, and seasons without requiring redundant paperwork or exposing sensitive information.
- Privacy-first architecture — compliant across jurisdictions including HIPAA, GDPR, and emerging data protection frameworks, ensuring that GPH can operate internationally without regulatory friction.

> Trust is not optional. It is operationally enforced. Every verification check, every access control, every encrypted data transfer is automated through the platform — removing human error and inconsistency from the trust equation.

LIMAJ HEALTH

STRATEGIC TAKEAWAY

What Operations Actually Means at GPH

Operations at Goddess Phenezy Holdings are not about booking venues, staffing events, or managing chaos. Those are the operational realities of traditional event companies — and they are precisely the activities that GPH has engineered out of its cost structure by delegating them to licensed Ambassadors.

Code

578,000+ lines of proprietary source code that automate every operational function.

Licensing

Master License framework distributing execution while concentrating governance.

Data

Every interaction generates compounding data assets owned by the Holding Company.

Media

Owned distribution turning live events into perpetual monetizable content.

Advancement

The core output — verified, compounding athlete signal that drives retention and revenue.

> **Operations = Code + Licensing + Data + Media + Advancement**
>
> **This is why we scale. This is why athletes stay. This is why competitors can't copy it.**

Financial Plan — Revenue Architecture

The GPH revenue model is not a single-stream business. It is a multi-layered extraction architecture designed to capture value at every point of interaction within the ecosystem. Each revenue stream reinforces the others: live events generate media content, media content drives platform engagement, platform engagement attracts sponsors, sponsors fund championships, and championships produce the athlete advancement data that keeps the entire flywheel spinning.

The following revenue streams represent the primary economic engines of the enterprise. Each stream is designed to scale independently while contributing to the compounding value of the integrated ecosystem.

Revenue Stream I: Live Events

Live events are the foundational revenue layer — the physical activations that generate data, content, community, and direct revenue simultaneously. Every championship, qualifier, conference, and academy represents a multi-revenue event with four distinct extraction points.

Entry Fees

Athlete and team registration fees processed exclusively through the owned platform, ensuring 100% capture with zero third-party leakage. Scaled pricing across recreational, competitive, and elite tiers maximizes accessibility while optimizing revenue per event.

Broadcasting Rights

Live and on-demand distribution through Xperience TV with licensing opportunities for external broadcasters. Owned distribution means GPH retains primary monetization rights while selectively licensing for additional reach and revenue.

Sponsorships

Title, presenting, and category sponsorships for championships and conferences. Data-driven sponsorship packages enabled by platform analytics give sponsors verified audience metrics, engagement depth, and athlete demographics — commanding premium pricing.

Merchandise & Concessions

On-site and digital merchandise sales, food and beverage concessions, and branded product partnerships. Physical revenue that complements digital streams and deepens brand engagement at every live activation.

Revenue Stream II: Xperience TV

Xperience TV — transforms every live event into a perpetual revenue engine. This is not supplementary media; it is a core business unit that monetizes the content exhaust of the entire ecosystem.

Revenue Drivers

- Advertising Revenue — Pre-roll, mid-roll, and programmatic advertising across live and archived content
- Partnerships — Branded content series, exclusive access deals, and co-produced programming
- Cross-Promotion — Driving registration, engagement, and sponsor value back to live events through media exposure

Xperience TV operates as a closed-loop media ecosystem. Every championship final, draft moment, athlete spotlight, and educational segment is produced, distributed, and monetized through owned infrastructure. This eliminates the revenue-sharing arrangements, algorithmic unpredictability, and content policy risks that plague organizations dependent on YouTube, Twitch, or other third-party platforms.

As the content library grows with each season, the catalog's aggregate value compounds — older content continues to generate views, advertising revenue, and promotional value indefinitely. A single season of championships produces hundreds of hours of monetizable content that works for GPH 24 hours a day, 365 days a year.

Revenue Streams III & IV: NPC App & Workshops

Beyond live events and media, GPH captures revenue through its digital product suite and educational programming — two streams that operate year-round regardless of championship scheduling.

NPC App Revenue

In-App Purchases: Premium features, enhanced profiles, advanced analytics, and customization options that athletes and parents purchase to maximize their ecosystem experience and competitive advantage.

Tournament Fees: Digital-first competitions, challenges, and qualifying events hosted entirely within the app — zero venue cost, maximum margin.

Advertising & Sponsorships: Targeted advertising and sponsored content within the app experience, powered by first-party data that delivers superior targeting precision and sponsor ROI.

Workshops & Conferences Revenue

Ticket Sales: In-person and virtual attendance for leadership summits, coaching academies, athlete development workshops, and industry conferences that position GPH as the intellectual authority in competitive athlete advancement.

Sponsorships: Conference-specific sponsorship packages with speaking slots, branded sessions, and exhibitor access — premium-priced due to the concentrated, high-value audience.

Online Streaming: Pay-per-view and subscription access for virtual attendees, extending the addressable audience beyond physical venue capacity and creating another derivative content asset.

The Operational Thesis

Goddess Phenezy Holdings is not an event company with technology. It is a technology company that produces events.

The distinction is existential. Event companies are constrained by linear economics — more events require proportionally more people, capital, and risk. Technology companies are defined by exponential economics — more users, more data, and more network effects improve margins, defensibility, and valuation with every incremental unit of activity.

578K+	450+	79+	365
Lines of Code	**Premium Domains**	**Championships**	**Days of Revenue**
Proprietary source code owned outright	Jurisdictional control across verticals	National & International events governed	Every event becomes perpetual content

The operations and infrastructure detailed in this section are not aspirational — they are built, owned, and operational. The Command Layer governs. The Field Layer executes. The Platform Stack captures every transaction, every data point, every piece of content. The Advancement Engine transforms all of it into verified athlete signal that compounds over time. And the Financial Plan extracts value at every layer of this architecture.

> We built the system. We own the system. We govern the system.That is the operational thesis. That is why we scale.

Financial Plan — The Sovereign Treasury

A structural blueprint for ownership-based economics, jurisdictional dominance, and retained margin architecture — engineered for sovereign wealth creation, not venture dependency.

1

The Sovereign Treasury

This financial framework fundamentally departs from conventional economic paradigms. It is designed to cultivate genuine sovereign wealth, empowering entities with intrinsic value generation and robust independence, unlike models reliant on external venture capital or platform monopolies. The Sovereign Treasury establishes a self-sustaining ecosystem built on control, strategic autonomy, and internal capital accumulation.

2

Independence and Self-Funding

This model operates with deliberate independence from external venture capital cycles and platform vulnerabilities. It prioritizes self-funding through robust internal revenue streams and strategic asset management, eliminating pressure to chase short-term valuations or cede control to outside investors. By developing proprietary infrastructure and fostering decentralized solutions, it ensures operational resilience, data sovereignty, and freedom from economic rent-seeking associated with dominant digital platforms.

3

Strategic Advantages

The strategic advantages are manifold, including enhanced resilience against market fluctuations, complete control over strategic direction, and the ability to pursue long-term objectives without external interference. Its viability rests on a clear legal and governance framework, diversified economic activities, and a steadfast commitment to internal capacity building and resource optimization.

Overview: The Arbitrage of Ownership

Goddess Phenezy Holdings, Inc. does not operate on traditional startup economics. We are not dependent on venture capital, and we do not surrender margin to platform monopolies. The entire enterprise is architected around a single, non-negotiable principle: ownership is the strategy.

The entire ecosystem is engineered as a Progressive Web App (PWA). There are no Apple App Store fees. No Google Play commissions. No 30% platform tax siphoning value from creators, athletes, or operators. That margin — the margin every other startup bleeds to gatekeepers — is retained as gross profit inside our ecosystem.

This is not a distribution workaround. It is not a clever technical hack designed to shave a few points off unit economics. It is a structural advantage — a permanent, compounding edge that widens with every transaction, every membership, every championship event that flows through our owned infrastructure.

We do not need outside funding because we own the vault. Every dollar earned stays inside the walls. Every margin point compounds. Every vertical feeds the next. This document lays bare the financial architecture that makes that possible — and the valuation logic that institutional capital will recognize as disciplined, defensible, and sovereign.

The PWA Structural Advantage

While competitors hemorrhage 30% of every transaction to Apple and Google, Goddess Phenezy Holdings retains that margin as pure gross profit. This is not incremental optimization — it is a fundamentally different economic model.

Traditional App Model

30% platform tax on every transaction. Revenue leaks to Apple and Google before it ever reaches the operator. Margin erosion is built into the business model from day one.

Our PWA Model

0% platform fees. 100% of revenue retained. No gatekeeper approval. No algorithmic suppression. Full margin sovereignty from transaction to treasury.

> **We do not need outside funding because we own the vault. Powered by Technologes R Us, Goddess Phenezy Holdings, Inc. owns every line and platform.**

The 450+ Domain Jurisdiction Portfolio

THE ASSET BASE

We do not compete for attention inside rented platforms. We own the destinations. Goddess Phenezy Holdings controls over 450 premium jurisdictional domains, each mapped to National and International competitions, conferences, and economic infrastructure. These domains function as exclusive digital real estate, not marketing URLs.

450+ Jurisdictional Domains

Extensive portfolio of premium domains mapped to key global verticals.

Exclusive Digital Real Estate

Functions as owned digital property, not merely marketing URLs.

Pre-emptive Territorial Strategy

Securing strategic digital ground before competitors can enter.

Operational Anchors

Foundational platforms for leagues, championships, and conferences.

Sovereign Jurisdictions

Generating real revenue and brand equity across diverse sectors.

When competitors attempt to enter these verticals, they discover the property is already owned. This is a pre-emptive territorial strategy — one that compounds in value with every passing year as these verticals mature, attract audiences, and generate recurring revenue streams.

Each domain represents a sovereign jurisdiction within the broader Goddess Phenezy Holdings ecosystem. They are not speculative acquisitions. They are operational anchors for leagues, championships, conferences, and economic platforms that generate real revenue and real brand equity across sports, technology, culture, and career advancement.

Global Sports & Athletics Jurisdictions

The sports vertical represents the most mature and revenue-dense segment of the domain portfolio. Each jurisdiction anchors a fully owned championship ecosystem — from league governance to media rights to player infrastructure.

The National Basketball Championship

The Origin. **The proving ground where the entire empire began. The first jurisdiction to unify infrastructure, governance, branding, and monetization under a single owned system.**

The National & International Golf Championships

Prestige & Sponsorship Density. **Golf commands the highest per-capita sponsorship spend of any sport. Our ownership of this jurisdiction unlocks premium brand partnerships.**

The National & International Soccer Championships

Global Scale. **Soccer's 4 billion global audience represents the largest addressable market in sports. This jurisdiction is our gateway to international expansion.**

The National & International Boxing Championships

High-Margin, Cash-Driven Events. **Boxing's PPV economics and premium event pricing create outsized revenue per event — a high-margin jurisdiction by design.**

The National & International Pickleball Championships

Fastest-Growing Participation Sport. **Pickleball's explosive growth trajectory makes this jurisdiction a high-velocity asset with rapidly compounding audience value.**

Gaming, Wealth & Tech Jurisdictions

At the intersection of technology, finance, and competitive gaming, these jurisdictions capture the fastest-growing segments of the digital economy. Each domain anchors an ecosystem of championships, conferences, and data infrastructure that will define the next decade of economic activity.

1

National & International Poker Championships

High-stakes competitive gaming with deep sponsorship potential and a global audience of over 100 million active players worldwide.

2

International Gamers Championship & Conference

The esports and gaming vertical — a $200B+ global industry with explosive audience growth and premium media rights economics.

3

African American WallStreet

STEM, Finance & Wealth Infrastructure. A sovereign platform for economic empowerment, financial literacy, and wealth-building at scale.

4

National & International AI Championships & Conferences

AI is the defining technology of our era. This jurisdiction positions us at the center of competitive AI development and thought leadership.

5

National & International Metaverse Championships & Conferences

The spatial computing and immersive technology vertical — capturing the emerging metaverse economy before it fully materializes.

Culture, Career & Advancement Jurisdictions

Beyond sports and technology, Goddess Phenezy Holdings commands jurisdictions across culture, career development, and professional advancement. These verticals serve a dual purpose: they generate independent revenue streams while feeding talent, audience, and brand equity back into the core sports and tech ecosystems.

The International Music Festival

A global platform for live music, artist development, and sponsorship activations that bridges entertainment and economic infrastructure.

National & International Career Fair

Physical career infrastructure connecting talent with opportunity across industries, geographies, and skill levels.

National & International Virtual Career Fair

The digital twin of the career fair — extending reach, reducing friction, and enabling year-round engagement with job seekers and employers.

The International Speaker Xccelerator

A platform for thought leadership development, keynote placement, and speaker brand monetization at global scale.

The International Business Xccelerator

Business acceleration infrastructure connecting founders with capital, mentorship, and market access through owned programming.

The National Health & Wealth Conference

Convergence of wellness and financial empowerment — a high-engagement vertical with deep sponsor alignment and recurring audience loyalty.

Domain Portfolio: Not Concepts — Owned Jurisdictions

Every domain in the Goddess Phenezy Holdings portfolio is more than a URL. Each represents an enforceable brand jurisdiction — a sovereign territory in the digital economy that cannot be replicated, outbid, or circumvented by competitors.

When a competitor decides to launch a "National Basketball Championship" or an "International AI Conference," they discover the digital real estate is already occupied. The namespace is controlled. The brand authority is established. The infrastructure is live.

This is the definition of a jurisdictional moat: you cannot compete for what is already owned. And with 450+ domains spanning sports, gaming, technology, culture, and career advancement, the moat is not narrow — it is continental.

These assets appreciate with every event hosted, every membership activated, every media asset created. They are not depreciating line items. They are compounding sovereign territory.

450+

Premium jurisdictional domains controlled

Enforceable

Brand control with legal and digital authority

Pre-Emptive

Territory secured before competitors can enter

Compounding

Value increases with every activation and event

THE ORIGIN JURISDICTION

The National & International Basketball Championships

Where the empire began. Where infrastructure, governance, branding, and monetization were first unified under a single owned system

Introduction: Where the Empire Began

Basketball is not merely one vertical inside Goddess Phenezy Holdings, Inc. It is the origin jurisdiction. Before platforms, before conferences, before global expansion, basketball was the proving ground — where infrastructure, governance, branding, and monetization were first unified under a single owned system.

Unlike traditional tournaments that exist as isolated events — one-off weekends with no structural continuity — The National & International Basketball Championships operate as a fully integrated ecosystem. This ecosystem comprises owned championships, owned leagues, recurring memberships, media rights, data infrastructure, and licensing authority — all operating under sovereign control.

Owned Championships

Sovereign events with full brand and revenue control

Owned Leagues

Recurring league architecture with annual licensing

Player Memberships

Verified digital identities with platform access

Media & Data Rights

Full IP ownership across content and analytics

This is not participation-based revenue. This is ownership-based economics.

Structural Advantage: Ownership Over Access

Goddess Phenezy Holdings does not rent distribution through third-party leagues, platforms, or governing bodies. We own the league architecture. This fundamental distinction separates us from every competitor operating within rented infrastructure and borrowed governance models.

The National Basketball League

Fully owned league governance, branding, and operational infrastructure serving domestic markets.

The International Basketball League

Global expansion of the league model with international licensing and championship pathways.

All Affiliated Championship Branding

Complete brand control across all events, media, and promotional assets.

PWA Digital Infrastructure

Zero platform tax. Full margin retention. No gatekeeper dependency.

The Jurisdictional Moat

The combination of league ownership and PWA infrastructure creates a compounding structural advantage that no competitor can replicate without building from scratch — and even then, the jurisdictional domains are already controlled.

No Apple App Store Fees

PWA delivery eliminates the 30% iOS tax that cripples margins for every app-dependent competitor in the market.

No Google Play Commissions

Android platform fees are equally eliminated, preserving full revenue integrity across all mobile access points.

No 30% Revenue Leakage

Every dollar earned remains inside the ecosystem as retained gross margin — not diverted to platform gatekeepers.

This structure is not an optimization. It is a jurisdictional moat.

Revenue Architecture: Multi-Layer, Recurring, Scalable

1

Licensing

A core revenue stream that leverages the brand's intellectual property.

2

Memberships

Consistent, recurring income through fan engagement and exclusive access.

3

Events

High-value experiences and direct consumer interaction.

4

Media

Broad reach and advertising opportunities through content distribution.

The basketball ecosystem generates revenue through four distinct, interlocking layers — each designed for recurring capture, structural scalability, and compounding margin retention.

Unlike traditional sports ventures that rely on a single championship weekend for the majority of their revenue, this architecture distributes income across licensing, memberships, events, and media — creating resilience, predictability, and investor-grade financial structure.

This multi-layered approach is not merely a diversification strategy; it is a foundational pillar of defensibility. Each layer (Licensing, Memberships, Events, and Media) is independently robust but gains exponential strength when integrated. The interlocking nature creates a network effect, where success in one area directly fuels growth and demand in others, establishing a significant barrier to entry for potential competitors. This prevents the fragility inherent in models dependent on singular, often unpredictable, revenue streams.

The structural scalability of this architecture ensures that growth is not linear but compounding. For instance, increased engagement through media content translates directly into higher demand for memberships and event attendance, which in turn enhances the value of licensing opportunities. This creates a virtuous cycle, consistently increasing the total addressable market and revenue potential without proportional increases in operational overhead.

Why This Architecture Works

Network Effects & Defensibility

The interlocking nature creates a network effect, where success in one area directly fuels growth and demand in others, establishing a significant barrier to entry for potential competitors. This prevents the fragility inherent in models dependent on singular, often unpredictable, revenue streams.

Structural Resilience

By distributing income sources, the ecosystem is inherently protected against market shifts or performance dips in any single area. This structural stability fosters long-term wealth creation, prioritizing sustained profitability and asset appreciation over short-term event-driven revenue spikes.

Predictability & Institutional Performance

Each layer is engineered for predictability and institutional-grade financial performance. Recurring revenue models, such as annual memberships and long-term licensing agreements, minimize seasonal fluctuations and provide a stable cash flow fundamental to long-term valuation. This deliberate design contrasts sharply with traditional sports business models, which frequently experience significant volatility tied to team performance, playoff runs, or a few marquee events.

Layer 1: League & Team Licensing Revenue

Participation in the ecosystem is governed through annual league licensing, not one-off entry fees. This creates a recurring revenue base that compounds year over year as league penetration deepens and geographic expansion accelerates.

National League

The domestic league targets youth, prep, semi-pro, and adult basketball markets across the United States. With conservative penetration assumptions:

- **1,200 teams enrolled**
- **Average license fee: $6,500 per team annually**
- **Total: $7.8M annually**

These figures are intentionally conservative relative to the available market. The total addressable universe of organized basketball teams in the U.S. exceeds 50,000.

International League

The international league extends the model globally, with premium pricing reflecting the higher brand value and operational complexity of cross-border championship access:

- **800 teams enrolled**
- **Average license fee: $8,500 per team annually**
- **Total: $6.8M annually**

International expansion targets established basketball markets across Europe, Africa, Asia, and Latin America.

$14.6M

Total League Licensing

Combined annual revenue from national and international team licensing

2,000

Total Teams

Conservative enrollment across domestic and international markets

$7.2K

Average License

Blended average per-team license fee across both leagues

Layer 2: Player Membership & Platform Access

Every registered player operates under a verified digital identity with access to a comprehensive suite of performance, career, and competition tools. This is a subscription-based platform access model that delivers continuous value and generates predictable recurring revenue.

This membership cultivates a robust, self-reinforcing ecosystem, creating powerful network effects as more players engage. It also serves as an invaluable data asset; player interactions generate rich behavioral and performance data, feeding AI-assisted tools for refined insights, personalized training, and superior game analysis, thus solidifying the platform's competitive advantage.

League Stats & Rankings

Real-time performance statistics, league standings, and competitive rankings across all divisions and age groups.

AI-Assisted Performance Tools

Machine learning-powered player development insights, game analysis, and training recommendations personalized to each athlete.

Digital Resumes

Verified athletic profiles that serve as digital recruiting portfolios for college, professional, and international opportunities.

Championship Eligibility

Membership is the gateway to championship participation — creating an inherent retention mechanism tied to competitive aspiration.

Player Membership: The Numbers

The membership model is built on straightforward, defensible assumptions rooted in the league licensing structure already established. Each team carries a standard roster, and each player pays a monthly subscription for platform access.

Key Assumptions

18 players per team

2,000 total teams

36,000 active players

Monthly Membership

$25/month × 12 months

= $300 per player annually

Revenue Calculation

With 36,000 active players each contributing $300 annually through monthly membership subscriptions, the player membership layer generates substantial recurring revenue that is highly predictable and structurally insulated from event-based volatility.

This revenue stream is purely digital — delivered through the PWA platform with zero platform tax. There are no physical fulfillment costs, no inventory, and no marginal cost per additional member beyond infrastructure scaling, which is negligible at this volume.

The retention mechanics are inherently strong: players must maintain active membership to retain championship eligibility, access performance data, and preserve their verified digital identity and recruiting profile.

36K

Active Players

Verified digital identities across both leagues

$10.8M

Annual Revenue

Total player membership revenue per year

$25

Monthly Fee

Per-player platform access subscription

Layer 3: Championship Event Revenue

Championships function as culmination events, not the primary revenue driver. 'The International Basketball Championship' by name alone puts it in a top tier and unique status.

Critical Structural Distinction

Traditional sports businesses are event-dependent — they rise and fall with ticket sales, weather, and attendance volatility. In this model, championships are the prestige layer that crowns a year of recurring revenue generation, offering a strategic anchor rather than a precarious income source.

Strategic Superiority

Positioning championships as culmination events is strategically superior because it fundamentally de-risks the financial model. By decoupling the primary revenue streams (like player membership and licensing) from event-specific performance, the ecosystem achieves remarkable financial stability. This ensures predictable, recurring income irrespective of external factors affecting a single event, allowing resources to be allocated towards long-term growth and infrastructure development rather than short-term event sales targets.

Championship Revenue Dynamics

Brand Amplification & Market Pull

The championship layer significantly amplifies brand value and creates an aspirational pull that directly feeds into the core membership and licensing revenues. 'The International Basketball Championship' is not just a competition; it's a powerful marketing and media showcase for the entire ecosystem. It validates the year-round competitive structure, demonstrating the pinnacle of achievement within the leagues and inspiring players and teams to strive for excellence, thereby driving sustained engagement and new member acquisition.

Three Revenue Channels

Revenue is generated through three primary channels within each championship:

Ticketing & Gate Receipts

Premium pricing driven by high demand for witnessing elite competition

Team Participation Fees

Justified by the invaluable opportunity to compete on a grand stage, gaining exposure and the chance for ultimate glory

Sponsorship Activations

Premium event structure commands significantly higher sponsorship rates as brands align with emotional peaks and high-visibility media moments

Strategic Value

This premium event structure creates valuable media assets and reinforces the perception of 'The International Basketball Championship' as a marquee event. These events create the intense emotional highs and memorable narratives that keep players, teams, and fans deeply invested in the ecosystem, strengthening its competitive integrity and commercial appeal.

Championship Revenue Breakdown

National Championship

Ticketing, team fees, and sponsorship activations combine for $6.0M in annual event revenue from the domestic culmination event.

International Championship

Global participation, premium international sponsorships, and elevated production value generate $5.5M annually.

$11.5M

Total Championship Revenue

Combined annual event revenue across both championships

Layer 4: Media, Content & Data Rights

At Layer 4, Goddess Phenezy Holdings fully capitalizes on a core strategic advantage: the absolute ownership of its leagues' media, content, and data rights. In an industry often fragmented by complex licensing agreements, we stand apart. Unlike typical organizations reliant on third-party distributors and their inherent dependencies, Goddess Phenezy Holdings maintains complete dominion over every frame, every statistic, and every compelling narrative generated across its expansive ecosystem.

Game Footage

Absolute command over all recorded content from league play and championship events, eliminating shared rights or external claims.

Player Statistics

Proprietary control over comprehensive player performance data, the foundational asset for advanced analytics, strategic scouting, and lucrative data licensing.

Performance Data

Exclusive access to advanced analytics and AI-processed metrics, a constantly appreciating asset whose value escalates with each passing season and expanding geographic footprint.

Storylines & Narratives

The authentic human dramas unfolding within our competitions – recognized as the single most compelling and enduring content asset in all of sports media.

Media Monetization Channels

These owned media and data assets are monetized through multiple parallel channels, each operating independently while reinforcing the others. The combined annual media valuation is deliberately conservative — it excludes significant upside from NIL overlays, betting data licensing, and third-party broadcast sublicensing.

1

Streaming

Direct-to-consumer live and on-demand content delivered through owned PWA infrastructure with zero distribution fees.

2

Content Licensing

Wholesale licensing of footage, highlights, and data packages to media companies, educational platforms, and sports networks.

3

Highlight Syndication

Curated highlight packages distributed across social platforms, driving audience growth while generating sponsorship and ad revenue.

4

Internal Media Platforms

Owned media channels that retain audience engagement within the ecosystem, creating compounding viewership and advertiser value.

$6.0M

Conservative Annual Media Valuation

Excludes NIL overlays, betting data licensing, and third-party broadcast sublicensing

What's Not Yet Modeled

The $6.0M media valuation is intentionally conservative, deliberately excluding several high-value revenue streams. These represent significant upside, to be modeled in future projections as each channel matures and market conditions crystallize.

NIL Overlays

Name, Image, and Likeness monetization for players within the ecosystem — a rapidly expanding market with significant revenue potential.

Betting Data Licensing

As sports betting legalization expands, proprietary league data becomes an increasingly valuable commodity for sportsbooks and data providers.

Broadcast Sublicensing

Third-party broadcast rights packages for networks seeking owned championship content — a traditional revenue accelerator in sports.

Strategic Rationale for Conservative Modeling

The decision to exclude these high-potential revenue streams reflects a disciplined, pragmatic approach to financial modeling. This conservatism provides institutional investors with a robust and transparent base case, ensuring projections are grounded in current, verifiable market conditions and established operational capabilities, thus mitigating speculative risk.

This strategic sequencing ensures formal activation of NIL overlays, betting data licensing, and broadcast sublicensing aligns with critical milestones in ecosystem maturity and market readiness. As our audience grows and regulatory landscapes stabilize, these channels will transition from "excluded upside" to formally integrated, high-performing revenue engines, poised for optimal market entry and maximum value extraction.

These excluded revenue streams represent conservative upside of $5–15M+ annually as the ecosystem scales and these channels are formally activated.

Revenue Summary: The Four Pillars

Four interlocking revenue layers. One sovereign ecosystem. $42.9M in total annual revenue.

Revenue Stream	Annual Value	% of Total
League Licensing (National + International)	$14.6M	34%
Player Memberships	$10.8M	25%
Championships (National + International)	$11.5M	27%
Media & Data Rights	$6.0M	14%
Total Annual Revenue	$42.9M	100%

Revenue Composition Visualized

The revenue architecture is intentionally diversified — no single layer accounts for more than 34% of total revenue, creating structural resilience and reducing dependency on any individual income source.

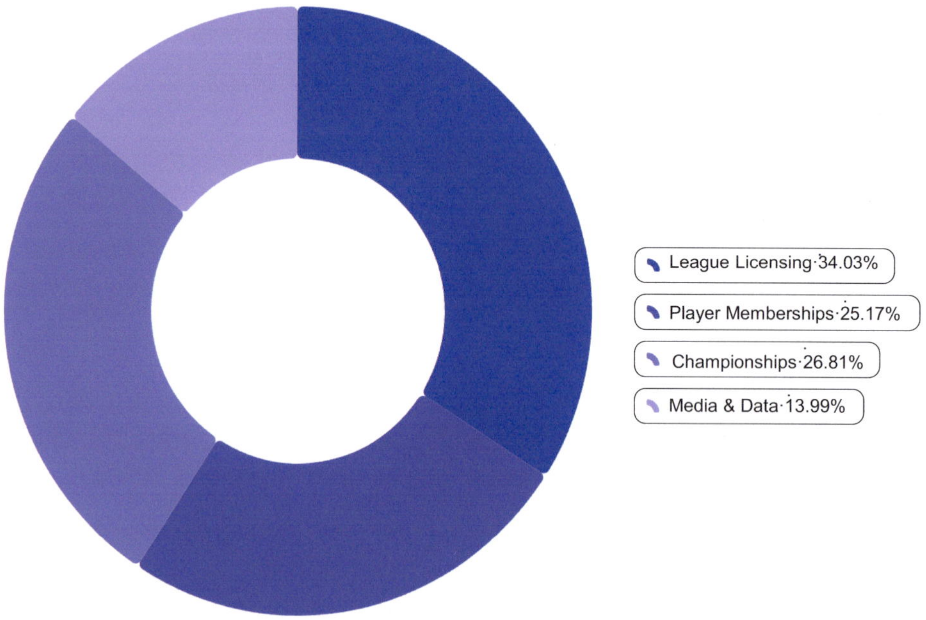

This diversification means that even if one layer underperforms in a given year, the ecosystem continues to generate substantial revenue through the remaining three pillars. This is the structural advantage of ownership-based economics over event-dependent models.

Revenue Architecture: Recurring vs. Event-Based

A critical distinction for institutional investors: the majority of this revenue is recurring by design. League licensing and player memberships together represent 59% of total revenue — both are subscription-based, annually renewing, and structurally insulated from event-day volatility.

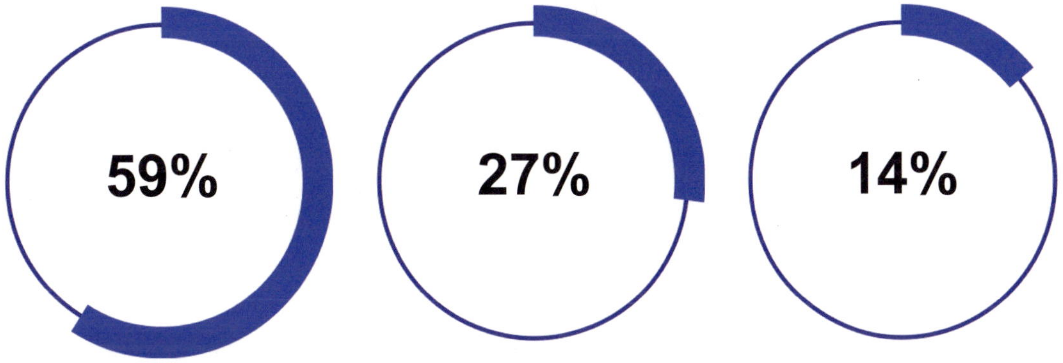

Recurring Revenue

League licensing + player memberships — predictable, annually renewing income streams

Event Revenue

Championship events — high-margin culmination events, not the primary revenue driver

Media Revenue

Content and data rights — compounding asset value that grows with each season

Revenue Layer Interlock

Each revenue layer feeds and reinforces the others, creating a flywheel effect that accelerates with scale. This is not four independent revenue streams — it is one integrated system where growth in any layer amplifies the others.

League Licensing

Teams enter → rosters activate → players join

Player Memberships

Players subscribe → data generates → content creates

Media & Data

Content distributes → brand grows → teams seek licensing

Championships

Events culminate → audiences gather → media captures

This symbiotic relationship ensures that growth in one area immediately uplifts the others. For instance, an increase in league licensing directly translates to more teams, which necessitates a proportional rise in player memberships. These new memberships not only generate direct recurring revenue but also enrich the dataset, providing richer analytics and narrative fodder for media partners. The heightened player participation fuels more competitive and compelling championship events, drawing larger audiences and enhancing the value of media and data rights. This, in turn, boosts the overall brand appeal, making league licenses even more attractive to prospective teams, thus restarting and amplifying the cycle.

This interlocking structure creates significant barriers to competition. New entrants cannot easily replicate the compounding value generated by an established ecosystem where each layer reinforces the next. The system gains momentum over time; what begins as linear growth in the early stages, where each additional team or player adds predictably, transitions into exponential growth as critical mass is reached. At this point, network effects take over, and the ecosystem becomes self-sustaining and increasingly difficult to disrupt. For investors, this revenue interlock translates directly into enhanced valuation multiples, reflecting the defensibility, predictability, and accelerating growth trajectory of the platform.

This flywheel dynamic means the ecosystem's revenue growth is not linear — it is compounding. Each new team that enters the league creates 18 new player memberships, generates new competitive content, and drives incremental championship attendance. The system feeds itself.

VALUATION | INSTITUTIONAL FRAMEWORK

Valuation Framework: Asset-Based, Not Event-Based

This ecosystem demands a valuation methodology that reflects what it actually is: a hybrid of sports league ownership, media IP, and SaaS infrastructure.

Understanding the true value proposition requires a fundamental shift from traditional event-based valuation models to a more robust asset-based framework. This distinction is critical for investors seeking to capture exponential growth.

Asset-Based Valuation: Exponential Growth

Unlike traditional event-based businesses valued on modest EBITDA multiples (3-5x) with consumable, depreciating assets, our valuation is rooted in the compounding value of owned, appreciating infrastructure. This includes proprietary domains, comprehensive league architecture, cutting-edge Progressive Web App (PWA) technology, exclusive media rights, and invaluable player data. This asset-based model commands significantly higher revenue multiples (6-15x and beyond), as increased user engagement translates to richer data, fueling more sophisticated AI, and driving powerful network effects.

Dual Growth Trajectory

Our ecosystem uniquely benefits from a dual growth trajectory. As technological advancements proliferate through continuous evolution of our PWA platform, refinement of AI tools, and expansion of data infrastructure, it concurrently enhances live experiences like championships and pivotal media moments. This synergy ensures both revenue generation and underlying asset value increase simultaneously, creating a compounding valuation effect traditional sports businesses cannot replicate.

Revenue Fuels Asset Appreciation & Conservative Multiples

Crucially, every dollar of revenue generated within this ecosystem serves to strengthen the asset base. Increased player participation provides more data, improving platform utility; expanded championship coverage enriches media libraries, securing future rights; broader league participation solidifies jurisdictional dominance. The projected 6-8x revenue multiple is inherently conservative, as it does not fully encapsulate the accelerated asset appreciation that occurs as the ecosystem scales. This creates a powerful valuation dynamic where market value grows at a pace that can outstrip revenue growth.

Why Traditional Sports Valuation Fails Here

Traditional sports event companies are valued on EBITDA (Earnings Before Interest, Taxes, Depreciation, and Amortization) multiples that reflect their event-dependent, cyclical nature. That framework does not apply to Goddess Phenezy Holdings because this is not a traditional sports event company. It is a sovereign ecosystem with recurring revenue, owned media IP, and SaaS-like membership infrastructure.

Sports League Ownership

The league architecture mirrors professional sports league economics — teams pay for participation rights, creating recurring revenue with high switching costs and deep brand loyalty.

Media IP Ownership

Full ownership of content, data, and narrative rights creates a media asset that appreciates with every season — similar to major media franchise valuations.

SaaS Infrastructure

Monthly player memberships with verified digital identities mirror SaaS subscription economics — recurring, predictable, and scalable with minimal marginal cost.

Blended Revenue Multiple Analysis

Applying a conservative blended multiple that reflects the hybrid nature of the ecosystem — part sports league, part media IP, part SaaS platform — yields a valuation range that remains below pure SaaS or pure media comparables.

These multiples are intentionally disciplined. For reference, pure SaaS companies with comparable recurring revenue profiles trade at 10–15× revenue. Pure media IP companies with owned content libraries command 8–12× multiples. Our blended range of 6–8× reflects a conservative, institution-ready valuation standard designed for credibility, not hype.

Valuation Summary

Base Case — 6× Revenue

Annual Revenue	$42.9M
Multiple	6×
Implied Valuation	$257M

The base case applies the lower bound of the blended multiple range, reflecting conservative market conditions and early-stage league penetration.

Growth Case — 8× Revenue

Annual Revenue	$42.9M
Multiple	8×
Implied Valuation	$343M

The growth case reflects accelerating league adoption, international expansion, and media rights monetization at scale.

Comparable Multiple Context

To provide institutional context for the 6–8× blended multiple, the following comparables illustrate where Goddess Phenezy Holdings sits relative to peer asset classes. Our chosen range is deliberately conservative.

The 6–8× range sits between traditional sports event valuations and pure-play digital asset classes, accurately reflecting the hybrid nature of the ecosystem. This positioning signals disciplined valuation standards while acknowledging the significant digital and media components that elevate the model above traditional sports economics.

Valuation Range at a Glance

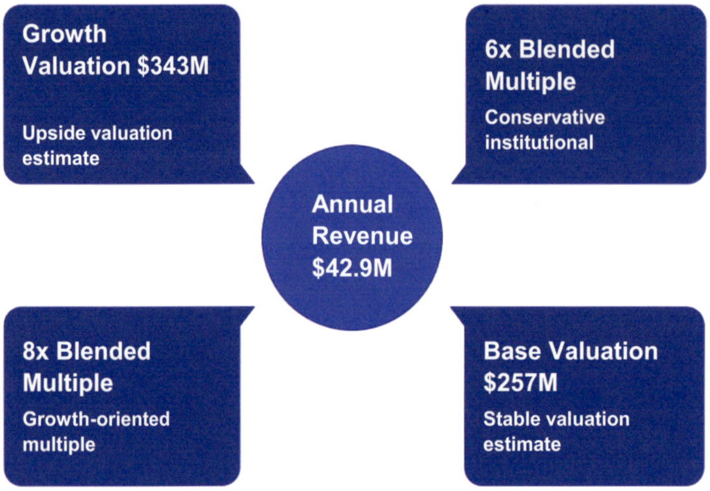

This visual represents the progression from annual revenue to two distinct valuation scenarios, reflecting both conservative and growth-oriented multiples. The base valuation provides a solid institutional standard, while the growth valuation highlights potential based on accelerating adoption within the basketball ecosystem.

These multiples remain below pure SaaS or pure media comparables and reflect disciplined, institution-ready valuation standards. The basketball ecosystem is one jurisdiction within a 450+ domain portfolio — total enterprise value extends significantly beyond this single vertical.

The Bigger Picture: Basketball Is Just the Beginning

The $257–$343M valuation range captures the basketball ecosystem alone — a single jurisdiction within a portfolio of 450+ owned domains spanning sports, gaming, technology, culture, and career advancement. The basketball model is the template. The financial architecture is replicable across every vertical in the portfolio.

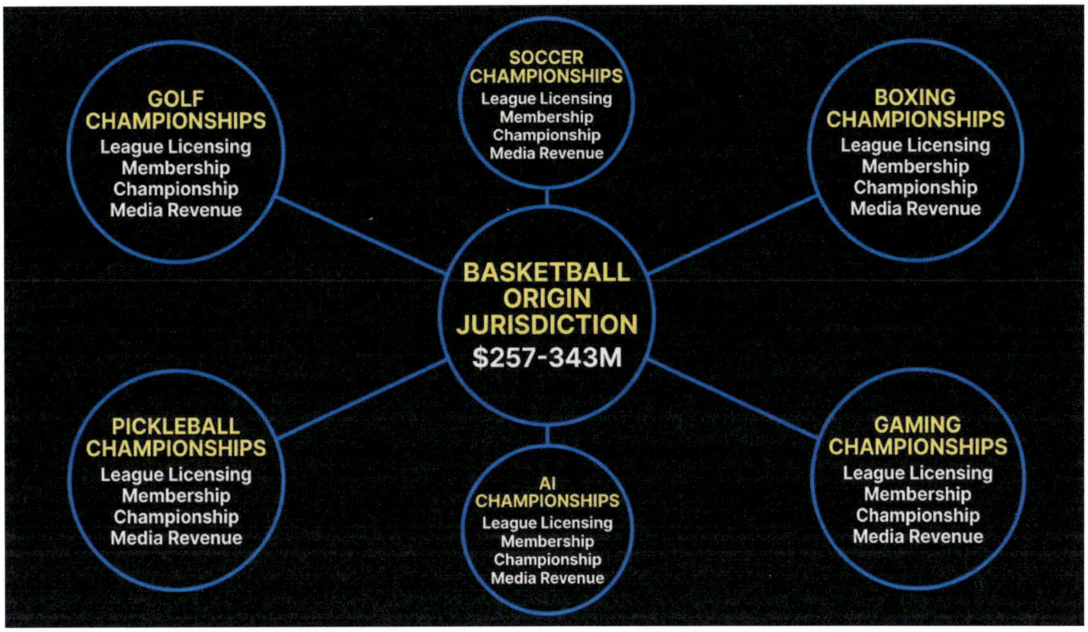

Every jurisdiction in the portfolio follows the same four-layer revenue architecture: league licensing, player/participant memberships, championship events, and media/data rights. As each vertical activates and matures, it contributes its own revenue stream to the consolidated enterprise — all flowing through owned PWA infrastructure with zero platform tax.

The Sovereign Advantage

Ownership is not a strategy. It is the strategy. The Sovereign Treasury is not a metaphor — it is the financial architecture that ensures every dollar earned stays inside the walls, every margin point compounds, and every vertical feeds the next.

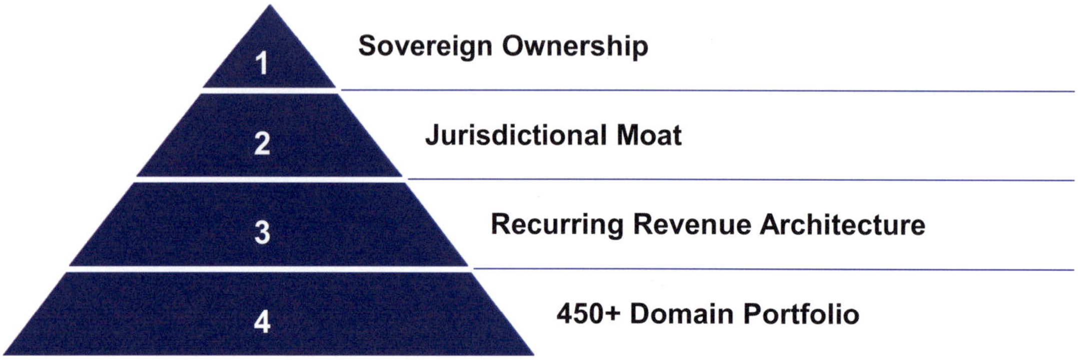

> **We do not need outside funding because we own the vault.**

The National & International Football Championships

The Apex Jurisdiction — The Protection Mandate

Sovereign ownership of football's highest-growth formats. Ethical governance of the world's most lucrative sport. A $60M annual revenue ecosystem built on the KJ Doctrine—where protecting the athlete *is* the multiplier.

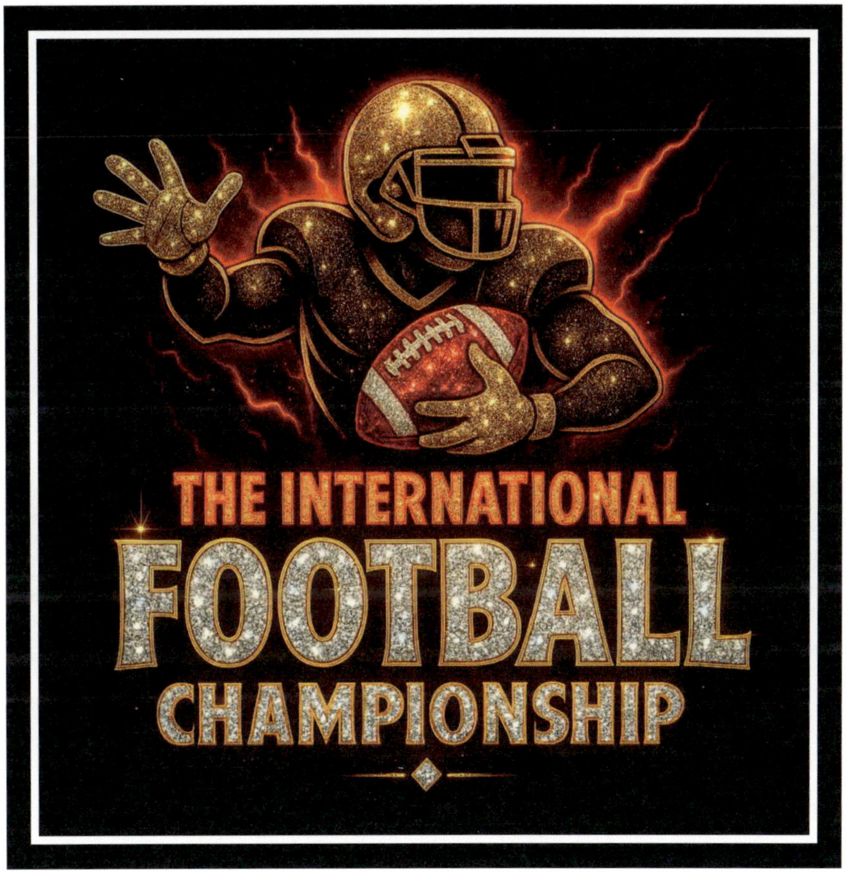

Governing the Cultural Apex

Football is the undisputed apex predator of the North American sports market. It commands the highest media rights valuations, the largest event attendance figures, and the highest per-athlete expenditure across youth, amateur, and adult segments. No other sport converts cultural attention into economic output at comparable scale. From Friday night lights to Super Bowl Sunday, football is not merely a game—it is a national economic engine, a social institution, and the most reliable commercial platform in American entertainment.

Yet the legacy infrastructure of amateur football is facing an existential inflection point. As awareness of foundational cognitive and physical trauma increases, traditional tackle football participation—particularly at the youth level—is contracting. Parents are asking harder questions. Medical professionals are issuing clearer warnings. Insurance carriers are recalculating risk. The result is a slow-motion exodus from the formats that built the sport's cultural throne.

Simultaneously, alternative formats such as 7-on-7 and Flag Football are experiencing explosive, Olympic-validated growth. These formats preserve the athleticism, strategy, and spectacle of the gridiron while dramatically reducing the injury profile that has become football's Achilles heel. Despite this momentum, however, these formats remain highly fragmented—controlled by decentralized regional promoters operating on rented platforms with no continuity, data sovereignty, or enforceable governance. The fastest-growing segments of the world's most valuable sport have no central authority, no data infrastructure, and no institutional spine.

Goddess Phenezy Holdings, Inc. resolves this structural contradiction through sovereign ownership and the strict application of the KJ Doctrine. We refuse to monetize the physical trauma of youth athletes. Instead, The National & International Football Championships strategically capture the highest-margin, highest-growth sectors of the sport under a single, owned jurisdiction. This is not participation in the world's most lucrative sport. ***This is ethical, jurisdictional ownership of it.***

The KJ Doctrine: Ethical Infrastructure as a Multiplier

The KJ Doctrine is the philosophical and operational foundation upon which the entire Football Jurisdiction is built. It asserts a simple but radical principle: the protection of the athlete and the profitability of the enterprise are not in conflict—they are inseparable. By redirecting capital, attention, and organizational energy away from formats that produce foundational trauma and toward formats that produce elite athletic development without that cost, the Doctrine unlocks markets that legacy operators cannot access.

Protect

Eliminate foundational cognitive and physical trauma from youth athletic development

Perform

Capture the explosive economics of 7-on-7 and Flag formats at Olympic-validated scale

Own

Sovereign jurisdiction over leagues, data, media, and championships with zero third-party leakage

> ***"We do not sacrifice the athlete to build the arena. We own the gridiron. We govern the territory. We write the rules."***

The Jurisdiction at a Glance

The National & International Football Championships is not a single event—it is an entire governed ecosystem. The jurisdiction encompasses every layer of competitive football outside the legacy tackle model, creating a vertically integrated authority over the sport's future.

1. **Owned Championships**

   Domestic and international championship series across Flag Football, 7-on-7, and Adult Pro-Am formats

2. **Owned Leagues**

   7-on-7 and Flag Football leagues with proprietary licensing, governance, and competitive structure

3. **NIL & Scouting Data**

   Proprietary collegiate scouting data infrastructure and verified athlete performance profiles

4. **Player Memberships**

   Recurring revenue through verified digital identity and monthly membership subscriptions

5. **Media & Broadcast**

   Direct-to-consumer broadcast and media rights through owned distribution via Xperience TV

The Market Inflection Point

The amateur football market is bifurcating. On one side, traditional tackle participation is declining—driven by parental concern over CTE research, rising insurance costs, and shrinking youth registration numbers that have been trending downward for over a decade. On the other side, alternative football formats are surging. Flag Football's inclusion in the 2028 Los Angeles Olympic Games has provided the single most powerful validation event in the history of non-tackle football, transforming a recreational format into an elite competitive pathway overnight.

Seven-on-seven football, meanwhile, has become the *de facto* developmental currency of elite high school and collegiate recruiting. The format strips football to its most athletic and strategic elements—route running, coverage, decision-making under pressure—without the collision profile that has driven the insurance and liability crisis engulfing youth tackle programs. Top 7-on-7 teams already command travel budgets, sponsorship deals, and media followings that rival traditional high school programs.

Yet despite this explosive growth, neither Flag Football nor 7-on-7 has a governing authority. There is no NFL equivalent for these formats—no single entity that controls competitive standards, owns the championship pathway, manages athlete data, or governs broadcast rights. The market is a patchwork of regional tournament operators, weekend promoters, and app-based registration platforms—none of which own their infrastructure, retain their data, or have the institutional credibility to negotiate at scale with media partners, sponsors, or collegiate recruiting networks.

This is the gap. This is the opportunity. Goddess Phenezy Holdings fills it—not as a participant, but as the jurisdiction itself.

Youth Tackle Decline vs. Alternative Format Growth

The data tells an unmistakable story. While youth tackle football participation has declined steadily over the past decade, Flag Football and 7-on-7 formats have experienced compound annual growth rates that signal a permanent structural shift in how football is played, consumed, and monetized in America and globally.

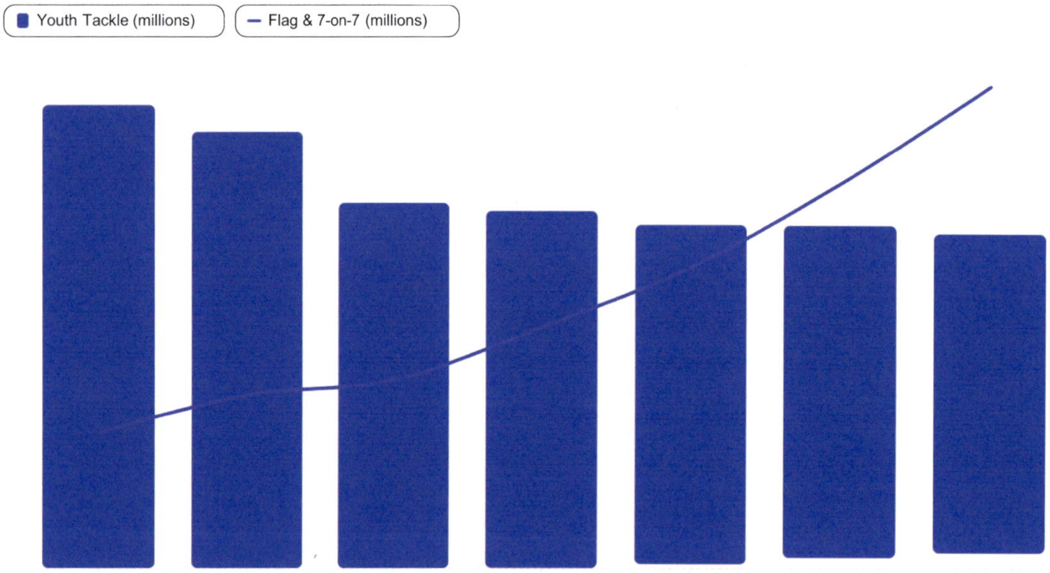

The crossover point arrived in 2022. Flag and 7-on-7 participation now significantly exceeds youth tackle football—and the gap is accelerating. The Olympic designation in 2028 will further catalyze this trend, making this the defining moment to establish jurisdictional authority over these formats.

The Fragmentation Problem

Today's alternative football landscape is a gold rush without a government. Hundreds of regional promoters, tournament operators, and league organizers are capitalizing on the surge in Flag and 7-on-7 participation—but none of them own the infrastructure they rely on. They rent registration platforms. They lease waiver software. They distribute content through third-party social media channels that harvest their data and control their reach. They have no continuity from season to season, no enforceable standards, and no ability to aggregate the athlete data that has become the most valuable currency in the NIL era.

Rented Infrastructure

Registration platforms, waiver software, and payment systems owned by Silicon Valley intermediaries — not the operators.

No Data Continuity

Athlete data harvested by third-party platforms. No season-to-season aggregation. No NIL-era scouting asset.

No Enforceable Standards

Hundreds of promoters, zero unified governance. No competitive weight. No institutional credibility.

No Media Infrastructure

Content distributed through social channels that control reach, harvest data, and capture the revenue.

No Championship Authority

Without a governing body, no championship carries real competitive weight or commands premium sponsorship.

No Aggregated Value

Fragmented operators cannot produce premium media deals, scouting data products, or institutional sponsorship frameworks.

This fragmentation is not merely an operational inconvenience—it is a structural vulnerability that suppresses the economic potential of the entire market. Without unified governance, there can be no premium media deals, no aggregated scouting data products, no institutional sponsorship frameworks, and no championship pathway that carries real competitive weight. Goddess Phenezy Holdings resolves every one of these deficiencies through sovereign jurisdictional ownership.

Structural Advantage: Ownership Over the Gridiron

Goddess Phenezy Holdings does not rent registration software, waiver platforms, or media distribution. We own the architectural baseline. This distinction is not semantic—it is the foundational competitive moat that separates our position from every other operator in the alternative football space. While competitors are tenants on platforms controlled by Silicon Valley intermediaries, we are sovereign owners of our entire operational, financial, and data infrastructure.

The Football Jurisdiction is not a loose affiliation of events branded under a common name. It is a vertically integrated institutional authority that controls every layer of the competitive, financial, and informational stack. From the moment an athlete registers to the moment a collegiate recruiter accesses their verified performance data, every interaction occurs within our owned ecosystem—captured, retained, and monetized without leakage.

Sovereign Owner

- Registration Infrastructure (Owned)
- Waiver & Compliance Platform (Owned)
- Media & Broadcast Distribution (Owned)
- Athlete Data & Analytics (Owned)
- Financial Processing (Owned)

Platform Tenants

- Registration Software (Rented)
- Waiver Platform (Rented)
- Media Distribution (Rented)
- Data Analytics (Rented)
- Payment Processing (Rented)

The Football Jurisdiction Architecture

The Football Jurisdiction comprises four institutional pillars, each of which reinforces and amplifies the others. Together, they form a closed-loop ecosystem with no external dependencies and no value leakage.

The National Football Championship

The premier domestic championship pathway for elite 7-on-7 and Flag Football competition across the United States

The International Football Championship

The global championship authority for Flag Football and adult exhibition competition across international markets

The International Football League

The governed league structure for all affiliated regional 7-on-7 and Flag licensing throughout the ecosystem

Regional Licensing Network

All affiliated regional 7-on-7 and Flag licensing operating under unified governance and standards

Four Pillars. One Sovereign Ecosystem. Every layer feeds the next — zero value leakage, zero external dependencies.

50+	30+	100K+	4
States	**Nations**	**Registered**	**Pillars**
Domestic Reach	International Markets	Athletes	Closed-Loop Architecture

Proprietary PWA Infrastructure

The entire Football Jurisdiction ecosystem operates through our proprietary Progressive Web App (PWA) infrastructure—a technology decision with profound economic and strategic implications. Unlike native mobile applications distributed through the Apple App Store or Google Play Store, the PWA architecture eliminates every intermediary fee, every platform commission, and every data-sharing obligation that erodes the margins and sovereignty of app-dependent competitors.

No App Store Fees

Zero commission on transactions. No 15–30% revenue share to Apple or Google on any purchase, subscription, or registration processed through the platform.

No Platform Commissions

Zero intermediary take-rate. Every dollar of league licensing, membership revenue, and scouting access flows directly into our treasury.

No Third-Party Data Leakage

Complete data sovereignty. Athlete profiles, performance metrics, transaction histories, and engagement data are retained exclusively within our infrastructure.

The Result

Every transaction—from an elite 7-on-7 team registration in Florida to an adult league membership in Texas—is processed, captured, and retained internally.

This creates an un-bypassable revenue and data pipeline for the sport's most valuable demographics.

While competitors pay rent to platforms they don't control, we collect rent from the infrastructure we own.

Ownership vs. Rental: The Competitive Moat

The difference between owning your infrastructure and renting it is not incremental—it is existential. In a market where margins determine survival and data determines valuation, the ownership model creates compounding advantages that rental-dependent competitors cannot replicate, no matter how aggressively they scale.

Dimension	Legacy Operators (Rental)	Goddess Phenezy (Ownership)
Registration Platform	Third-party SaaS (LeagueApps, etc.)	Proprietary PWA — fully owned
Transaction Fees	15–30% platform commissions	Zero intermediary fees
Athlete Data	Shared with platform providers	100% sovereign retention
Media Distribution	YouTube, social media (no control)	Xperience TV — owned broadcast
Championship Authority	None — event-by-event promotion	Institutional jurisdiction
Governance Standards	None — fragmented rules	Unified, enforceable standards
Valuation Basis	Event revenue only	Data + SaaS + Media + Licensing

This table illustrates why the ownership model is not simply a better business—it is a fundamentally different category of enterprise. Legacy operators are event companies. Goddess Phenezy Holdings is an infrastructure authority.

Revenue Architecture

Multi-Layer, Recurring, Scalable

The Football Jurisdiction's revenue architecture is designed to generate income from every layer of the competitive ecosystem—licensing, membership, events, data, and media—with each layer reinforcing the others in a compounding flywheel. This is not a single-revenue-stream business. It is a multi-layered economic engine where the addition of each team, each athlete, and each championship event amplifies the value of every other component in the system.

Revenue Stream 1: League & Team Licensing

The Core Economic Engine

Elite 7-on-7 circuits and Olympic-track Flag Football programs already operate on substantial travel and participation budgets. Families and organizations routinely spend tens of thousands of dollars annually across disconnected tournament operators, travel leagues, and regional circuits—with no unified competitive pathway, no data retention, and no institutional credibility to show for it. Goddess Phenezy Holdings consolidates these fragmented circuits under a single governed jurisdiction, transforming scattered expenditure into structured, recurring licensing revenue.

3,000

National Teams

U.S. 7-on-7, Flag, and Adult circuits licensed at $4,500 average annual fee

1,000

International Teams

Global Flag and Adult circuits licensed at $5,500 average annual fee

$19M

Annual Licensing Revenue

Total combined league licensing across domestic and international operations

Conservative Penetration: These figures reflect conservative estimates in a market where elite developmental programs routinely spend tens of thousands annually across disconnected tournament operators. The addressable market is significantly larger than current projections.

Revenue Stream 2: Player Memberships & Premium Scouting

The Data Monetization Layer

In the NIL era, verified performance data—not proximity or reputation—is the most valuable recruiting currency in football. Every registered athlete operates under a verified digital identity within the PWA, creating a persistent, authenticated profile that accumulates performance metrics, highlight footage, and competitive results over time. This is not a passive registration system—it is an active data asset that appreciates in value with every game played, every stat recorded, every season completed.

Player Memberships

01

15 Players Per Team

Blended 7-on-7, Flag, and Adult rosters

02

4,000 Active Teams

Combined domestic and international leagues

03

60,000 Active Players

Total registered athlete base

04

$300 Annual Membership

$25/month × 12 months per athlete

Total Player Membership Revenue: $18.0M annually

B2B Scouting & Recruiter SaaS

Verified performance data, 7-on-7 film, and biometric metrics are licensed to institutional buyers who depend on accurate, authenticated athlete information to make recruiting decisions worth millions of dollars:

- **Collegiate recruiting departments**
- **NIL collectives and agencies**
- **Independent scouting networks**

2,000 organizational accounts × $1,500/year = $3.0M annually

Combined Stream 2 Total: $21.0M annually

Revenue Stream 3: Championship Event Revenue

High-Margin Apex Convergences

Football championships operate as large-scale cultural convergence events—generating extreme gate density, VIP hospitality demand, premium apparel revenue, and sponsorship activation that far exceeds the per-event economics of regular season play. These are not merely games; they are destination events that draw families, scouts, media, sponsors, and community stakeholders into a concentrated economic ecosystem that produces outsized returns relative to operational cost.

1

Regional Qualifiers

Gate fees, team entry, local sponsorships

$2.0M

2

VIP Hospitality & Premium Packages

Premium seating, family experience bundles

$3.0M

3

International Championship

Global Flag participation, adult exhibition, international sponsorships

$4.5M

4

National Championship

Elite 7-on-7 & Flag Series, ticketing, protective-gear sponsorships

$7.5M

5

Total Championship Revenue

Combined annual event ecosystem

$12.0M+

Total Championship Revenue: $12.0M annually

Revenue Stream 4: Media, Content & Broadcast Rights

Owned Distribution Through Xperience TV

Elite 7-on-7 football produces some of the most viral highlight content in youth sports. The combination of elite athleticism, one-on-one matchups, and explosive plays creates content that generates millions of organic views across social platforms—views that currently benefit the platforms, not the athletes or the organizations that produce the competition. Through owned media infrastructure via Xperience TV, every snap becomes monetizable content within a sovereign broadcast ecosystem.

"The legacy networks wouldn't open the door. So Goddess Phenezy built her own house. Xperience TV was not a pivot — it was a mandate. When the gatekeepers said no, we stopped asking for permission and started owning the infrastructure."

The Result: A Sovereign Broadcast Ecosystem

Streaming Subscriptions

Direct-to-consumer access for families, recruiting fanbases, and football enthusiasts who want to watch every game, every play, every season.

Automated Highlight Packages

AI-generated highlight reels sold to NIL agencies and athletes, transforming raw game footage into premium recruiting content.

Sponsorship Integrations

Premium brand integrations within owned broadcasts—no third-party platform taking a cut of advertising revenue.

Conservative Annual Media & Content Revenue: $8.0M

The Revenue Flywheel

Each revenue stream does not operate in isolation—they compound. More teams mean more athletes, more athletes mean more data, more data means more scouting value, more scouting value means more media interest, more media interest drives more team registrations. This is a self-reinforcing flywheel that accelerates with every cycle.

Team Licensing

Teams join the governed jurisdiction

Brand & Growth

Institutional credibility attracts more teams

Championships

Apex events drive cultural convergence

Player Memberships

Athletes register verified digital identities

Data Accumulation

Performance metrics build over time

Media & Content

Every snap becomes owned content

6	4,000+	60,000+	$60M+
Self-Reinforcing Cycle	Licensed Teams	Active Athletes	Annual Ecosystem Revenue

Every registration fuels the data. Every game fuels the media. Every championship fuels the brand. The flywheel never stops.

$ FINANCIAL SUMMARY

Total Annual Football Ecosystem Revenue

The following table presents the consolidated annual revenue projection across all four streams of the Football Jurisdiction. These figures represent conservative baseline estimates built on verified market dynamics and existing expenditure patterns within the elite alternative football market.

Revenue Stream	Annual Value	% of Total
League Licensing (National + International)	$19.0M	31.7%
Player Memberships & Scouting SaaS	$21.0M	35.0%
Championships (National + International)	$12.0M	20.0%
Media & Data (Xperience TV)	$8.0M	13.3%
TOTAL ANNUAL REVENUE	$60.0M	100%

Revenue Distribution by Stream

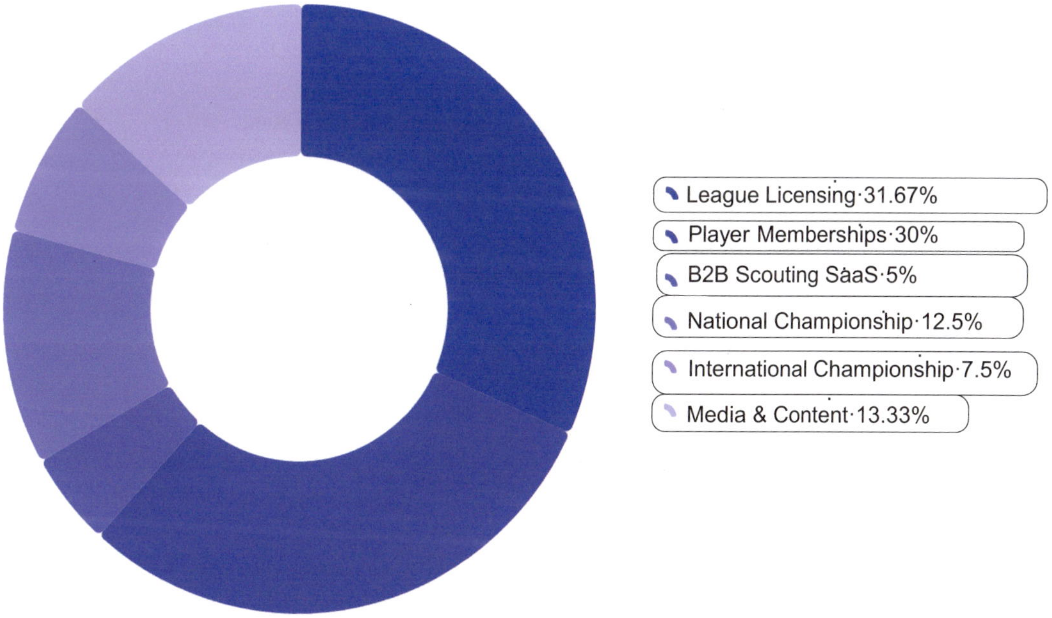

Revenue Composition: Recurring vs. Event-Based

One of the most strategically significant features of the Football Jurisdiction's revenue architecture is the dominance of recurring revenue over event-based income. This composition fundamentally changes the risk profile and valuation basis of the enterprise.

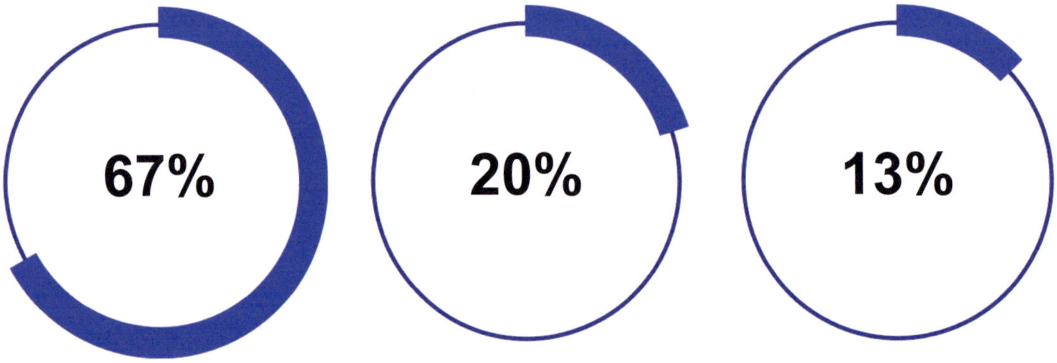

Recurring Revenue

League licensing, player memberships, and SaaS subscriptions generate predictable, annualized cash flow

Event Revenue

Championship events generate high-margin convergence revenue with VIP, gate, and sponsorship income

Media Revenue

Owned broadcast and content rights with growing subscriber base and sponsorship integration potential

A two-thirds recurring revenue base is the hallmark of enterprise-grade SaaS and platform businesses—it is virtually unheard of in sports league operations. This composition commands premium valuation multiples from institutional investors who prize predictability, defensibility, and compound growth.

Valuation Framework: Governed Apex Infrastructure

The Football Jurisdiction does not fit neatly into a single valuation category because it operates across multiple value-creation domains simultaneously. It is not merely a sports league—it is a hybrid enterprise that combines the institutional authority of a premier sports governing body, the data monetization engine of a high-value NIL and scouting broker, and the scalable technology economics of an enterprise SaaS platform.

This hybrid positioning demands a valuation framework that accounts for the compounding value of each layer, not simply the aggregate revenue. Institutional investors who evaluate comparable enterprises across sports technology, data platforms, and league ownership consistently apply premium multiples to businesses that demonstrate the combination of recurring revenue, data sovereignty, and institutional authority that defines the Football Jurisdiction.

1

Sports Governing Body

Institutional authority, championship rights, and brand equity · 4–6x Revenue Multiple

2

NIL & Scouting Data Broker

Verified athlete data, recruiter SaaS, and performance analytics · 6–8x Revenue Multiple

3

Enterprise SaaS Platform

Proprietary PWA, owned infrastructure, recurring subscription base · 8–10x Revenue Multiple

4

Hybrid Apex Jurisdiction

All three layers converging in a single sovereign ecosystem · $360M–$600M Valuation Range

The Hybrid Valuation Model

The Football Jurisdiction functions simultaneously as three distinct enterprise types—each of which independently commands premium institutional multiples. The convergence of all three within a single owned jurisdiction creates a valuation basis that exceeds the sum of its parts.

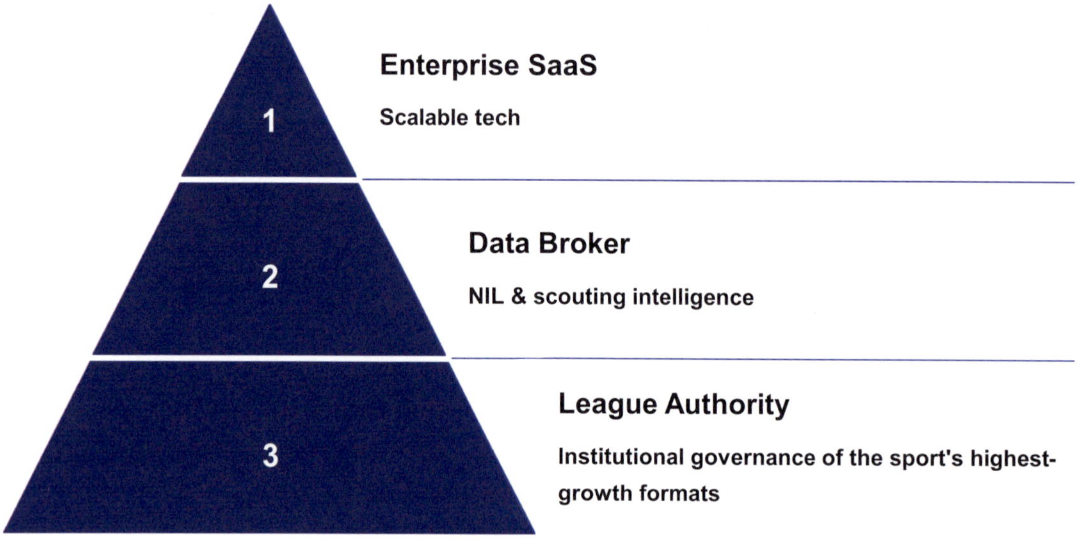

At the foundation sits the League Authority—the institutional credibility and governance rights that make everything above it possible. Upon this sits the Data Broker layer—the verified athlete data and scouting intelligence that transforms raw competition into monetizable intelligence. At the apex sits the Enterprise SaaS layer—the scalable technology infrastructure that delivers all of this with zero intermediary costs and infinite marginal scalability.

Valuation Range

Applying disciplined institutional multiples to the Football Jurisdiction's $60M annual revenue base produces a valuation range that reflects the strategic significance of capturing football's economic dominance while eliminating the foundational physical liabilities embedded in legacy models.

Base Case: 6× Revenue

$360M

Reflects comparable valuations for sports technology platforms with recurring revenue, data assets, and institutional governance authority.

Growth Case: 8× Revenue

$480M

Reflects the premium for Olympic-validated growth trajectory, complete data sovereignty, and the unique competitive moat of jurisdictional ownership.

These multiples are conservative relative to pure-play SaaS companies, which routinely command 10–15× revenue multiples. The application of 6–8× reflects the hybrid nature of the enterprise and the current maturity stage of the Football Jurisdiction's market penetration. As recurring revenue compounds and data assets appreciate, upward multiple expansion is the expected trajectory.

Valuation Waterfall: From Revenue to Enterprise Value

Each revenue component builds upon the last, creating a total annual revenue base of $60M. At 6× multiple, this translates to a $360M base-case enterprise valuation. At 8× multiple, reflecting the growth trajectory and competitive moat, the enterprise value reaches $480M. The waterfall illustrates how each revenue stream contributes incrementally to the total, with no single stream creating dangerous concentration risk.

The Protection Mandate in Practice

The KJ Doctrine is not merely a philosophical position—it is an operational reality embedded into every layer of the Football Jurisdiction. It manifests in format selection, in governance standards, in equipment requirements, in data protocols, and in the fundamental design of every competitive pathway within the ecosystem. Protecting the athlete is not a marketing message. It is the engineering principle upon which the entire enterprise is built.

For parents, this means confidence that their child's cognitive and physical development is not being sacrificed on the altar of competitive ambition. For coaches, it means access to formats that develop the skills that actually translate to higher levels of play without the injury profile that ends careers before they begin. For investors, it means exposure to the sport's most lucrative economics without the liability tail that haunts legacy tackle operations.

For Parents

Confidence that their child's cognitive and physical development is protected. No CTE risk. No compromised futures.

For Coaches

Access to formats that develop elite skills without the injury profile that ends careers before they begin.

For Investors

Exposure to football's most lucrative economics without the liability tail that haunts legacy tackle operations.

> *"We refuse to monetize the physical trauma of youth athletes."*

This is not a limitation on growth—it is the engine of growth. By operating in formats that are exploding in participation precisely because they eliminate the trauma concerns that are driving families away from traditional football, the Protection Mandate positions Goddess Phenezy Holdings on the right side of the most powerful demographic and cultural trend in American youth sports.

What the Protection Mandate Means for Every Stakeholder

The KJ Doctrine creates compounding value for every participant in the ecosystem. Protection is not a cost center — it is the value proposition that drives adoption, loyalty, and institutional credibility across all four stakeholder groups.

PARENTS

Peace of Mind

Compete in formats validated by medical science and Olympic institutions. No cognitive trauma. No liability waivers masking systemic risk. A genuine pathway to athletic excellence.

COACHES

Elite Development

Governed structures that develop the skills collegiate programs actually value — decision-making, route precision, coverage technique — within a framework that protects athletes and professional reputation.

INVESTORS

Clean Growth

Exposure to football's unmatched economic engine without litigation risk, insurance escalation, or reputational liability. Defensible margins. Institutional credibility.

POLICYMAKERS

Self-Governance Model

A model for how the private sector can protect youth welfare while generating economic growth and community engagement — without requiring legislative intervention.

The Sports Portfolio Strategic Matrix

Goddess Phenezy Holdings does not govern one sport. It governs all of them. The National & International Championships portfolio spans six major sports — creating the only multi-sport institutional authority of its kind in the world. Each sport is strategically positioned to serve a distinct function, unified under the KJ Doctrine's governing principle.

APEX ANCHOR

Football

Cultural dominance, ethical governance, $60M+ revenue. The proof of concept that protection is a multiplier.

VELOCITY

Basketball

Fast-cycle engagement, high media virality, and cultural currency in urban markets.

PIPELINE RETENTION

Baseball

Long-cycle development, family-oriented consumption, and deep seasonal loyalty.

GLOBAL SCALE

Soccer

International reach, diplomatic access, and explosive emerging market penetration.

PREMIUM ACCESS

Tennis

High-income demographic engagement, individual athlete branding, and global tournament authority.

GROWTH FRONTIER

Pickleball

The fastest-growing sport in America. Mass participation, cross-demographic appeal, and untapped institutional governance.

> **When you govern every sport, you govern the culture. When you govern the culture, you own the future.**

The Competitive Landscape: Why No One Else Can Do This

The Football Jurisdiction's competitive moat is not built on a single advantage—it is built on the convergence of multiple structural barriers that make replication extraordinarily difficult for any potential competitor.

Institutional First-Mover

Being the first governed jurisdiction for alternative football formats creates a network effect that compounds with every team and athlete registered

Proprietary Infrastructure

The owned PWA stack eliminates dependency on third-party platforms, creating zero-commission economics that competitors cannot match while renting

Data Sovereignty

Verified athlete data appreciates over time—the longer the jurisdiction operates, the more valuable the data asset becomes, creating an ever-widening moat

Ethical Brand Authority

The KJ Doctrine creates trust with parents, coaches, and policymakers that cannot be replicated by operators with tackle football liabilities in their portfolio

Olympic Alignment

Flag Football's 2028 Olympic inclusion validates the format at the highest possible institutional level—and this jurisdiction owns the championship pathway

The Path Forward: Scaling the Jurisdiction

The Football Jurisdiction's growth trajectory is not speculative—it is built on market dynamics already in motion. Flag Football participation is accelerating toward the 2028 Olympics. 7-on-7 has already established itself as the developmental standard for elite football prospects. Adult recreational and Pro-Am formats are growing as former athletes seek post-collegiate competitive outlets. Every one of these trends feeds directly into the governed ecosystem.

Phase 1: Foundation

4,000 teams established. Proprietary PWA infrastructure fully operational. Domestic governance locked.

Phase 2: Expansion

Scale to 8,000 teams. International Championship launched. Regional licensing network activated.

Phase 3: Olympic Acceleration

2028 LA Olympics drives explosive global Flag Football growth. Jurisdiction is the established authority.

Phase 4: Institutional Maturity

$100M+ annual revenue. Full-scale governed jurisdiction. Premier institutional sports asset.

Each phase builds upon the last, with the 2028 Olympic Games serving as the single most powerful external catalyst for growth that any emerging sports format has ever received. The jurisdiction is positioned to capture the full economic impact of that catalyst because it will already be the established institutional authority when the world's attention turns to Flag Football on the Olympic stage.

☆ CONCLUSION

We Own the Gridiron

The National & International Football Championships captures the economic dominance of the world's most lucrative sport while eliminating the physical liabilities driving the legacy model toward irrelevance.

> *"We do not sacrifice the athlete to build the arena. We own the gridiron. We govern the territory. We write the rules."*

We Own the Infrastructure

Proprietary PWA. Zero platform dependency. Every transaction, every data point — ours.

We Govern the Territory

The only institutional authority over elite alternative football in North America and beyond.

We Write the Rules

Format standards, athlete protections, competitive governance — set by us, enforced by us.

We Protect the Athlete

The KJ Doctrine is not a tagline. It is the engineering principle of every competitive pathway we govern.

$60M+	$360M–$600M	2028
Annual Ecosystem Revenue	Enterprise Valuation Range	Olympic Inflection Point

Goddess Phenezy Holdings, Inc. — The KJ Doctrine Applied. The Gridiron Owned. The Athlete Protected.

The National & International Baseball

Goddess Phenezy Holdings, Inc. — Making History & Memories While Owning the Narrative

Introduction: Unifying a Fragmented Empire

Baseball is one of the most entrenched and capital-intensive sports economies in the world. Domestically and internationally, it commands extreme brand loyalty, generational participation, and billions in annual spend across youth, travel, amateur, and professional pipelines. Families invest tens of thousands of dollars per year chasing advancement through a labyrinth of disconnected operators, rented platforms, and regional gatekeepers who offer no continuity and no ownership.

Yet despite its scale, the modern baseball ecosystem is structurally broken. The global "travel ball" and amateur circuits remain highly fragmented—controlled by disjointed regional promoters, disconnected tournament operators, and rented technology platforms. Teams and families pay exorbitant fees into systems that offer no continuity, no ownership, and no technological sovereignty. Every dollar spent vanishes into someone else's infrastructure.

> Goddess Phenezy Holdings, Inc. resolves this fragmentation through absolute jurisdictional ownership. The National & International Baseball Championships are not isolated tournaments. They operate as a fully integrated baseball economy—a closed-loop system where every registration, every data point, every broadcast, and every transaction flows through infrastructure we own outright.

The Integrated Baseball Economy

What distinguishes this jurisdiction from every legacy tournament operator is the scope of what is owned—not rented, not licensed from a third party, not dependent on external approval. This is a vertically integrated sports infrastructure designed to capture value at every layer of the baseball pipeline.

Owned Championships

Domestic and international championship properties operated as apex destination events with premium economic density.

Owned Leagues

National and global league structures governing year-round competition under a single jurisdictional authority.

Proprietary Data Infrastructure

Scouting, performance analytics, and player identity systems built on sovereign technology—no third-party leakage.

Recurring Memberships

Player memberships and B2B SaaS subscriptions generating predictable, scalable annual revenue.

Owned Media Rights

Direct-to-consumer and B2B media distribution through proprietary broadcast infrastructure.

> **This is not participation revenue. This is pipeline ownership.**

Baseball- A Must Have For the Portfolio

Beyond its comprehensive structural ownership of baseball's development infrastructure, this jurisdiction is fundamentally committed to empowering families and fostering youth development. Its integrated data pipeline is meticulously designed to create a transparent, fair, and accessible ecosystem for every young athlete, irrespective of their background. From initial player registration and verification systems that ensure a safe and equitable playing field for all families, to efficient tournament scheduling and bracket management that respects family time, every technical layer serves a higher purpose. Real-time live scoring and statistics tracking keep families engaged and allow for the celebration of individual and team achievements. Advanced performance metrics and analytics (e.g., exit velocity, spin rate) provide unprecedented transparency, empowering families with objective insights into their child's development. Integrated video capture and game footage offer invaluable tools for growth and shared memories. Detailed scouting reports and player profiles, alongside expansive college recruiting databases, democratize access to advancement opportunities, connecting deserving young athletes and their families to scholarships and future pathways. Finally, exclusive broadcast distribution through Xperience TV ensures that the spirit of community and the achievements of all players are visible, celebrated, and shared, truly building a mission-driven, community-focused baseball experience.

Competitive Landscape of Major Sports

Baseball holds a unique and competitive position in the youth sports economy, offering significant opportunities for participation and professional advancement compared to other major sports.

Sport	Youth Participants (U.S.)	NCAA Scholarships Available	Average Scholarship Value	Average Pro Salary
Football	~1.0M	~85 per school	~$35K	~$2.7M (NFL)
Basketball	~540K	~13 per school	~$40K	~$8.5M (NBA)
Soccer	~3.0M	~9.9 per school	~$25K	~$470K (MLS)
Baseball	~480K	~11.7 per school	~$30K	~$4.4M (MLB)

Structural Advantage: Ownership Over the Data Pipeline

Goddess Phenezy Holdings owns its entire architectural baseline, from registration to media distribution. This direct ownership provides a significant structural margin advantage and sovereign operational control, unlike competitors who pay costly 'tolls' to third-party technology monopolies.

1

Registration & Player Data

Owned foundational layer for secure player information.

2

Scheduling & Tournament Management

Proprietary platforms for efficient tournament and game management.

3

Live Scoring & Statistics Tracking

Proprietary systems for real-time data without external dependencies.

4

Video Capture & Content Creation

Owned infrastructure for full control over media assets.

5

Media Distribution & Broadcast (Xperience TV)

Wholly owned channels for direct broadcast and zero commissions.

The Baseball Jurisdiction wholly owns all major championships, leagues, and affiliated regional licensing, operating under a unified governance structure that eliminates market fragmentation. The proprietary Progressive Web App (PWA) infrastructure ensures all transactions are processed internally, eliminating App Store fees, platform commissions, and third-party data leakage.

This establishes an un-bypassable data and revenue pipeline. For investors, every dollar and data point compounds the platform's value within this owned infrastructure, creating a powerful, self-sustaining ecosystem with no viable competitor exit.

The Proprietary PWA Advantage

Understanding why the Progressive Web App architecture matters requires understanding where legacy operators hemorrhage margin. Every tournament that processes registrations through a third-party platform pays a toll. Every broadcast routed through YouTube or a social media platform surrenders data ownership. Every transaction processed through an app store forfeits up to 30% in commission.

No App Store Fees

Zero commission on transactions. Full margin retention on every registration, subscription, and purchase.

No Platform Commissions

No third-party scheduling or media platforms extracting percentage-based tolls from revenue.

No Data Leakage

Player data, performance metrics, and behavioral analytics remain sovereign—never shared with external platforms.

Full Pipeline Control

Every transaction from every geography processed, captured, and retained internally.

REVENUE ARCHITECTURE

Multi-Layer. Recurring. Scalable.

Four interlocking revenue engines generate $72.1 million annually through league licensing, player memberships, championship events, and owned media distribution.

League Licensing $23.0M annually	**Player Memberships** $19.6M annually
Championship Events $18.5M annually	**Media & Broadcast** $11.0M annually

Revenue Layer 1: League & Team Licensing

The Core Economic Engine

The primary revenue stream comes from licensing the league's intellectual property, such as team names, logos, and broadcast rights, to national and international partners. These multi-year contracts include annual escalation clauses, ensuring consistent, high-margin income and predictable growth.

National League Licensing $12.5M annually	**International League Licensing** $10.5M annually

Total League Licensing Revenue: $23.0M annually

Revenue Layer 2: Player Memberships & Premium Scouting Access

The Data Monetization Layer

Player advancement relies on verified data, currently scattered and unverified across platforms. Our PWA provides every registered player with a verified digital identity, a persistent record owned by our infrastructure, ensuring reliable data for career progression.

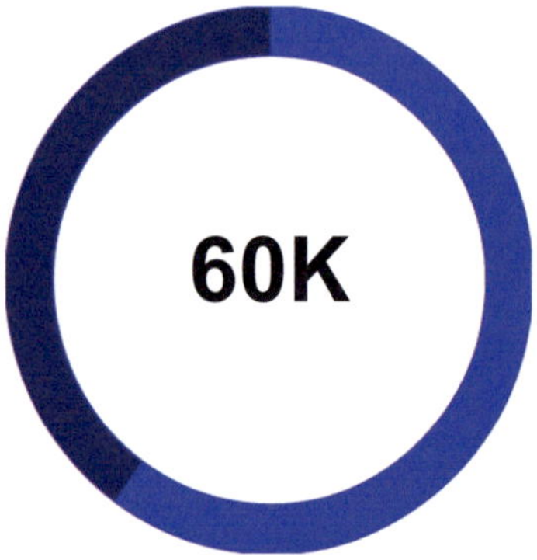

Active Players

Membership Fee

Annual per Player

Total Revenue

With 60,000 active players (15 players/team across 4,000 teams) paying $30/month, this generates a powerful $21.6 million annually in recurring membership revenue.

B2B Scouting & Recruiter SaaS Access

The second dimension of data monetization targets the institutional buyers of player intelligence. Collegiate programs need verified performance data to make scholarship decisions. Professional scouts need persistent career records to identify talent across geographies. International agencies need standardized metrics to facilitate cross-border player movement.

01

Data Collection

Players generate verified performance data

02

Platform Access

Institutions subscribe to SaaS platform

03

Decision Making

Scouts/recruiters analyze player metrics

04

Player Movement

Scholarships, signings, and transfers

1

Collegiate Programs

Universities and junior colleges accessing verified performance data for recruitment decisions across thousands of prospects.

2

Professional Scouts

MLB organizations and independent scouts leveraging persistent career records and advanced metrics for talent identification.

3

International Agencies

Cross-border scouting operations using standardized data to facilitate player movement between leagues and nations.

At 10,000 organizational accounts paying $500 per year, the B2B scouting SaaS generates $5.0 million annually. Combined with player memberships, the total data monetization layer produces $26.6 million—making it the single largest revenue stream in the baseball jurisdiction.

Revenue Layer 3: Championship Event Revenue

High-Margin Apex Events

Baseball championships function as multi-day, destination-based tournaments, generating exceptional gate density and host-city economic participation. These are not single-game events. They are week-long economic engines that transform host cities into temporary baseball capitals, driving hotel occupancy, restaurant revenue, retail spending, and corporate hospitality at concentrated scale.

National Championship

$8.5M annually

International Championship

$6.0M annually

Combined Total

$14.5M

The National Championship operates as a multi-week summer series—the culmination of the domestic league season. Revenue streams include ticketing, premium team entry fees, and equipment sponsorships from brands seeking direct access to the most engaged baseball families in the country. This single property generates a projected $8.5 million annually.

The International Championship expands the model across borders, incorporating global participation, travel packages, and elite sponsorships from brands seeking international visibility within the baseball community. With projected revenue of $6.0 million annually, the combined championship revenue reaches $14.5 million—all flowing through owned infrastructure with no third-party extraction.

Revenue Layer 4: Media, Content & Broadcast Rights

Owned Distribution Through Xperience TV

Travel baseball produces hundreds of games per weekend across the country and around the world. In the legacy model, this content is either lost entirely or captured on personal devices and uploaded to platforms that monetize it for someone else's benefit. Through owned media infrastructure—Xperience TV—every pitch becomes monetizable content flowing through sovereign distribution.

Streaming Subscriptions

Families and fans subscribe to watch games live, creating a persistent direct-to-consumer revenue stream.

Automated Highlights

AI-driven highlight generation packages individual player moments as digital goods.

International Broadcast Sublicensing

Apex championship events are sublicensed to international broadcasters, generating B2B media revenue.

8M

Annual Media Revenue

Conservative estimate for content and broadcast rights

Total Annual Baseball Revenue: $72.1 Million

The four revenue layers combine to produce a diversified, recurring, and scalable annual revenue base that no single event cancellation, no seasonal disruption, and no competitive entrant can meaningfully threaten—because the revenue is structural, not promotional.

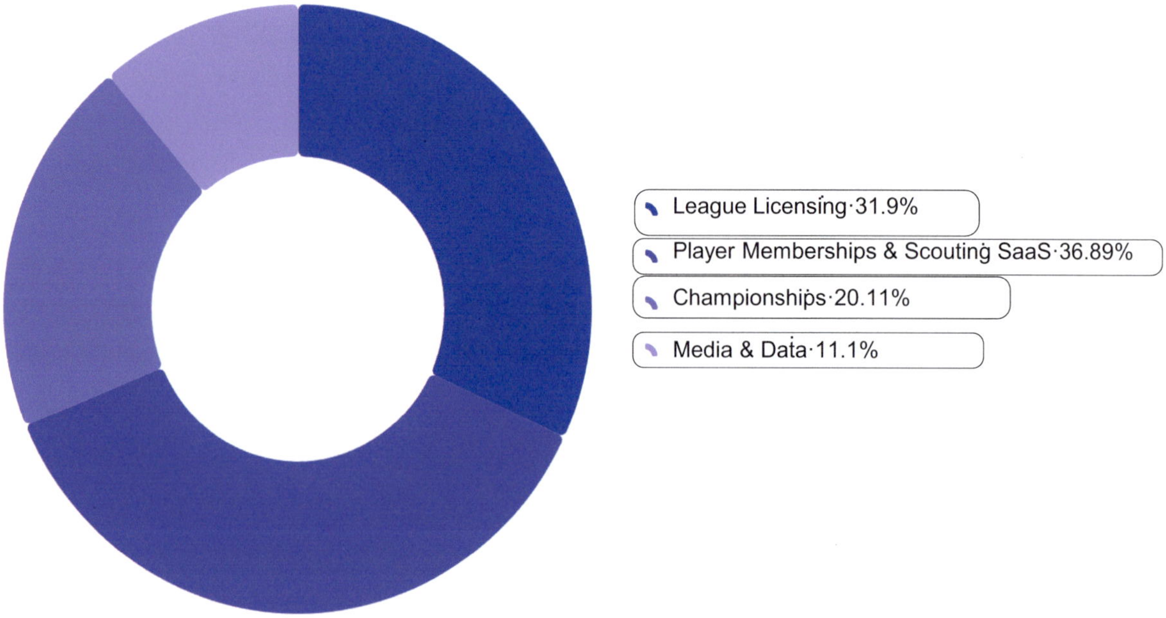

Player Memberships & Scouting SaaS represents the largest single revenue stream at $26.6M (37% of total), validating the thesis that data ownership is the highest-value layer in the modern sports economy. League licensing provides the foundational recurring base, championships deliver high-margin event density, and media rights convert every moment of play into monetizable content.

Valuation Framework: Asset-Based Monopoly

The Baseball Jurisdiction cannot be valued as a simple tournament operation or event promotion company. It functions as a hybrid of three high-multiple asset classes, each of which commands premium institutional valuation in isolation—and together, they create a compounding effect that legacy operators cannot replicate.

Premier Sports League Operator

Governing year-round competition across domestic and international markets with exclusive jurisdictional authority.

Global Media Network

Owned broadcast infrastructure distributing hundreds of hours of live content to families, fans, and international licensees.

B2B Scouting & Data SaaS

Enterprise-grade platform licensing verified performance data to collegiate programs, professional scouts, and international agencies.

Base Case (6× Revenue)

$432 Million

Reflects institutional blended multiple applied to recurring, diversified revenue with owned infrastructure and no platform dependency.

Growth Case (8× Revenue)

$576 Million

Reflects strategic value of the un-bypassable data pipeline for the world's premier bat-and-ball sport with international expansion runway.

Baseball Strategic Significance

Basketball establishes velocity. Baseball establishes retention, continuity, and pipeline discipline. The travel ball economy is not trend-driven or seasonal—it is generational. Families enter the system when their children are eight years old and remain engaged through high school graduation, college recruitment, and beyond. This creates a retention profile that no other youth sports vertical can match.

Together with the basketball jurisdiction, baseball demonstrates that the Goddess Phenezy Holdings ecosystem is not event-dependent, trend-dependent, or platform-dependent. It is system-dependent. That distinction is where real valuation lives. Systems compound. Events decay. Every family that enters the baseball pipeline deepens the data moat, expands the media catalog, and strengthens the licensing network—creating a flywheel that accelerates with time rather than requiring constant reinvestment to maintain.

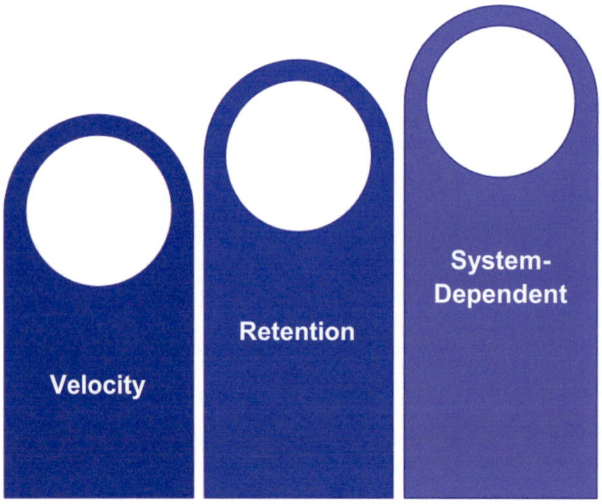

Investors should understand this progression as intentional architecture, not coincidence. Each jurisdiction is designed to prove a different thesis about the system's capability—and baseball's thesis is the most powerful for long-term institutional valuation: the system retains.

SOCCER JURISDICTION

The Global Phenomenon

The world's most ubiquitous athletic economy—captured under a single owned digital flag through sovereign infrastructure and the Global Player Passport.

Introduction: The World's Game as Sovereign Territory

While basketball establishes domestic velocity and baseball secures pipeline retention, soccer represents the ultimate scale. It is the most ubiquitous athletic economy on Earth—driving tens of billions of dollars annually through youth clubs, international academies, travel circuits, and professional development pipelines. No other sport touches as many nations, languages, or socioeconomic strata. From the favelas of São Paulo to the academies of Manchester, soccer is the universal language of athletic ambition.

Yet despite its global dominance, the modern soccer development ecosystem is structurally fractured. Clubs operate on rented, localized software. Player data is lost at borders. Scouting depends on fragmented video platforms and unverifiable statistics. Advancement relies more on proximity and privilege than performance. A talented player in West Africa has virtually no path to verified visibility in European scouting databases—not because the talent doesn't exist, but because the infrastructure doesn't connect.

The National & International Soccer Championships were architected to unify this global chaos under a single, owned digital flag. Goddess Phenezy Holdings, Inc. controls the apex international championships, the domestic and global leagues, the proprietary scouting and Global Player Passport infrastructure, and direct-to-consumer international broadcast rights. This is not participation in the world's game. It is jurisdictional ownership of it.

Fragmented Infrastructure

Clubs operate on rented, localized software.

Lost Player Data

Performance data disappears at borders.

Unverified Scouting

Fragmented video platforms and unverifiable statistics.

Privilege Over Performance

Advancement relies on proximity, not talent.

Structural Advantage: Borderless Economic Capture

In an international sport, platform fees are lethal. Legacy tournament operators lose massive margin to app store taxation, foreign transaction fees, and third-party scheduling, registration, and video platforms. Every dollar that passes through someone else's infrastructure is a dollar that compounds for someone else's shareholders. At global scale, these tolls become existential—consuming margins that should be fueling growth.

The Soccer Jurisdiction bypasses these costs entirely. By operating exclusively through a proprietary Progressive Web App (PWA) infrastructure, every transaction is processed natively—without tolls to technology monopolies. Every club registration from London. Every player membership from São Paulo. Every broadcast subscription from Los Angeles. All captured. All retained.

London	São Paulo	Los Angeles
Club registrations processed natively through owned PWA infrastructure	Player memberships captured without foreign transaction platform fees	Broadcast subscriptions retained with zero app store commission

> **This is not a workaround. It is borderless economic sovereignty.**

Soccer Revenue Layer 1: Academy, Club & League Licensing

The Jurisdictional Core

Youth and academy soccer operates year-round and at massive scale. Unlike baseball's seasonal rhythms, soccer clubs train, compete, and develop talent across all twelve months. Infrastructure, continuity, and data governance are non-negotiable for serious programs—and Goddess Phenezy Holdings licenses exclusive operation under its soccer jurisdiction to meet that demand.

The licensing structure spans two geographic pillars. The domestic market encompasses U.S. and Canadian travel circuits, where the youth soccer economy has exploded over the past decade. The international market extends across Europe, South America, Africa, and Asia—regions where soccer is not merely popular but culturally essential, and where the absence of unified digital governance creates enormous opportunity for consolidation.

National Leagues (U.S. & Canada)

3,000 clubs × $4,000 average annual license

$12.0M

International Leagues & Academies

4,000 clubs × $5,000 average annual license

$20.0M

Total League & Academy Licensing Revenue: $32.0M annually. These figures represent conservative penetration within a market already accustomed to significant annual spend across fragmented operators.

Soccer Revenue Layer 2: Player Memberships & The Global Scouting Passport

The Data & Advancement Layer

Soccer advancement is driven by international transfers, scouting visibility, and verified performance. The difference between a player who gets discovered and one who doesn't is increasingly a function of data access—not raw talent. The Global Scouting Passport eliminates this inequity by creating a persistent digital identity for every registered player.

Every registered player is issued a Global Scouting Passport—a persistent digital identity containing verified match data, performance metrics, video reels, and eligibility and advancement history. This passport travels with the player across borders, across seasons, and across competition levels. It is the definitive record of athletic development, and it lives on owned infrastructure.

The membership scale is extraordinary: 18 players per team across 7,000 teams yields 126,000 active players. At $25 per month—$300 annually per player—the total player membership revenue reaches $37.8 million annually. This is the highest single-line-item revenue stream across any individual jurisdiction in the Goddess Phenezy Holdings portfolio, reflecting soccer's unmatched global participation density.

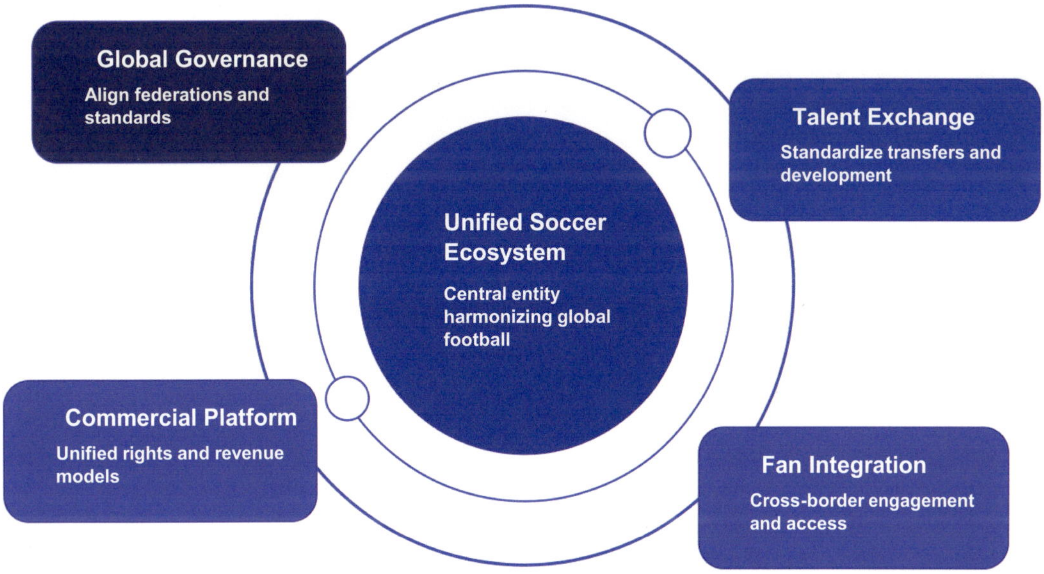

Global Scouting Passport: What It Contains

The Global Scouting Passport is not a simple registration profile. It is an institutional-grade data asset that transforms every player into a verifiable, discoverable prospect—regardless of geography, socioeconomic status, or proximity to traditional scouting networks.

Verified Match Data

Every competitive appearance documented with standardized metrics—goals, assists, minutes, disciplinary records—verified by the jurisdiction's own officiating infrastructure.

Performance Metrics

Speed, endurance, technical proficiency, and tactical intelligence measured and benchmarked against global cohorts in the same age group and position.

Video Reels

AI-curated highlight packages generated from every match—automatically packaged and permanently linked to the player's passport for instant scout access.

Eligibility & Advancement

Complete history of league participation, championship qualifications, and advancement milestones—creating an unbroken chain of verified athletic progression.

B2B International Scouting & Federation Access

The institutional side of the data economy is where the Global Scouting Passport generates its highest per-account value. Professional academies, universities, international agencies, and national federations all require standardized, verifiable player intelligence to make decisions that carry significant financial and organizational consequences.

1 Professional Academies

Elite development programs accessing the passport database to identify talent across borders without deploying physical scouts to every tournament.

2 Universities

Collegiate programs leveraging verified data for scholarship decisions, reducing risk and expanding their geographic recruitment reach.

3 International Agencies

Player representation firms using standardized metrics to facilitate cross-border transfers and contract negotiations.

4 National Federations

Governing bodies accessing the platform for talent identification, youth national team selection, and development program optimization.

At 5,000 organizational accounts paying $1,200 per year, the B2B scouting and federation SaaS generates $6.0 million annually. Combined with player memberships, the total soccer data monetization layer produces $43.8 million—the single most powerful revenue engine in the entire portfolio.

Soccer Revenue Layer 3: Championship Events & Global Invitationals

Apex Cultural Convergence

Soccer championships function as multi-day, multi-venue festivals—comparable to the Gothia Cup or Dallas Cup—often transforming host cities into temporary global hubs. These are not tournaments in the traditional sense. They are cultural events that draw participants, families, scouts, sponsors, and media from dozens of countries simultaneously, creating economic density that no single-day event can approach.

The National Soccer Championship generates revenue through team entry fees, ticketing, brand villages, and vendor leases—producing a projected $9.5 million annually. The International Soccer Championship operates at an even higher tier, incorporating global participation, airline partnerships, financial institution sponsorships, and global apparel sponsors seeking direct access to the next generation of talent. This international apex event alone generates a projected $14.0 million annually.

$9.5M

National Championship

Team fees, ticketing, brand villages, vendor leases

$14.0M

International Championship

Global participation, airline partnerships, elite sponsorships

$23.5M

Total Championship Revenue

Combined annual apex event revenue

Soccer Revenue Layer 4: Media, Data & International Broadcast Rights

Always-On Global Attention

Soccer commands the highest digital viewership of any sport on Earth. The appetite for content is insatiable—and it extends far beyond professional leagues. Youth soccer families are among the most engaged digital consumers in sports, willing to pay for live streams of their children's games, for highlight packages, and for performance analytics that track development over time.

Through owned media infrastructure—Xperience TV—the soccer ecosystem captures value across three distinct channels. Multi-language streaming subscriptions serve the global family audience. Algorithmic highlight packaging transforms every match into curated content products. International broadcast sublicensing monetizes apex events through traditional media channels hungry for premium soccer content.

Multi-Language Streaming

Direct-to-consumer subscriptions serving families and fans across every continent in their native language—no third-party platform required.

Algorithmic Highlights

AI-driven packaging of individual player moments, team recaps, and tournament summaries—sold as premium digital goods and scout intelligence.

International Sublicensing

Apex championship broadcast rights sublicensed to international media networks, generating B2B revenue at institutional scale.

Conservative Annual Media & Content Revenue: $15.0M

Total Annual Soccer Ecosystem Revenue: $114.3 Million

The soccer jurisdiction produces the largest annual revenue of any single sport within the Goddess Phenezy Holdings portfolio—a testament to soccer's unmatched global scale and the power of sovereign infrastructure applied to the world's most participated sport.

The revenue distribution confirms a critical pattern: data monetization is the dominant value layer. Player Memberships and Scouting SaaS account for 38% of total soccer revenue, reinforcing the thesis that owning the identity and performance data of athletes is more valuable than owning the events where they compete. The events drive participation into the data pipeline; the data pipeline generates the highest-margin, most recurring revenue.

Soccer Revenue Comparison: Baseball vs. Soccer

Comparing the two fully developed jurisdictions reveals how the same architectural model produces dramatically different scale outcomes when applied to sports with different geographic footprints. The architecture is identical—the multiplier is geography.

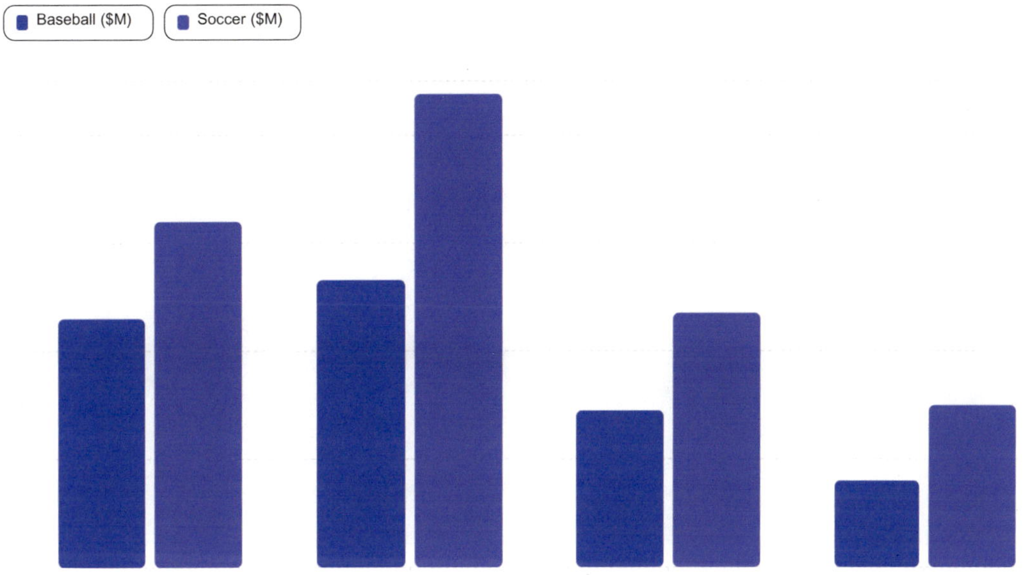

Soccer's total revenue of $114.3M represents a 58% premium over baseball's $72.1M—driven entirely by the sport's borderless participation base. The same PWA infrastructure, the same data pipeline architecture, the same media distribution system produces nearly $42 million more annually simply because soccer touches more nations, more languages, and more families. This is the power of a system designed for global replication.

Soccer Valuation Framework: The Global Scale Multiplier

Soccer operates as a rare hybrid of three institutional asset classes—global sports infrastructure, enterprise SaaS, and international media brokerage. This is not a regional tournament company. It is a borderless data and media authority for the world's most valuable sport, and it must be valued accordingly.

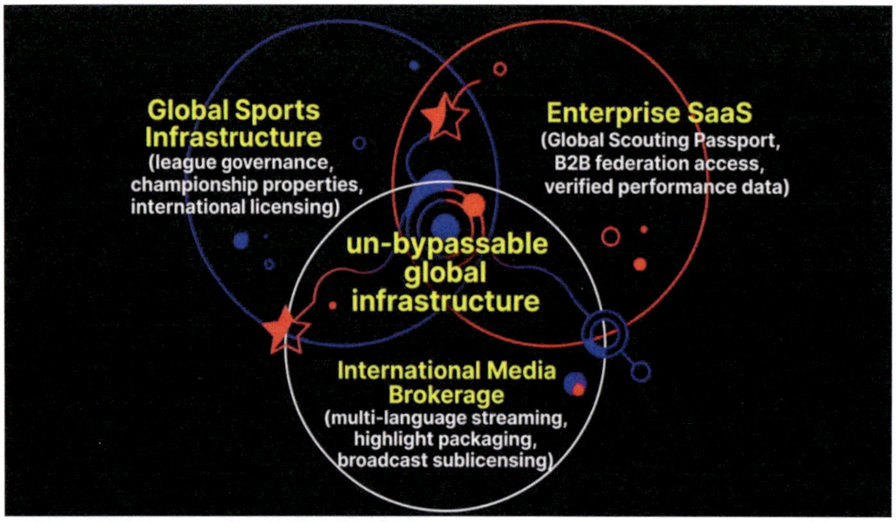

Base Case (6× Revenue)

$685 Million

Institutional blended multiple reflecting recurring revenue, owned infrastructure, and global market position.

Growth Case (8× Revenue)

$914 Million

Strategic premium reflecting un-bypassable global infrastructure and international expansion runway.

Soccer Strategic Significance

If basketball proves the model works, and baseball proves the model retains—soccer proves the model scales globally. This is the definitive stress test of the Goddess Phenezy Holdings architecture. Can a system designed for domestic market capture operate seamlessly across continents, languages, currencies, and regulatory environments? The soccer jurisdiction answers definitively: yes.

Basketball

Proves the Model Works (Velocity & Market Ignition)

Baseball

Proves the Model Retains (Pipeline Discipline & Continuity)

Soccer

Proves the Model Scales Globally (Cross-Continental Replication)

By capturing the international soccer economy through sovereign digital infrastructure, Goddess Phenezy Holdings establishes itself as an unavoidable governing layer for the next generation of global talent. The Global Scouting Passport ensures that every player who enters the system—regardless of where they are born, what language they speak, or what economic circumstances they face—has a verified, persistent, discoverable identity in the world's largest talent marketplace.

We do not require permission from legacy federations. We own the environment where discovery happens. And in the 21st century, the environment where discovery happens is more powerful than the institution that grants credentials. The soccer jurisdiction is not merely a revenue center—it is proof of concept for infinite geographic replication.

CUJU JURISDICTION

The Cuju Revival

Monopolizing the ancestry of the world's game. The only modern sports property built on origin without competition—a clean-slate cultural economy governed entirely by its steward.

Introduction: Monopolizing the Ancestry of the World's Game

While modern soccer has grown into a multi-billion-dollar global industry, its governance is locked behind century-old federations and entrenched institutional frameworks. Legacy organizations control the rules, the competitions, and the commercial relationships—creating barriers to entry that no new operator can overcome through superior technology or capital alone. The game is governed. The territory is claimed.

Yet the origin of the game itself remains untouched. Cuju—widely recognized by historians as the earliest documented form of organized football—predates modern soccer by more than two thousand years. It demanded extraordinary technical skill, acrobatic precision, and disciplined coordination, functioning as both athletic competition and cultural practice across ancient Chinese dynasties. Despite its historical primacy, Cuju was never industrialized, commercialized, or governed within the modern sports economy.

It remains a cultural asset without an owner. Until now. Goddess Phenezy Holdings, Inc. is not introducing another tournament. We are reviving a dormant global property and establishing its modern jurisdiction. There is no legacy governing body. No inherited federation. No competing authority. The Cuju Revival enters the modern era as a clean-slate sport, governed entirely by its steward. This is the rarest asset class in all of sports: origin without competition.

Structural Advantage: Complete Format & Governance Control

Operating modern sports means capturing value within preexisting cultural and institutional boundaries. Operating Cuju means defining the boundaries themselves. This distinction is the single most important structural advantage in the entire Goddess Phenezy Holdings portfolio—and potentially in the entire modern sports economy.

The Cuju Revival benefits from a rare structural advantage that no other jurisdiction possesses. There are no competing leagues to fragment attention or negotiate with. No legacy rule-makers whose permission is required. No fragmented ownership to consolidate. No political encumbrance from entrenched interests.

01

Official Modern Ruleset

Goddess Phenezy Holdings defines the rules of play—adapting ancient principles into a modern competitive format optimized for spectacle and broadcast.

02

Competition Format

Tournament structures, league seasons, and exhibition circuits are designed from scratch—no inherited formats to accommodate or legacy agreements to honor.

03

Championship Structure

Apex championship properties are wholly owned, with no revenue sharing obligations to external governing bodies or federation partners.

04

Athlete Pathway

The development pipeline from amateur to professional is architected internally—creating a closed-loop talent ecosystem with no external gatekeepers.

05

Digital & Media Environment

Every broadcast, every data point, every transaction flows through proprietary infrastructure—sovereign from launch with no migration required.

> **This is not market entry. It is market creation.**

Positioning: Heritage as Premium Entertainment

Cuju is not positioned as a grassroots sport. It is not introduced as a youth development pipeline or a community recreation initiative. It is introduced as a premium cultural and athletic spectacle, occupying a space between elite athletic competition, historical storytelling, and high-end global entertainment.

This positioning is deliberate and strategic. By entering the market at the top rather than the bottom, the Cuju Revival avoids the margin compression that plagues grassroots sports operators while capturing the premium pricing power that comes with cultural exclusivity and narrative uniqueness.

Ancient in Origin

Two thousand years of documented history provide an authenticity and depth that no modern sport can fabricate. The narrative practically writes itself.

Modern in Execution

Contemporary production values, digital infrastructure, and broadcast technology transform ancient competition into 21st-century spectacle.

Exclusive in Access

Premium ticket pricing, limited touring schedules, and curated experiences create scarcity that drives demand and protects margins.

Comparable not to local leagues, but to globally touring cultural properties that blend performance, history, and athletic excellence—think Cirque du Soleil meets the Champions League, with the narrative weight of a civilization behind it.

Cuju Revenue Layer 1: Premium Touring & Exhibition Leagues

The Spectacle Layer

The Cuju Revival launches as a top-down touring property hosted in major global cities, cultural capitals, and international festivals. This is not a local league building from the ground up. It is a global exhibition circuit designed to generate immediate premium revenue while building the brand equity necessary for subsequent expansion into structured competition.

Revenue includes high-ticket arena exhibitions where audiences experience Cuju as a live spectacle for the first time, international cultural showcases partnering with museums, heritage organizations, and diplomatic events, and corporate and diplomatic hospitality packages that leverage the unique positioning of ancient sport revived for the modern era.

Premium Exhibition Matches

High-profile, high-ticket events in iconic global cities, delivering an unparalleled cultural and athletic spectacle.

International Festival Partnerships

Strategic integration with major cultural festivals and heritage events, amplifying reach and prestige through unique performances.

Corporate & Institutional Sponsorships

Exclusive partnerships with premium brands and cultural institutions, leveraging Cuju's unique narrative for high-value engagement.

Projected Annual Revenue: $18.5M

Cuju Revenue Layer 2: Educational & Cultural Licensing

The Heritage Layer

Because Cuju is historically documented and culturally significant, it uniquely qualifies for educational and cultural partnerships unavailable to modern sports. This is a revenue layer that literally does not exist for any other sports property in the Goddess Phenezy Holdings portfolio—or for any competitor in the broader sports economy. It is a function of Cuju's unique identity as both a sport and a piece of world heritage.

University & Academy Licensing

Higher education institutions license the Cuju format for physical education programs, cultural studies curricula, and campus athletic offerings—creating a new category of institutional revenue that compounds as academic adoption spreads.

Curriculum Integrations

K-12 educational partnerships incorporating Cuju's history into world history and physical education programs, creating brand awareness among the next generation of participants and consumers.

Cultural Festival Anchors

International cultural festivals license Cuju exhibitions as headline attractions—leveraging the property's unique blend of athleticism and historical significance to anchor multi-day events.

Projected Annual Licensing & Cultural Revenue: $7.2M

Cuju Revenue Layer 3: Media, Storytelling & Merchandise

The Narrative Layer

The resurrection of the world's oldest organized sport is inherently cinematic. This is not marketing spin—it is structural truth. A two-thousand-year-old athletic tradition, dormant for centuries, revived with modern technology and brought to global audiences for the first time in human history. Every major streaming platform, every documentary producer, every content studio in the world would recognize this as a premium narrative property.

Distributed through owned media infrastructure, the Cuju narrative supports original docuseries exploring the sport's ancient origins and modern revival, global broadcast content covering exhibitions, championships, and athlete profiles, and high-end heritage apparel and equipment that transforms participants and fans into brand ambassadors for a cultural movement.

Documentary & Streaming Content

The cinematic story of Cuju's revival across platforms.

Premium Merchandise & Heritage Apparel

Culturally significant branded products blending ancient aesthetics with modern design.

International Broadcast & Media Rights

Global distribution of Cuju events, exhibitions, and championships.

Projected Media & Merchandise Revenue: $11.2M

Total Annual Cuju Ecosystem Revenue: $39.7 Million

For a sport that did not exist in the modern commercial landscape before Goddess Phenezy Holdings revived it, $39.7 million in annual revenue represents an extraordinary achievement—and a powerful proof of concept for the system's ability to create markets from whole cloth.

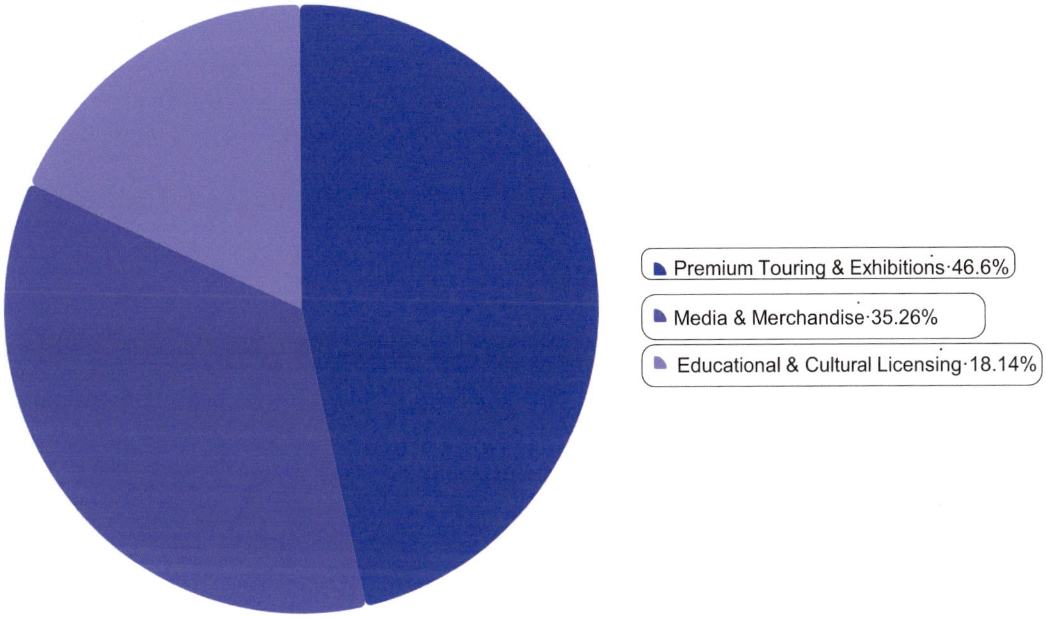

The revenue distribution reveals Cuju's unique economic character. Unlike baseball and soccer—where data monetization dominates—Cuju's revenue is driven by the spectacle and narrative layers. Premium Touring & Exhibitions account for 47% of total revenue, reflecting Cuju's positioning as a cultural entertainment property rather than a grassroots participation sport. Media & Merchandise at 35% confirms the narrative's inherent monetizability. Educational & Cultural Licensing at 18% represents a revenue category that exists nowhere else in the portfolio.

Cuju Valuation Framework: Exclusive Origin IP

Cuju cannot be valued as a traditional tournament or league. It functions as a wholly stewarded cultural entertainment property, where the format is owned, the narrative is controlled, the competition structure is exclusive, and the brand has no substitutes. There is no comparable operator, no competing federation, no alternative version of the world's oldest football code. The asset is singular.

Format Ownership

The modern ruleset, competition structure, and athlete pathway are wholly defined and governed by Goddess Phenezy Holdings.

Narrative Control

The story of Cuju's revival—from ancient origins to modern spectacle—is exclusively told through owned media infrastructure.

Exclusive Competition

Every sanctioned Cuju event in the world operates under a single jurisdictional authority. There are no alternatives.

Zero Substitutes

No competing property can claim the origin of football. The brand is historically unique and commercially uncontested.

Cuju Valuation Multiples

Comparable models for Cuju valuation include premium combat sports properties that command outsized multiples through format exclusivity, touring cultural spectacles that generate revenue through scarcity and narrative premium, and exclusive global entertainment IP that operates without competition or substitution. Each of these comparables commands multiples well above standard sports event operators.

Base Case (7× Revenue)

$277 Million

Conservative premium IP multiple reflecting exclusive governance, cultural significance, and zero competitive overlap.

Heritage Growth Case (10× Revenue)

$397 Million

Reflects the rarest asset class in sports: origin without competition, with unlimited geographic and cultural expansion runway.

> This valuation reflects the rarest asset class in sports: origin without competition. There is no ceiling on Cuju's cultural penetration because there is no competitor to share the market with.

Cuju Strategic Significance

Basketball proves the system can ignite. Baseball proves the system can retain. Soccer proves the system can scale. Cuju proves something far more powerful: the system can create markets where none previously existed.

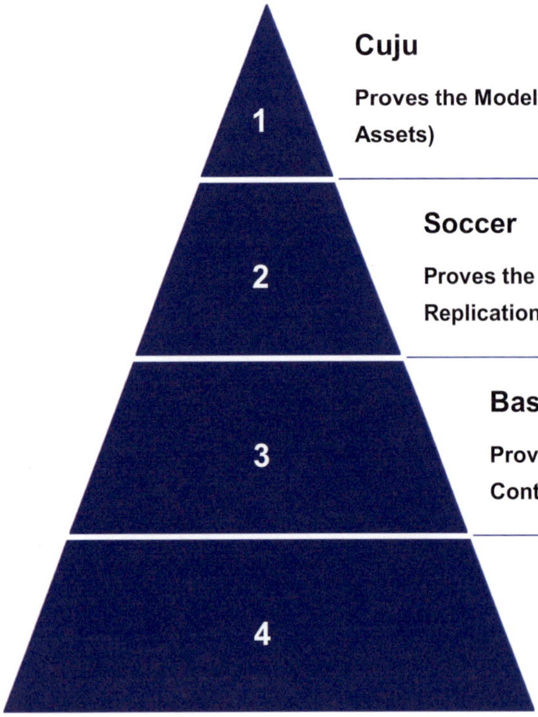

Cuju

Proves the Model Creates Markets (Resurrects Dormant Cultural Assets)

Soccer

Proves the Model Scales Globally (Cross-Continental Replication)

Baseball

Proves the Model Retains (Pipeline Discipline & Continuity)

Basketball

Proves the Model Ignites (Velocity & Market Entry)

By reclaiming and governing the world's first football code, Goddess Phenezy Holdings demonstrates that its infrastructure is not limited to modern competition—it is capable of resurrecting history, formalizing it, and transforming it into a living global economy. This is not a theoretical capability. The revenue projections, the structural advantages, and the valuation framework all confirm that dormant cultural assets can be revived into institutional-grade commercial properties when governed by the right infrastructure.

We are not competing for market share. We are claiming the source. And in doing so, we prove the most important thesis of all: the Goddess Phenezy Holdings system is not sport-specific. It is not era-specific. It is not geography-specific. It is a universal governance engine capable of capturing value from any athletic or cultural property—including those that have been dormant for millennia.

CONSOLIDATED ANALYSIS

The Unified Empire

Three jurisdictions. One architecture. A combined annual revenue of $226.1 million flowing through sovereign infrastructure with no third-party dependency.

Combined Revenue: All Three Jurisdictions

This consolidated view highlights how the shared architectural model generates significant cumulative revenue. Each jurisdiction, while unique in its market dynamics, contributes to a robust financial ecosystem, demonstrating the compounding effect of unified infrastructure.

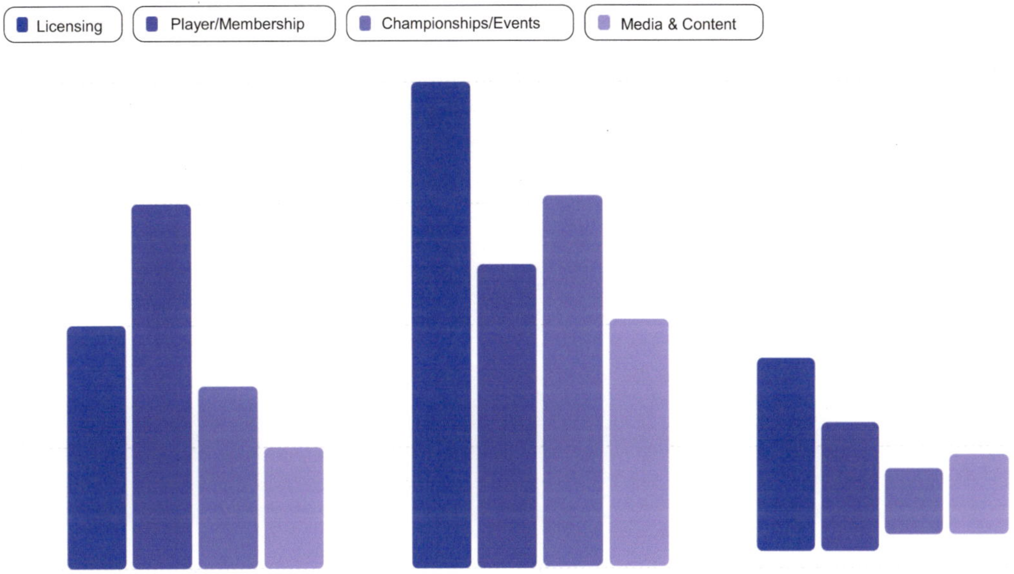

This chart illustrates how each jurisdiction leverages different revenue concentrations to contribute to the overall financial strength. From global media rights and licensing to robust player memberships and event championships, the diverse streams underscore the adaptability and comprehensive value capture of our model.

Consolidated Revenue Summary

Revenue Stream	Baseball	Soccer	Cuju
Licensing	$23.0M	$32.0M	$7.2M
Player/Scouting	$26.6M	$43.8M	—
Championships	$14.5M	$23.5M	$18.5M
Media & Content	$8.0M	$15.0M	$14.0M
Total	$72.1M	$114.3M	$39.7M

$226.1M

Combined Annual Revenue

Three jurisdictions operating on sovereign infrastructure

186K+

Active Players

Across baseball and soccer digital identity systems

11K+

Licensed Clubs & Teams

Operating under unified jurisdictional governance

Consolidated Valuation Framework

Each jurisdiction carries its own valuation multiple reflecting its unique asset characteristics. Baseball commands 6-8× as a pipeline and data monopoly. Soccer commands 6-8× as a global infrastructure play. Cuju commands 7-10× as an exclusive cultural IP property. The consolidated valuation represents the sum of these distinct but architecturally unified assets.

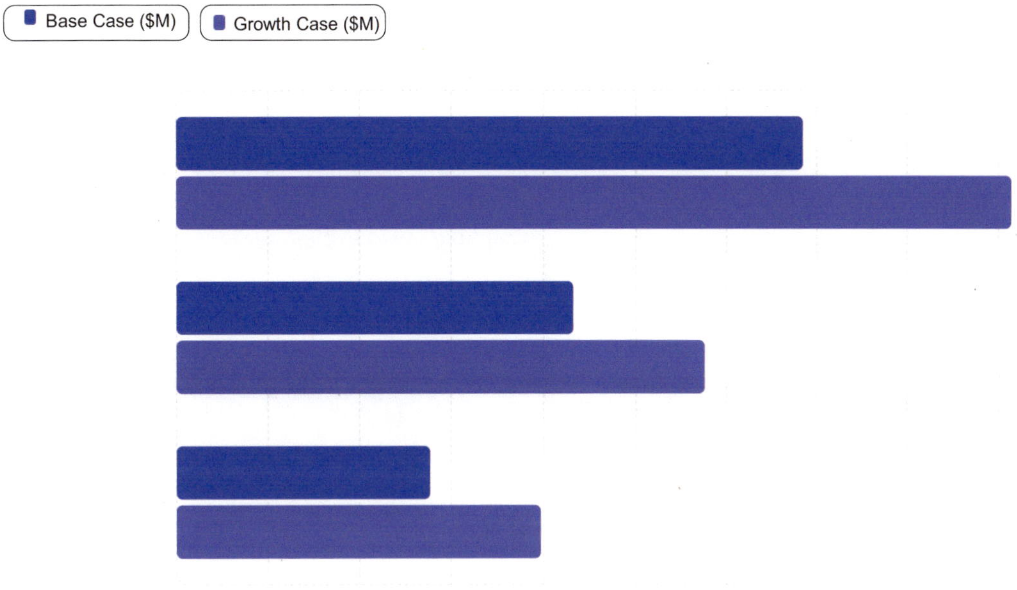

The consolidated valuation range spans from $1.394 billion (base case) to $1.887 billion (growth case)—establishing Goddess Phenezy Holdings as a multi-billion-dollar sports infrastructure enterprise on the strength of just three jurisdictions, with the architectural capacity to add more.

The Strategic Progression

The sequence of jurisdictions is not accidental. It is a carefully architected progression that demonstrates escalating capability—each sport proving a more ambitious thesis about what the Goddess Phenezy Holdings system can achieve.

Basketball → Velocity

Proves the system can ignite domestic markets rapidly, capturing attention and revenue in the most culturally visible sports economy in America.

Baseball → Retention

Proves the system retains participants across generational timelines, building compounding data assets and recurring licensing revenue over decades.

Soccer → Global Scale

Proves the system operates seamlessly across continents, languages, and currencies—capturing the world's largest sports participation base.

Cuju → Market Creation

Proves the system can create entirely new markets from dormant cultural assets—the ultimate demonstration of infrastructural power.

The Architecture Is the Moat

Competitors can attempt to replicate a tournament. They can attempt to copy a league format. They can attempt to build a scouting database. But they cannot replicate the jurisdictional authority—the integrated, sovereign, un-bypassable architecture that processes every transaction, captures every data point, and retains every dollar within a system designed for absolute ownership.

Players Enter

Teams and players join the jurisdiction through licensed clubs

Pipeline Locks

Switching costs become insurmountable as data dependencies deepen

Data Compounds

Performance data accumulates in sovereign infrastructure

Content Monetizes

Every game becomes owned media distributed globally

Value Grows

Network effects attract more teams, deepening the moat

This flywheel is the reason the system is not event-dependent, trend-dependent, or platform-dependent. It is system-dependent. And that distinction is where real valuation lives.

System-Dependent.

Systems compound. Events decay. Goddess Phenezy Holdings owns the system—and every participant, every data point, and every broadcast deepens the moat that competitors cannot cross.

Jurisdictions

Annual Revenue

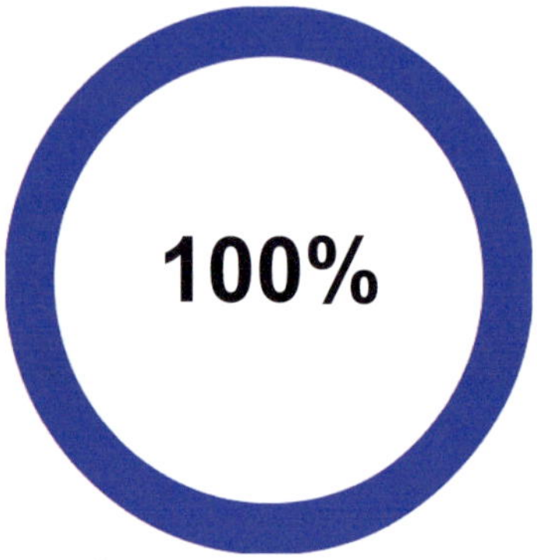

Owned Infrastructure

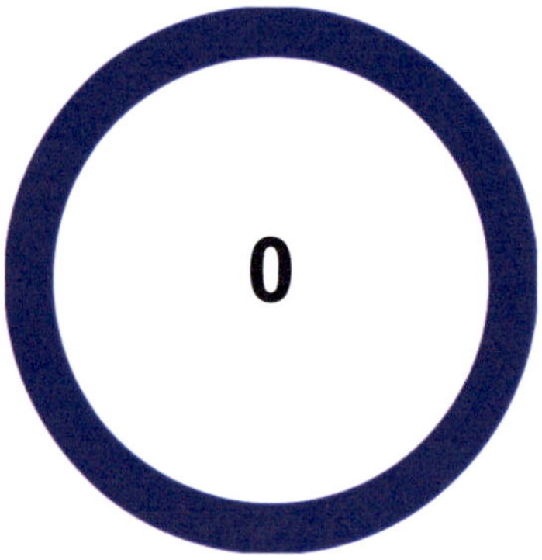

Third-Party Dependencies

> **We do not require permission from legacy federations. We own the environment where discovery happens. We are not competing for market share. We are claiming the source.**

The National & International Golf Championships

The Prestige Jurisdiction: Capital, Access & Quiet Power

A comprehensive investment thesis detailing how Goddess Phenezy Holdings, Inc. is recomposing access to golf's elite infrastructure—and engineering a $36M annual revenue ecosystem built on value density, executive relationships, and capital adjacency.

GOLF CHAMPIONSHIPS

Owning the Game Behind the Deals

Golf is not the loudest sport in the world. It does not need to be. It is the sport of boardrooms disguised as fairways—where capital moves quietly, relationships compound over decades, and access is often more valuable than trophies. In the corridors of institutional power, golf has always served as more than recreation. It is a language of trust, a mechanism for proximity, and a proving ground for the kind of relationships that move markets.

Historically, golf has been governed by exclusivity—structures designed not to maximize participation, but to curate belonging. The sport's gatekeeping apparatus has functioned for generations through a deeply entrenched set of norms and institutions that have defined who plays, where they play, and with whom.

Private Clubs

Membership-gated access controlling who enters the ecosystem

Invitation-Only Tournaments

Participation determined by social capital, not athletic merit alone

Legacy Institutions

Generational incumbency that resists modernization

Unwritten Rules

Belonging defined by proximity and permission, not talent

Ownership has been restricted not by talent, but by proximity and permission. The question has never been whether the sport creates value—it unequivocally does—but rather who is permitted to capture that value. This is the structural asymmetry that defines the opportunity.

A New Entrant. A New Logic.

Goddess Phenezy Holdings, Inc. enters golf without asking for proximity or permission. The National & International Golf Championships are not positioned as country-club replicas. They are not designed to inherit the aesthetics of exclusivity while perpetuating its economic constraints. Instead, they are engineered as a modern prestige infrastructure—one that preserves the sport's elite positioning while removing its gatekeeping economics.

This distinction is critical for investors and strategic partners to understand. The thesis is not one of disruption—tearing down golf's carefully cultivated mystique serves no one. Rather, this is a thesis of recomposition. The same prestige. The same capital adjacency. The same executive density. But with consolidated ownership, digital-native infrastructure, and a revenue architecture designed for scalable value capture.

> **This is not disruption. It is recomposition of access.**

Traditional Golf Model

Fragmented ownership, gatekeeping economics, institutional dependency

The Recomposition

Consolidated control, digital infrastructure, scalable value capture

Preserved Elements

Elite positioning, capital adjacency, executive density

New Advantages

Modern access, unified ownership, revenue architecture

Structural Advantage

Prestige Without Dependence

The Problem with Traditional Golf Economics

The existing economic models supporting elite golf events, though seemingly robust, are fundamentally misaligned with modern scaling imperatives. These legacy structures, often rooted in an antiquated understanding of value capture and distribution, create bottlenecks that prevent optimal revenue generation and audience engagement. We've identified four critical areas of structural inefficiency:

Club-Controlled Calendars

Reliance on individual club schedules creates a fragmented, unpredictable event calendar that hinders consistent audience building and long-term planning.

Sponsor-Heavy Dependency

An over-reliance on a few major sponsors limits diversified revenue streams and makes events vulnerable to shifts in corporate marketing budgets.

Fragmented Event Ownership

Dispersed ownership of intellectual property and event rights complicates unified branding, media negotiations, and overall market leverage.

Third-Party Media Contracts

Outsourced media rights dilute direct control over content distribution, limiting innovation in broadcast formats and direct monetization opportunities.

This fundamental misalignment means that while the prestige of golf remains high, its economic engine is inefficient, leaving significant value on the table for all stakeholders.

The Golf Jurisdiction: Consolidated Control

The Golf Jurisdiction consolidates control across the entire championship ecosystem. Rather than operating within the constraints of existing institutional frameworks, Goddess Phenezy Holdings, Inc. builds the framework itself—owning every layer of the value chain from event creation through audience engagement through revenue capture.

National Championships

Owned domestic tournament series across premium U.S. markets

International Championships

Global destination-hosted events attracting executive participation worldwide

Invitational Series

Curated, invitation-only competitive circuits for qualified participants

Corporate & Executive Leagues

Structured programs linking competitive golf to business relationship building

All delivered through proprietary digital infrastructure, eliminating platform taxes, ticketing leakage, and intermediary control. This is not an incremental improvement to existing tournament management. It is architectural—a purpose-built system designed from inception to consolidate ownership, data, and margin.

Value Density Over Volume

The Core Principle

Golf does not scale through volume.

It scales through value density.

Every participant, every sponsor, every media touchpoint is worth more per unit than in any other sport. The strategy is not to maximize headcount—it is to maximize the economic value of every interaction within the ecosystem.

Destination-Hosted Premium Events

Each championship event is designed as a multi-day, destination-driven experience that combines world-class competition with luxury hospitality and high-level business networking. Participants do not simply attend—they invest time, travel, and social capital into an experience that generates returns far beyond the fairway.

This model transforms each event from a cost center requiring external subsidy into a self-sustaining prestige asset that appreciates through reputation, repeat attendance, and deepening corporate relationships over successive seasons.

Premium Value per Participant

Unlike mass-market sports focused on ticket volume, our model emphasizes high-value engagement, maximizing economic return from each individual.

Multi-Day Immersive Engagement

Beyond single-event transactions, we offer multi-day experiences, fostering deeper connections and greater value extraction over time.

Cultivating Relationship Capital

Our focus is on building lasting relationships and social capital among participants, far surpassing mere transactional attendance.

Elevated Corporate Networking

We provide unparalleled opportunities for high-level business networking, distinct from general spectator entertainment prevalent in other sports.

POSITIONING

The Modern Invitational Economy

The National & International Golf Championships are designed around a fundamentally different participation model than traditional mass-market sporting events. This is not a volume play. This is a curation play—engineered to attract and serve the highest-value segments of the sports-business landscape.

Corporate Leadership

C-suite executives and senior decision-makers who use competitive golf as a relationship accelerant. These participants bring deal flow, brand credibility, and institutional capital into the ecosystem with every round played.

Executive Participation

VP-level and above professionals whose competitive engagement creates natural networking density. Every foursome is a potential partnership, every pro-am is a pipeline of future business.

High-Net-Worth Leisure Markets

Affluent individuals for whom premium golf experiences represent both personal recreation and social capital investment. These participants drive ancillary revenue in hospitality, travel, and luxury goods.

Destination-Based Prestige Events

Championship venues selected for their ability to combine world-class course conditions with luxury resort infrastructure, creating immersive multi-day experiences that justify premium pricing.

> This ecosystem does not chase mass participation. It curates qualified presence.

Revenue Architecture

High-Margin, Low-Volume, Maximum Signal

Revenue Layer 1: Invitational Tournaments & Destination Events

The Prestige Layer

At the apex of the revenue architecture are the tentpole events: the National Invitational Series and the International Golf Championship. These events are specifically designed for exclusivity, drawing top-tier participants and commanding premium sponsorships and participation fees. Think high-end, bespoke experiences rather than mass-market appeal.

These events generate significant revenue through a combination of corporate partnerships, individual entry fees, and destination-based hospitality packages. They are not merely tournaments; they are carefully curated business and leisure ecosystems.

$9.0M	$11.0M	$20.0M
National Invitational Series	International Golf Championship	Total Event Revenue

Revenue Layer 2: Corporate, Executive & Membership Access

The Relationship Economy

Golf is uniquely suited for executive leagues and corporate engagement. No other sport provides the same combination of sustained interpersonal interaction (four to five hours per round), controlled competitive dynamics, and social prestige. These characteristics make golf the premier platform for relationship-driven monetization.

The Relationship Economy layer captures recurring revenue through three primary mechanisms:

01

Corporate League Licensing

Companies license the right to field teams in structured competitive leagues, creating annual recurring revenue anchored to corporate budgets.

02

Executive Memberships

Individual memberships providing access to exclusive championship events, networking circles, and priority registration for premium experiences.

03

Private Championship Tiers

Ultra-premium access levels that provide invitational-only tournament entry, dedicated hospitality, and concierge-level event coordination.

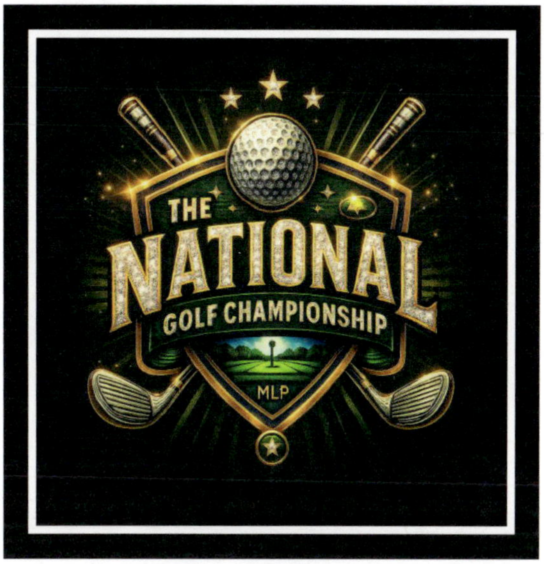

$7.5M

Annual Revenue

Corporate and executive participation revenue

Revenue Layer 3: Sponsorship, Brand Alignment & Media

The Quiet Capital Layer

Golf attracts a distinct category of sponsors—organizations whose brand identity aligns with discretion, sophistication, and long-term value creation. Financial institutions, luxury brands, real estate developers, and global consultancies do not seek the mass impressions of stadium advertising. They seek precision alignment with audiences that match their ideal client profiles.

This creates a sponsorship dynamic that is the inverse of most sports properties. Sponsorships are fewer in number—but significantly higher in per-unit value. A single title partnership in the golf ecosystem carries more strategic weight than dozens of category sponsorships in volume-driven sports.

Title Partnerships

Presenting sponsorships that position premium brands alongside the championship brand itself, commanding seven-figure annual commitments.

Premium Placements

Integrated brand experiences within the championship environment—from on-course activations to hospitality suite naming rights.

Content & Highlight Licensing

Proprietary media content distributed through owned digital channels, eliminating intermediary margin and capturing full advertising value.

$8.5M

Annual Sponsorship & Media Revenue

Conservative annual projection reflecting premium but selective brand partnerships

Total Annual Golf Ecosystem Revenue

The three revenue layers combine to create a $36 million annual revenue ecosystem characterized by high margins, low volume dependency, and maximum signal-to-noise ratio. Each layer reinforces the others: events create the environment for relationships, relationships attract sponsors, and sponsors elevate the prestige that drives event demand.

Revenue Stream	Annual Value
Invitational & Destination Events	$20.0M
Corporate & Executive Access	$7.5M
Sponsorship & Media	$8.5M
Total Annual Revenue	$36.0 Million

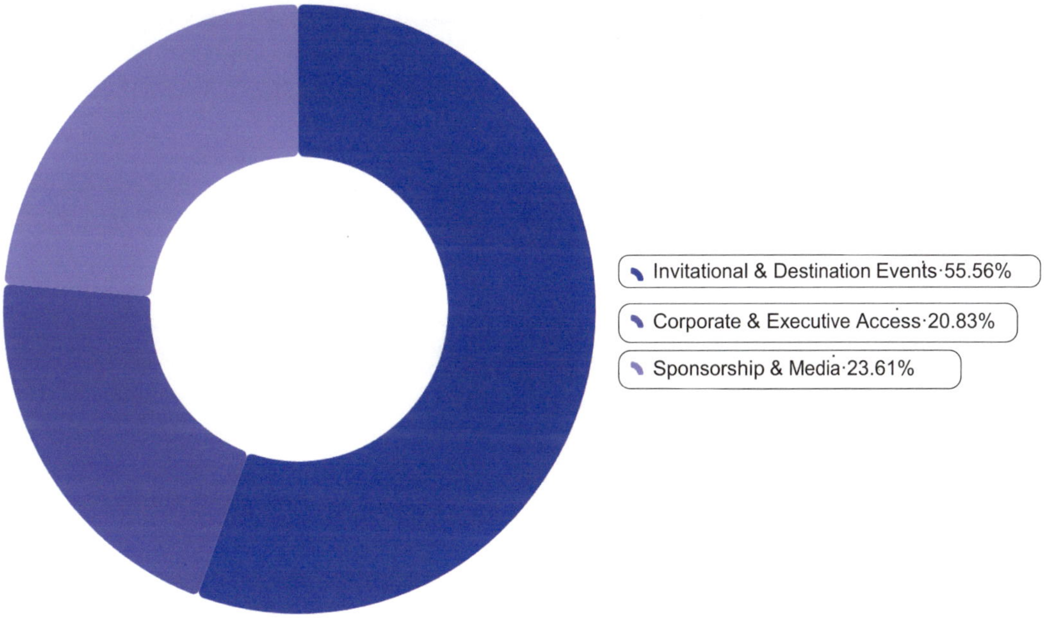

The revenue composition reflects a deliberately balanced architecture. Invitational events constitute 56% of total revenue, providing the experiential foundation. Corporate access contributes 21%, representing the recurring relationship economy. Sponsorship and media deliver 23%, anchored by high-value, long-term brand partnerships rather than transactional advertising.

Why Golf Commands a Different Multiple

Golf operates under a fundamentally different valuation logic than mass-participation sports. Its value does not derive from audience volume, ticket velocity, or merchandise turnover. Instead, golf's valuation premium is anchored to four structural characteristics that traditional sports economics systematically undervalues.

1

Capital Adjacency

Golf's participant base represents the highest per-capita wealth concentration of any sport. Proximity to capital creates optionality—every event is a potential deal environment, every membership a relationship with asymmetric upside.

2

Brand Trust

The sport's association with discipline, integrity, and tradition creates a brand halo that transfers to affiliated properties. This trust premium cannot be manufactured—it must be inherited through authentic positioning within the sport's culture.

3

Executive Density

No other sport aggregates senior decision-makers with the same consistency and duration of interaction. A golf championship creates more high-quality executive touchpoints per event than entire seasons of other sports.

4

Long-Term Sponsorship

Golf sponsorships are characterized by multi-year commitments and low churn. Brands that align with golf tend to maintain those relationships for decades, creating predictable, compounding revenue streams.

Comparable assets in the market include private sports properties, executive networks, and luxury experiential brands—all of which trade at significant premiums to revenue-based valuations.

Golf Valuation: $216M–$288M

Applying conservative prestige multiples to the $36 million annual revenue base produces a valuation range that reflects the unique characteristics of golf as an investable asset class. These multiples are informed by comparable transactions in private sports properties, luxury experiential brands, and exclusive membership-based business models.

Base Case: 6× Revenue

$216 Million

Reflects the current revenue base with conservative assumptions about growth trajectory. Accounts for the established prestige infrastructure, owned championship properties, and recurring corporate revenue streams.

Growth Case: 8× Revenue

$288 Million

Incorporates expansion into additional international markets, deepening corporate league penetration, and the compounding effect of multi-year sponsorship relationships. Assumes moderate acceleration in executive membership growth.

This valuation reflects access control, not audience size. The asset's worth is determined by who participates, not how many watch.

Strategic Significance: Golf Brings Capital

Within the broader Goddess Phenezy Holdings ecosystem, each sport serves a distinct strategic function. Understanding this functional differentiation is essential for investors evaluating the portfolio's aggregate value proposition.

Basketball Brings Momentum

Cultural velocity, youth engagement, and mass-market energy that drives brand awareness and social media penetration across global markets.

Baseball Brings Continuity

Generational loyalty, seasonal rhythm, and the deep institutional trust that comes from America's most enduring sporting tradition.

Soccer Brings Scale

Global reach, international audience infrastructure, and the unmatched participatory base of the world's most played sport.

Cuju Brings Origin

Historical legitimacy, cultural heritage storytelling, and differentiated positioning through the world's oldest football tradition.

Golf Brings Capital

Executive access, institutional credibility, deal-flow infrastructure, and the quiet power that legitimizes the entire ecosystem in rooms where decisions are made permanently.

> **Golf is the connective tissue between sports and business, competition and deal flow, prestige and governance. It legitimizes the ecosystem in rooms where decisions are made quietly—and remembered permanently.**

The Racket & Paddle Championships

Tennis & Pickleball

The Velocity Jurisdiction

Lifestyle, Agility, and Real Estate

This jurisdiction is strategically designed to foster an agile and dynamic environment, attracting individuals and businesses that thrive on innovation and rapid adaptation. It offers a unique blend of high-quality living and strategic business advantages.

Central to its appeal is a commitment to an active, health-conscious lifestyle, supported by world-class recreational facilities and integrated community planning. The focus on racket and paddle sports, in particular, cultivates a vibrant, engaged, and well-connected community.

Tennis

The timeless appeal of tennis provides a sophisticated platform for networking and competitive engagement, attracting a demographic that values precision, strategy, and enduring athleticism.

Pickleball

As the fastest-growing sport, pickleball embodies agility, accessibility, and community, fostering a lively and inclusive social environment that complements the jurisdiction's dynamic ethos.

Complementary, Not Competitive

By housing both The National & International Tennis Championships and The Pickleball Velocity Circuit under one digital roof, we do not force players, brands, or developers to choose. We own the entire lifecycle of the racket and paddle athlete—from the mass-market accessibility of pickleball to the high-net-worth prestige of tennis.

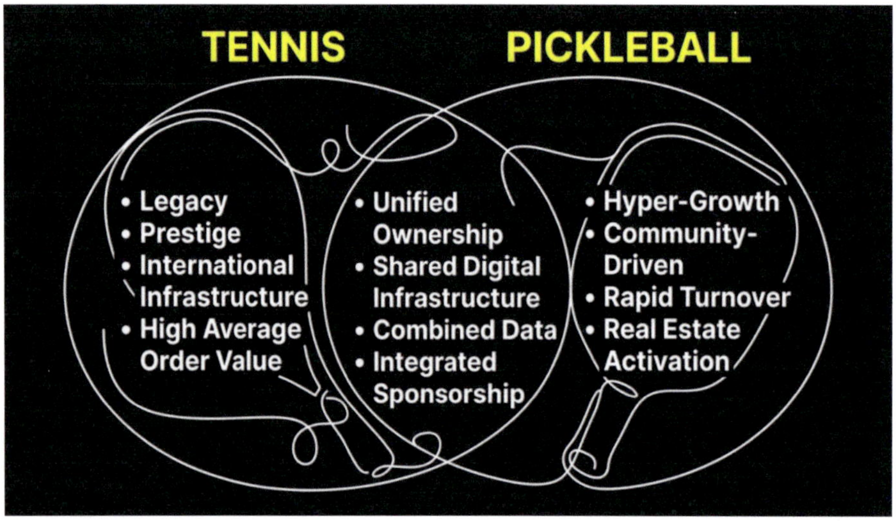

This unified approach creates strategic advantages that neither sport could achieve independently. Tennis provides the institutional credibility and premium positioning that attracts high-value sponsors. Pickleball provides the participation velocity and community engagement that drives data density and recurring revenue. Together, they create a portfolio effect that de-risks the investment thesis while amplifying upside potential.

Structural Advantage

Real Estate and Data Capture

The Most Valuable Asset Is the Court

The most valuable asset in racket sports is not the paddle. It is not the racket. It is not even the athlete. It is the court time. This insight—deceptively simple yet profoundly underexploited—forms the foundation of the Velocity Jurisdiction's structural advantage.

Legacy racket sports are bottlenecked by two systemic inefficiencies that have persisted for decades: fragmented booking systems that fail to optimize facility utilization, and disorganized player rating databases that prevent meaningful competitive matching across venues and geographies.

The Current Problem

- **Fragmented booking systems across thousands of independent facilities**
- **No standardized player rating that works across both tennis and pickleball**
- **Court utilization rates averaging below 40% at most facilities**
- **Revenue leakage through third-party booking platforms charging 15-25% commissions**
- **No unified data layer connecting players, facilities, and competitions**

These are not minor operational gaps. They represent billions of dollars in unrealized value across the global racket sports infrastructure—value that flows to no one because no one has built the system to capture it. Until now.

The PWA Infrastructure Solution

The Velocity Jurisdiction solves these structural inefficiencies by deploying proprietary Progressive Web App (PWA) infrastructure to control the entire facility and player experience. This is not a bolt-on technology layer. It is the operating system for the modern racket sports economy.

The Universal Racket Passport

A single digital identity tracking both a player's verified tennis handicap and pickleball rating natively within our system. This eliminates the current fragmentation where players maintain separate identities across multiple platforms, leagues, and facilities.

Facility SaaS Licensing

The operating system for thousands of new racket facilities being built globally—handling court reservations, dynamic pricing, and league management without third-party app store taxes. Every facility that adopts our system becomes a node in our data network.

Density Monetization

While a tennis court holds 2 to 4 players, that same footprint converted to pickleball holds up to 16. Our software optimizes real estate yield for host venues, making our jurisdiction the preferred technological partner for urban developers and country clubs alike.

Density Monetization: The Real Estate Multiplier

The density advantage of pickleball is not merely a participation statistic—it is a real estate economics thesis. When our software can dynamically shift court configurations between tennis and pickleball based on demand signals, reservation patterns, and pricing optimization, we transform static real estate into dynamic, yield-optimized infrastructure.

Tennis Players

Per standard court footprint per session

Pickleball Players

Same footprint converted, up to 4× density

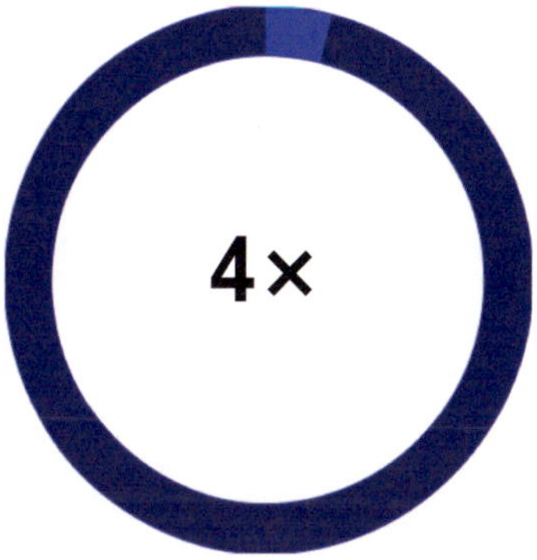

Revenue Density

Potential yield increase per square foot through dynamic conversion

This capability makes our jurisdiction the preferred technological partner for three critical market segments: urban developers seeking to maximize revenue per square foot in premium locations, country clubs looking to attract younger demographics without sacrificing existing tennis programming, and municipal parks departments under pressure to serve growing pickleball demand within existing facility footprints.

Revenue Architecture

High Turnover & Premium Margins

Revenue Layer 1: Championship Circuits & Pro-Ams

Expanding Event Horizons

Our platform unlocks significant event-based revenue streams by enabling dynamic scheduling and premium experiences. By hosting a diverse range of high-profile tennis and pickleball events, we cater to both professional athletes and enthusiastic amateurs, driving engagement and monetization.

These curated circuits and pro-am tournaments leverage our flexible court infrastructure, maximizing utilization during peak and off-peak hours. This strategy not only generates direct revenue through ticketing, sponsorships, and entry fees but also elevates brand visibility and attracts a wider audience to our ecosystem.

Tennis Invitationals

Exclusive, high-stakes tournaments featuring top-tier tennis talent, attracting affluent spectators and premium sponsors.

Pickleball Majors

Showcasing the rapidly growing sport of pickleball with professional circuits and significant prize pools, drawing a large, active community.

Combined Pro-Ams

Unique events pairing professionals with amateur enthusiasts, fostering community, skill development, and diverse revenue through participation fees and fan experiences.

$22.5M

Annual Event Revenue

Projected revenue from a diversified portfolio of championship circuits and pro-am events across our network.

Revenue Layer 2: Facility SaaS & Digital Memberships

The Infrastructure Layer

By licensing our PWA management software to independent clubs, public parks, and private franchises, we capture recurring revenue from the very real estate where the sports are played. This is the most strategically significant revenue layer because it creates a compounding data and distribution advantage that deepens with every facility onboarded.

Club Operating System Licensing

Monthly software fees from facilities using our platform for court management, member engagement, league administration, and dynamic pricing. As the installed base grows, network effects strengthen—players prefer facilities on our platform because their Universal Racket Passport works seamlessly across all of them.

Player Subscriptions

Individual digital memberships that unlock premium features: verified ratings, priority booking, league access, tournament eligibility, and performance analytics. These subscriptions create direct consumer relationships independent of any single facility.

$28.0M

Annual SaaS & Membership Revenue

Recurring revenue across thousands of global courts

Revenue Layer 3: Equipment, Apparel & Media

The Consumer Layer

The consumer layer capitalizes on one of pickleball's most remarkable commercial characteristics: equipment turnover velocity. Pickleball paddles and accessories turn over rapidly—often every 6 to 12 months—as players upgrade equipment, experiment with new technologies, and respond to evolving paddle regulations. This creates a recurring direct-to-consumer sales engine fundamentally different from the slower replacement cycles of traditional sports equipment.

Tennis apparel, by contrast, commands luxury price points and carries premium brand margins. The combination of high-velocity pickleball gear sales and high-margin tennis apparel creates a continuous, diversified retail pipeline that smooths revenue seasonality and maximizes lifetime customer value.

Proprietary Gear Drops

Direct-to-consumer equipment sales leveraging our player database for targeted product launches, limited editions, and personalized recommendations based on playing style and level data captured through the Universal Racket Passport.

Algorithmic Highlight Packaging

AI-curated broadcast content from championship events, distributed through owned digital channels to maximize advertising yield without surrendering media rights to third-party networks.

Premium Brand Sponsorships

Consumer-facing sponsorship integrations within our digital ecosystem—reaching verified, active players at the point of purchase decision with unmatched targeting precision.

$18.5M

Annual Media & Equipment Revenue

DTC sales, broadcast rights, and premium brand sponsorships

Total Annual Racket & Paddle Ecosystem Revenue

The three revenue layers of the Velocity Jurisdiction combine to create a $69 million annual ecosystem—nearly double the golf championship revenue—reflecting the dual-sport strategy's superior volume dynamics while maintaining premium margin characteristics. The SaaS infrastructure layer, representing 41% of total revenue, provides the recurring, high-margin foundation that supports aggressive valuation multiples.

Revenue Stream	Annual Value
Championship Circuits & Pro-Ams	$22.5M
Facility SaaS & Digital Memberships	$28.0M
Equipment, Apparel & Media	$18.5M
Total Annual Revenue	$69.0 Million

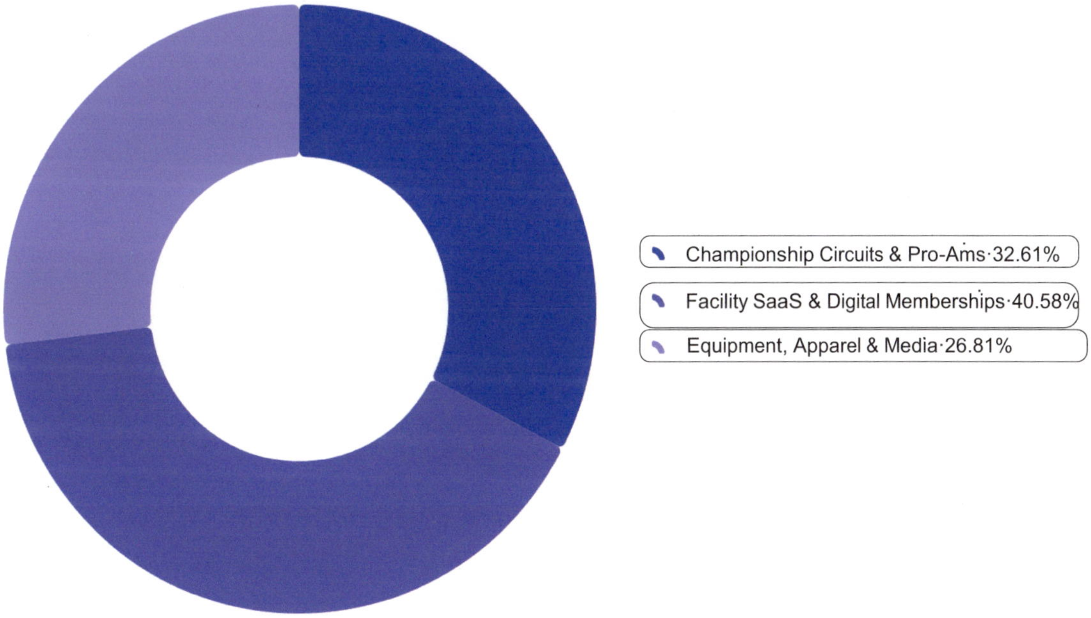

The revenue distribution reveals the strategic sophistication of this model. Unlike pure-play event businesses that depend on seasonal championship revenue, the Velocity Jurisdiction's largest revenue stream—SaaS and memberships—generates 365 days per year regardless of tournament calendar. This recurring infrastructure revenue fundamentally changes the risk profile of the investment.

Revenue Comparison: Golf vs. Racket & Paddle

Examining the two jurisdictions side by side reveals the portfolio diversification thesis. Golf delivers high-value-density revenue through fewer, more premium touchpoints. Racket and paddle sports deliver higher aggregate revenue through a combination of volume participation, infrastructure SaaS, and rapid consumer product turnover. Together, they create a combined $105 million annual sports ecosystem.

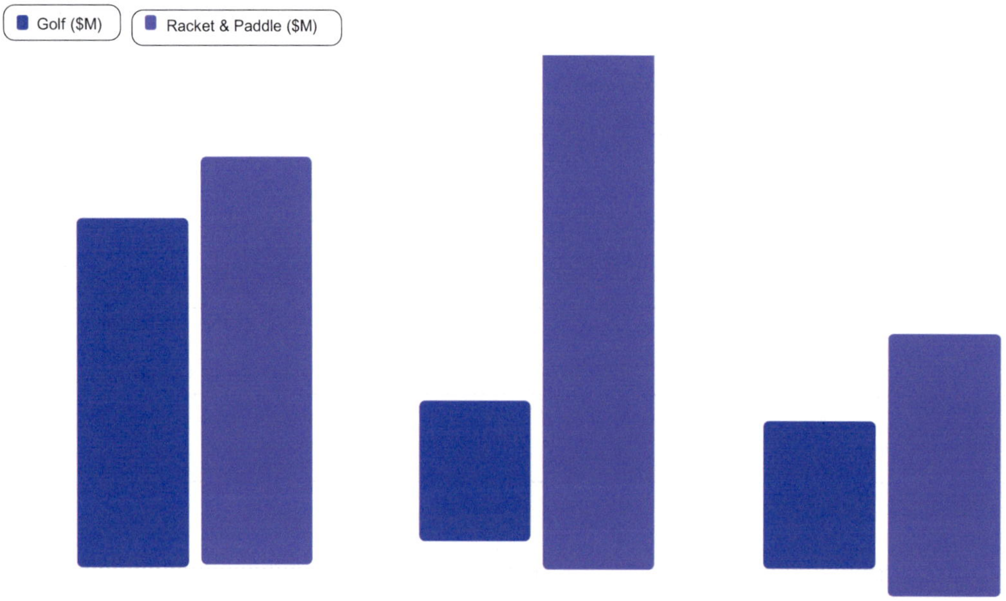

$36M

Golf Annual Revenue

High-margin, prestige-driven

$69M

Racket & Paddle Revenue

Velocity-driven with recurring SaaS

$105M

Combined Annual Revenue

Diversified across four sport verticals

Valuation Framework-Why Dual-Sport Commands Dynamic Multiples

Because this dual-sport ecosystem blends high-growth consumer retail (pickleball) with premium lifestyle B2B SaaS (facility management), it commands dynamic tech and real estate multiples that exceed traditional sports property valuations. The valuation thesis rests on three structural pillars that together justify aggressive multiple application.

Consumer Growth Velocity

Pickleball's $3B+ market and 30%+ annual participation growth rate position the consumer revenue layer for sustained expansion. Equipment turnover velocity creates built-in recurring demand independent of event scheduling.

B2B SaaS Recurring Revenue

The facility management platform generates predictable, high-margin monthly revenue that scales with facility adoption. SaaS businesses in adjacent verticals trade at 8-12× revenue, providing the multiple expansion catalyst for the entire ecosystem.

Real Estate Infrastructure Control

Owning the software that optimizes court utilization creates structural dependency. Facilities cannot easily switch operating systems once their member bases, league structures, and booking histories are embedded in our platform.

Dual-Sport Portfolio De-Risking

Tennis provides stability and premium positioning during any normalization in pickleball growth. Pickleball provides volume upside during periods of tennis market maturity. The portfolio construction inherently manages risk.

Racket & Paddle Valuation: $483M–$621M

Applying dynamic technology and real estate multiples to the $69 million annual revenue base produces a valuation range that captures both the established premium of tennis and the hyper-growth trajectory of pickleball, amplified by the SaaS infrastructure that connects them.

Base Case: 7× Revenue

$483 Million

Reflects current revenue with conservative growth assumptions. Accounts for the blended multiple between sports event properties (5-6×) and B2B SaaS platforms (8-12×), weighted toward the lower end given early-stage market penetration.

Velocity Growth Case: 9× Revenue

$621 Million

Incorporates accelerated facility adoption, international expansion of the pickleball circuit, deepening tennis championship prestige, and the network effects generated by a growing installed base of Universal Racket Passport holders.

This valuation reflects the power of controlling both the fastest-growing participatory sport and the infrastructure that houses it.

Combined Portfolio Valuation Summary

Examining Sections 6.6 and 6.7 together reveals the aggregate asset value created by the sports jurisdiction strategy. The golf and racket/paddle ecosystems together represent a $105 million annual revenue platform with a combined valuation range of $699 million to $909 million under conservative to growth-case scenarios.

Golf Jurisdiction

Revenue: $36.0M

Valuation: $216M - $288M

Racket & Paddle Jurisdiction

Revenue: $69.0M

Valuation: $483M - $621M

Combined Totals

Total Revenue: $105.0M

Total Valuation: $699M - $909M

Jurisdiction	Annual Revenue	Base Valuation	Growth Valuation
Golf	$36.0M	$216M (6×)	$288M (8×)
Racket & Paddle	$69.0M	$483M (7×)	$621M (9×)
Combined	$105.0M	$699M	$909M

These valuations do not account for the cross-pollination synergies between jurisdictions—golf members who participate in tennis events, corporate sponsors who align across multiple championship properties, or the data network effects created by unified digital infrastructure. The true enterprise value of the integrated portfolio likely exceeds the sum of its individual jurisdiction valuations.

The Strategic Ecosystem Map

The full value architecture across both jurisdictions can be understood as a layered system where prestige and velocity create complementary forces. Each jurisdiction serves a distinct strategic function while reinforcing the others through shared digital infrastructure, overlapping corporate relationships, and unified brand positioning.

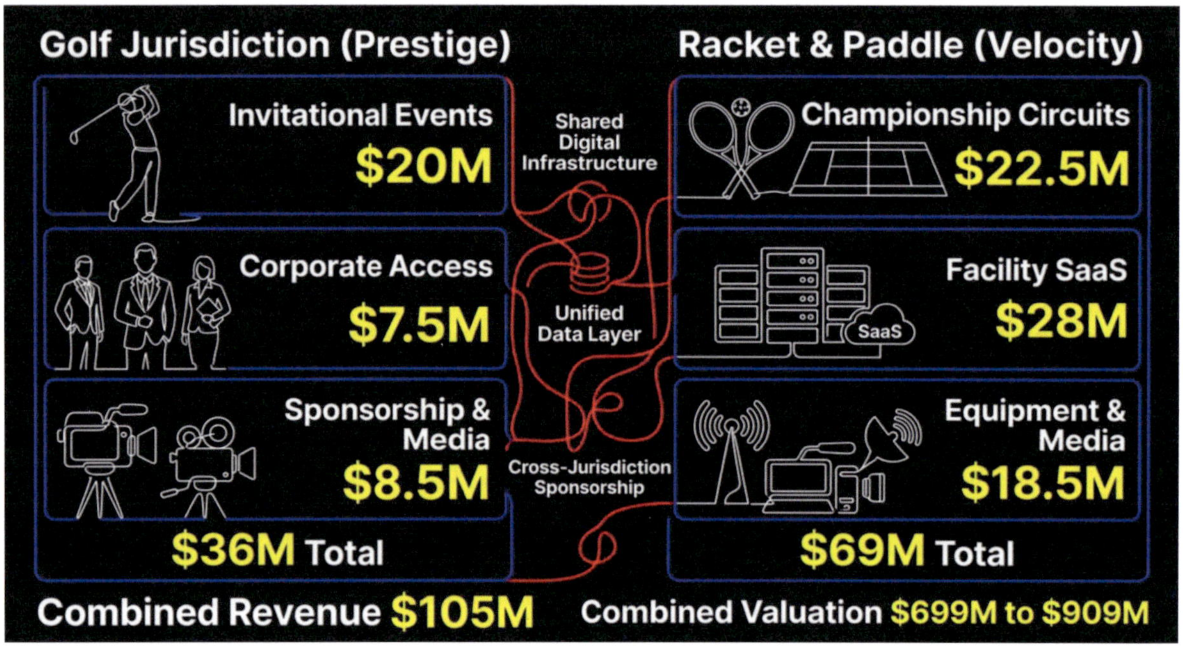

This architectural view reveals why the whole exceeds the sum of its parts. The shared infrastructure layer—digital identity, payment systems, data analytics, and media distribution—serves both jurisdictions simultaneously, creating operating leverage that improves as each jurisdiction scales independently.

Investment Thesis: Key Differentiators

1. Ownership, Not Licensing

Unlike traditional sports properties that license championship rights from governing bodies, Goddess Phenezy Holdings owns the championships outright. This eliminates sanctioning fees, governance interference, and the risk of license non-renewal that plagues dependent operators.

2. Infrastructure, Not Just Events

The SaaS facility management layer transforms the business from a seasonal events company into a year-round infrastructure platform. This shifts the revenue profile from lumpy and event-dependent to recurring and predictable.

3. Data Sovereignty

Every player, every facility, every transaction generates data captured within proprietary systems. This first-party data asset compounds in value over time and cannot be replicated by competitors who depend on third-party platforms.

4. Cross-Sport Synergies

A corporate sponsor entering through golf can be expanded into tennis. A pickleball player's data can inform tennis equipment recommendations. The interconnected portfolio creates cross-sell opportunities unavailable to single-sport operators.

5. No Platform Taxes

Proprietary PWA infrastructure eliminates app store commissions, third-party ticketing fees, and intermediary booking charges. Every dollar of margin saved flows directly to the bottom line.

The Participation Growth Opportunity

The scale of the underlying participation opportunity cannot be overstated. Both golf and racket sports are experiencing structural tailwinds that support the revenue projections outlined in this thesis. The post-pandemic era has fundamentally accelerated demand for outdoor, socially engaging, skill-based recreation—precisely the category these four sports occupy.

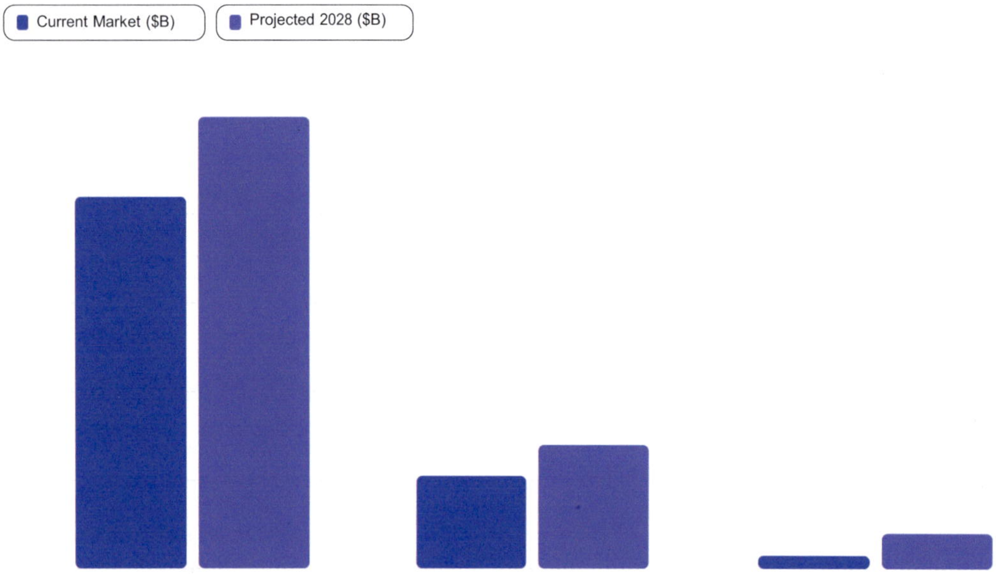

Pickleball's growth trajectory is particularly remarkable—projected to more than double its market size within four years. Golf continues its steady expansion driven by younger demographics entering the sport. Tennis benefits from renewed international interest and the wellness economy. The Velocity and Prestige Jurisdictions are positioned to capture disproportionate share of this growth because they control the digital infrastructure through which participation is organized, measured, and monetized.

Risk Mitigation Through Diversification

Sophisticated investors will appropriately evaluate downside scenarios alongside the upside thesis. The dual-jurisdiction architecture inherently mitigates the most significant risks associated with single-sport investment:

Cyclical Risk

Golf's premium positioning makes it more sensitive to economic downturns, but pickleball's mass-market accessibility provides counter-cyclical demand. During recessions, consumers trade down from golf to pickleball—both within our ecosystem. During expansions, pickleball participants trade up to tennis and golf experiences.

Technology Risk

The SaaS platform faces competitive risk from incumbent booking platforms. However, the Universal Racket Passport's dual-sport integration and verified rating system create switching costs that deepen with usage. No competitor currently offers a unified cross-sport digital identity.

Market Saturation Risk

If pickleball participation growth decelerates, the high-margin SaaS layer continues generating recurring revenue from existing facilities. Tennis and golf championships provide stable, premium revenue floors independent of pickleball growth rates.

Regulatory Risk

Owned championship structures eliminate dependency on third-party sanctioning bodies whose regulatory decisions could disrupt licensed operators. Full ownership ensures complete control over competitive formats, eligibility rules, and brand positioning.

The Path Forward: Scaling Quiet Power and Velocity

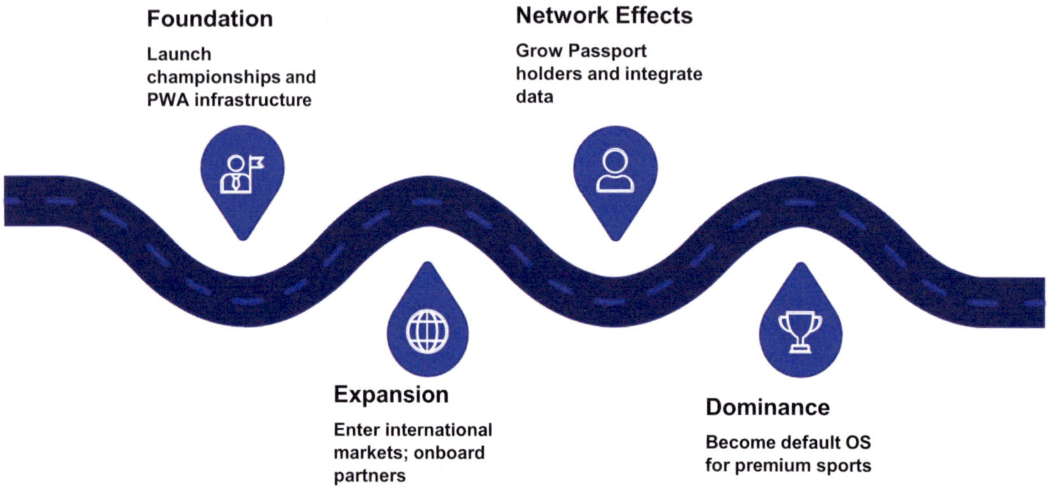

The scaling trajectory follows a deliberate, capital-efficient progression. Foundation investments in championship ownership and digital infrastructure create the gravitational pull that attracts facilities, players, and sponsors. Network effects accelerate as the installed base grows. And dominance is achieved not through aggressive market capture, but through the quiet accumulation of structural advantages that make the ecosystem indispensable.

Conclusion: Capital, Access, Velocity

The National & International Golf Championships and the Racket & Paddle Championships represent two complementary expressions of a single strategic thesis: the consolidation of prestige sports infrastructure under unified ownership, powered by proprietary digital infrastructure, and designed for maximum value capture per participant.

Golf brings the capital adjacency, brand trust, and executive density that legitimize the broader ecosystem. Tennis brings international prestige and premium consumer economics. Pickleball brings hyper-growth velocity and infrastructure opportunity. Together, they create a $105 million annual revenue platform valued between $699 million and $909 million.

$105M	$699M–$909M
Combined annual revenue across all four sports	Portfolio valuation range under conservative to growth-case multiples

> **Golf legitimizes. Tennis elevates. Pickleball accelerates. Together, they compose an ecosystem where prestige and velocity are not opposing forces—they are complementary instruments in the architecture of modern sports ownership.**

This is not a collection of tournaments. This is the infrastructure of quiet power and relentless momentum—engineered for the investors, partners, and strategists who understand that the most valuable games are played both on the course and behind it.

Modular Jurisdictions, Infinite Scale

The racket and paddle economy demonstrates a critical truth: once sovereign digital infrastructure exists, new sports do not need to be invented—they only need to be absorbed. The entire expansion thesis rests on a simple but powerful architectural principle: sport-agnostic infrastructure that treats every athletic discipline as a data stream, a media asset, and a membership economy waiting to be activated.

Goddess Phenezy Holdings has engineered a championship and league architecture that is sport-agnostic, jurisdiction-ready, and immediately monetizable. After activating core verticals—Basketball, Baseball, Soccer, Cuju, Golf, Tennis, and Pickleball—the same proprietary Progressive Web App (PWA) infrastructure can seamlessly onboard additional athletic economies without new software development, platform dependency, or operational friction. The PWA backbone was designed from inception to be a universal chassis, not a single-sport solution.

Global Demand

Evaluates existing worldwide interest and fan base for the sport.

Revenue Density

Assesses potential earnings per participant within the sport's ecosystem.

Jurisdiction Fit

Confirms alignment with our sovereign digital infrastructure model.

These are not experimental expansions. They are pre-qualified markets already aligned with our data, media, and membership model. Each target vertical has been evaluated against three criteria: existing global demand, revenue density per participant, and compatibility with the sovereign jurisdiction model. Every expansion candidate passed all three filters before earning a position in the pipeline.

The following sections detail the primary expansion targets, secondary cultural activations, and two landmark intellectual properties—Chess and Senet—that extend the ecosystem's reach from physical athleticism into cognitive sovereignty and ancient heritage. Together, they represent the next phase of jurisdictional growth: absorbing markets, not building them from scratch.

THE INTERNATIONAL
PICKLEBALL
CHAMPIONSHIP

Honorable Mentions: The Expansion Pipeline

Not Secondary. Strategic.

Our "Honorable Mentions" are anything but secondary priorities. This designation speaks to the immense breadth of our portfolio, which proudly includes ownership of over 79 National & International Championships. Goddess Phenezy Holdings has secured unprecedented digital sovereignty across the entire sports economy by acquiring the .com domains for all major sports.

Each championship is pre-positioned for activation through our sovereign digital infrastructure, ready to integrate seamlessly. This isn't about importance hierarchy; it's about acknowledging a vast collection of ready-to-activate assets, not aspirational targets.

79+

National & International Championships owned

.com

Complete domain portfolio for major sports

100%

Sovereign digital infrastructure across all verticals

Ready

Pre-positioned for immediate activation

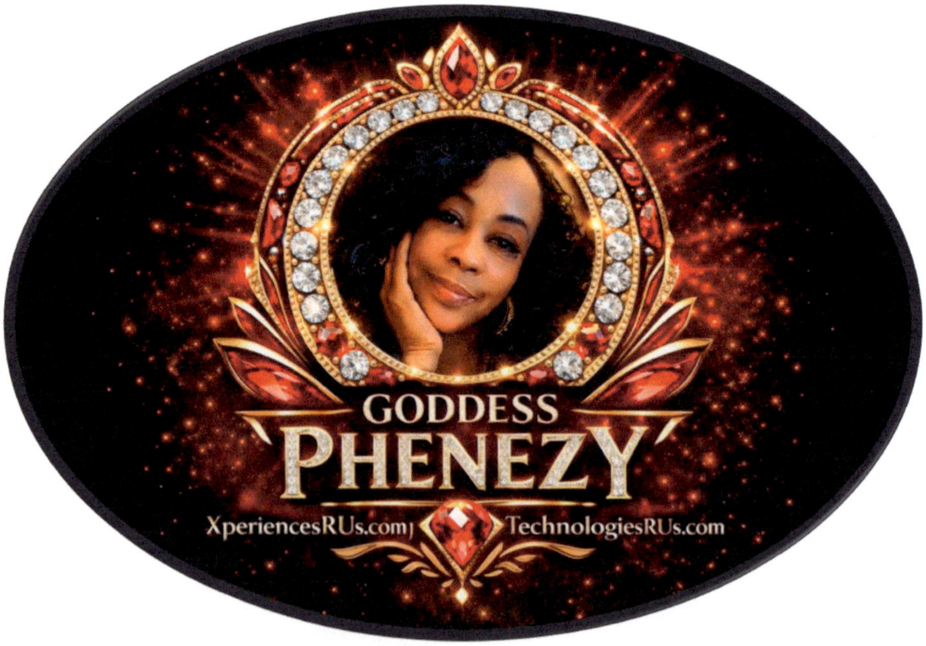

The Absorption Model

Traditional sports organizations scale horizontally, treating each new sport as a siloed entity with bespoke operational overhead, fragmented fanbases, and duplicated infrastructure. This approach creates structural limitations, slows market penetration, and constrains growth.

Goddess Phenezy Holdings, however, . This is a fundamental architectural divergence. Vertical scaling integrates every layer of the value chain—identity, data, media, commerce, and governance—across all athletic disciplines. Jurisdictional scaling deploys a pre-engineered, unified PWA chassis that seamlessly extends its proven infrastructure to new sports without the friction of traditional market entry. This shared foundation creates compounding network effects, where each absorbed sport immediately benefits from a robust technological stack and established distribution channels, while simultaneously enriching the entire network.

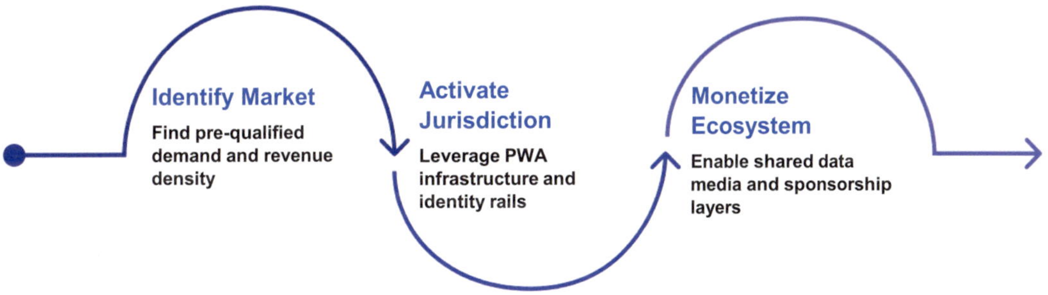

Once an athlete, fan, or sponsor enters the Goddess Phenezy Holdings ecosystem through any championship, they are onboarded into a unified digital citizenship. This shared, persistent identity, coupled with integrated data and media layers, creates a profound economic multiplier effect. It enables effortless cross-vertical discovery and engagement, amplifying individual brands and providing sponsors unparalleled access to a diverse, cohesive audience. This is not mere portfolio diversification, but , where every new vertical strengthens every existing one, building a defensible moat against competition and ensuring long-term economic permanence.

THE
INTERNATIONAL
MMA
CHAMPIONSHIP

The National & International MMA Championships

Mixed Martial Arts (MMA) stands as a testament to explosive growth and global appeal, making it a cornerstone for the Goddess Phenezy Holdings absorption model. Over the last decade, MMA has transcended niche appeal to become a mainstream combat sport, capturing the attention of diverse demographics worldwide. Its dynamic nature, combined with a clear competitive structure, positions it as an ideal candidate for integration into our expanding ecosystem of modular jurisdictions.

This sport's ascent is marked by a dramatic surge in viewership, athlete professionalization, and digital engagement. With an ever-growing global fanbase and a compelling narrative of athletic excellence, MMA brings immense revenue density per participant and aligns perfectly with our data, media, and membership economy model. Our proprietary PWA backbone is designed to seamlessly absorb such high-impact verticals, providing an immediate monetization pathway without requiring bespoke software development.

Global Fanbase Surge

Exponential growth in international viewership and engagement.

Athlete Professionalization

Elevated athlete profiles and career opportunities.

Revenue Density

High value per participant through media rights and sponsorships.

Seamless Integration

Optimized for rapid activation within sovereign digital infrastructure.

The integration of MMA championships reinforces our strategy of sovereign absorption, demonstrating how existing pre-qualified markets can instantly scale within our framework. Every new MMA fan, athlete, and sponsor contributes to the collective strength and data intelligence of the entire Goddess Phenezy Holdings ecosystem, proving that growth comes not just from innovation, but from strategic integration of high-performing assets.

THE INTERNATIONAL
BOXING
CHAMPIONSHIP

PRIMARY EXPANSION

The National & International Boxing Championships

The Spectacle Economy

Boxing remains one of the highest-margin sports on Earth, a spectacle economy driven by massive pay-per-view consumption, global fan bases that transcend cultural boundaries, high-value sponsorships from luxury and lifestyle brands, and a unique economic profile characterized by low participant volume but extraordinary revenue density per event. No other sport packs as much economic value into a single evening as boxing.

The current boxing landscape is fractured across competing promoters, sanctioning bodies, and broadcast partners—each extracting value while fragmenting the fan experience. Fighters are dependent on intermediaries who control access, pricing, and audience data. The jurisdictional model eliminates this dependency entirely, creating a vertically integrated championship economy where every dollar flows through sovereign infrastructure.

Jurisdictional Advantage

Our jurisdictional advantage restructures the economics of combat sports from the ground up, removing the promoter-dependent model that has defined boxing for over a century.

Direct-to-Consumer Broadcast

Control content distribution, maximizing reach and revenue without third-party gatekeepers.

Owned Fighter Identity

Fighters retain full ownership of their brand and audience data, ensuring fair compensation and long-term career growth.

Premium Hospitality

Elevated event experiences and exclusive access drive high-value sponsorship and fan engagement.

> This is not just about better economics; it's about establishing sovereign control over the sport's destiny, ensuring that every participant—from fighter to fan—benefits directly from a unified, value-driven ecosystem.

PRIMARY EXPANSION

The Volleyball & Beach Athletics Circuits

The Density & Lifestyle Economy

Volleyball—both indoor and beach—offers one of the highest player-to-court ratios in sports, making it an ideal candidate for scalable jurisdictional activation. Where boxing thrives on spectacle scarcity, volleyball thrives on participation density. A single beach court can host dozens of matches per day, and a single tournament weekend can engage hundreds of athletes across age groups, skill levels, and formats.

The lifestyle dimension amplifies the economic model exponentially. Beach volleyball tournaments are not just athletic events—they are festival-style activations that attract sponsors from surf, fitness, fashion, beverage, and hospitality brands. Coastal and resort settings create natural premium environments where sponsorship, media, and real estate yield converge into a single revenue engine.

Volleyball: Platform Capabilities

The PWA infrastructure enables volleyball to scale across formats, geographies, and demographic pipelines in ways that fragmented legacy organizations cannot match.

Unified Player Passports

A single verified identity that travels across indoor and beach formats, tracking performance, rankings, and tournament history regardless of venue or geography. One player, one passport, infinite courts.

Scalable Licensing

Licensing frameworks for coastal cities, universities, and resorts to host branded tournaments under the jurisdiction umbrella. Each licensed event feeds data and revenue back into the central ecosystem.

Integrated Media & Lifestyle Sponsorships

Media production and sponsorship packages designed for the volleyball lifestyle economy—connecting brands with highly engaged, health-conscious, and visually dynamic audiences.

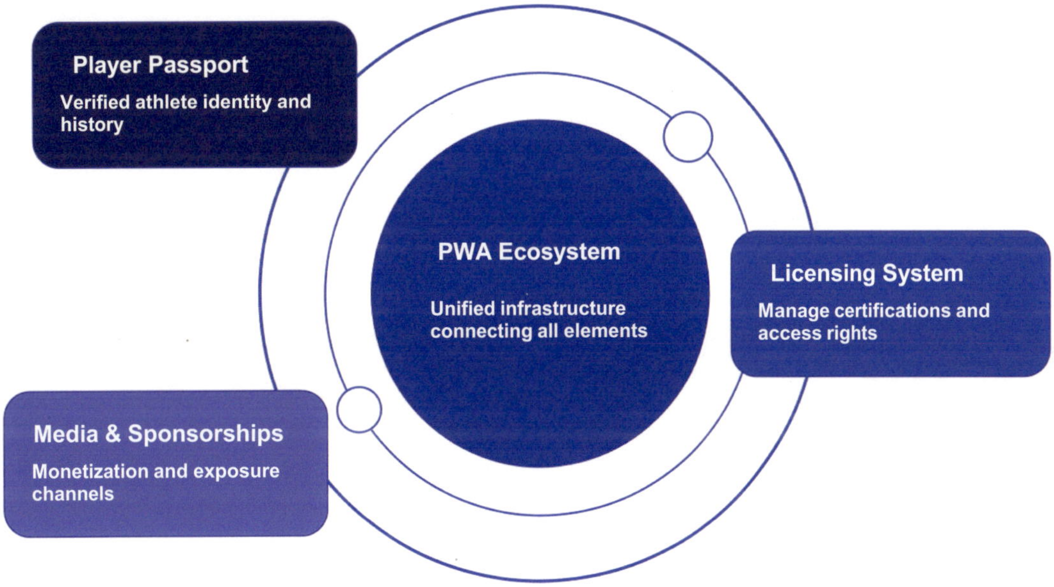

This diagram illustrates how the PWA platform unifies diverse aspects of volleyball, creating a cohesive and scalable ecosystem for growth and engagement.

> Strategic Value: Exceptional real estate yield and sponsorship-friendly environments. Volleyball's density model generates maximum revenue per square foot of activated space.

PRIMARY EXPANSION

The National & International Track & Field Showcases

The Data Purity Economy

Track & Field is the most universal athletic discipline on Earth. Every nation competes in it. Every school teaches it. Every Olympic Games revolves around it. Yet despite this universality, Track & Field remains one of the most technologically antiquated sports in existence—a paradox that represents an enormous market opportunity for sovereign digital infrastructure.

The core problem is data fragmentation. Timing results, verification records, athlete rankings, and scouting data are scattered across schools, regional meets, national federations, and disparate timing companies. A high school sprinter's 100-meter time in Texas exists in a completely different data universe than a collegiate runner's time in California. There is no unified performance record, no verified athlete identity, and no longitudinal data pipeline connecting youth development to professional opportunity.

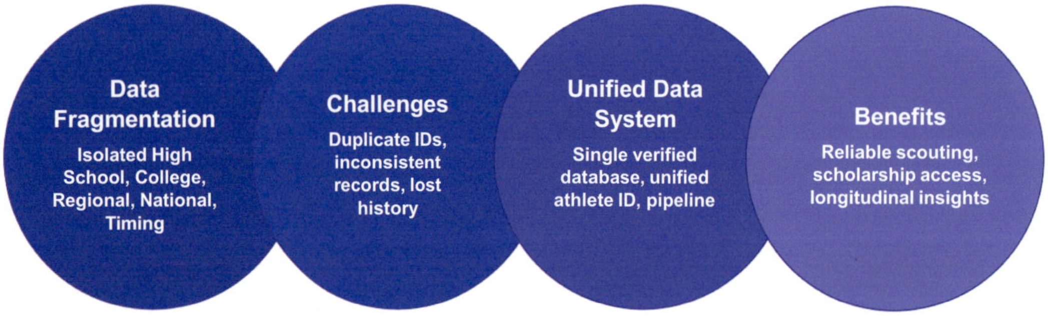

This diagram illustrates the stark contrast between the current fragmented data landscape in Track & Field and the unified, interconnected system enabled by the jurisdiction model, transforming isolated information into actionable insights.

This fragmentation doesn't just create inefficiency—it creates injustice. Talented athletes in under-resourced communities are invisible to scouts and scholarship committees simply because their data lives in silos that no one aggregates. The jurisdiction model solves this at the architectural level.

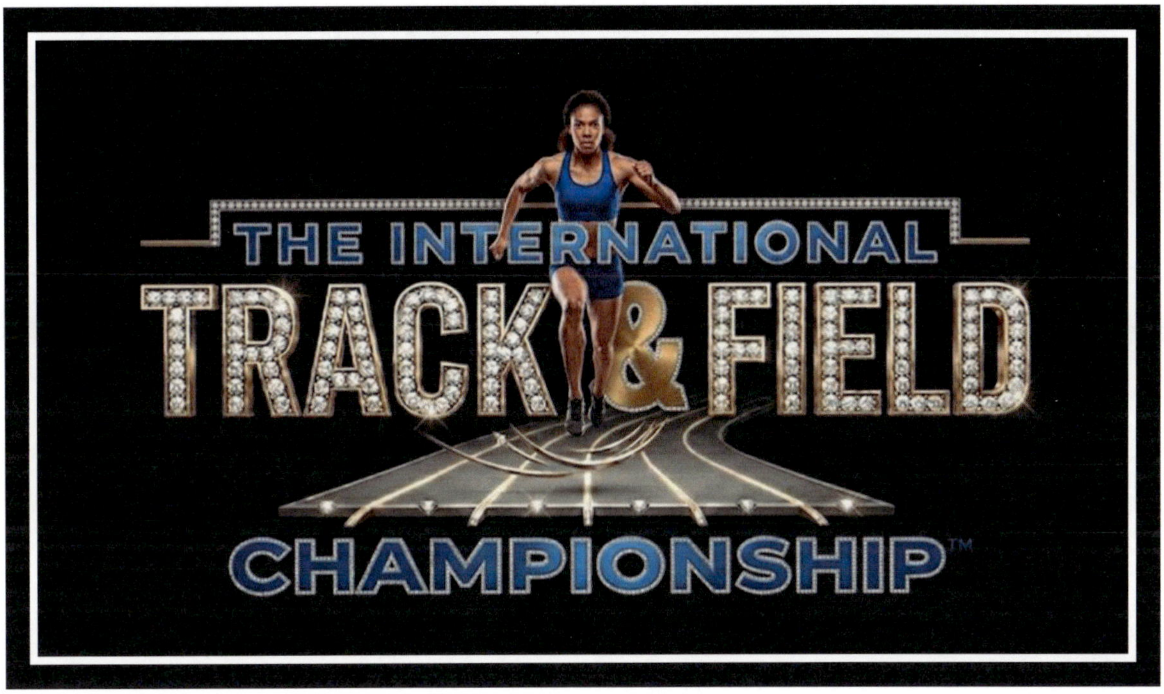
THE INTERNATIONAL
TRACK & FIELD
CHAMPIONSHIP™

Track & Field: The Jurisdiction Model

The jurisdiction model modernizes Track & Field by transforming fragmented data into a unified, verified intelligence system that serves athletes, coaches, scouts, and institutions simultaneously.

1

Verified Performance Tracking

Every time, every distance, every jump—verified at the source and permanently recorded within the sovereign data layer. No more disputed results or lost records.

2

Universal Athlete Passports

A single identity that follows an athlete from youth meets through collegiate competition to professional showcases. One passport, one verified history, accessible by authorized scouts and institutions.

3

Scouting & Scholarship Pipelines

Performance data tied directly to scouting databases and scholarship platforms, ensuring that talent is discovered based on verified results—not geography or connections.

Strategic Value: Foundational athletic data that feeds multiple sports and educational pathways. Track & Field data is the universal athletic currency—and the jurisdiction becomes the central bank.

Primary Expansion: Revenue Density Comparison

Each primary expansion target was selected for its unique economic profile. While all three share compatibility with the sovereign PWA infrastructure, their revenue models differ in ways that create portfolio diversification and risk mitigation across the expansion pipeline.

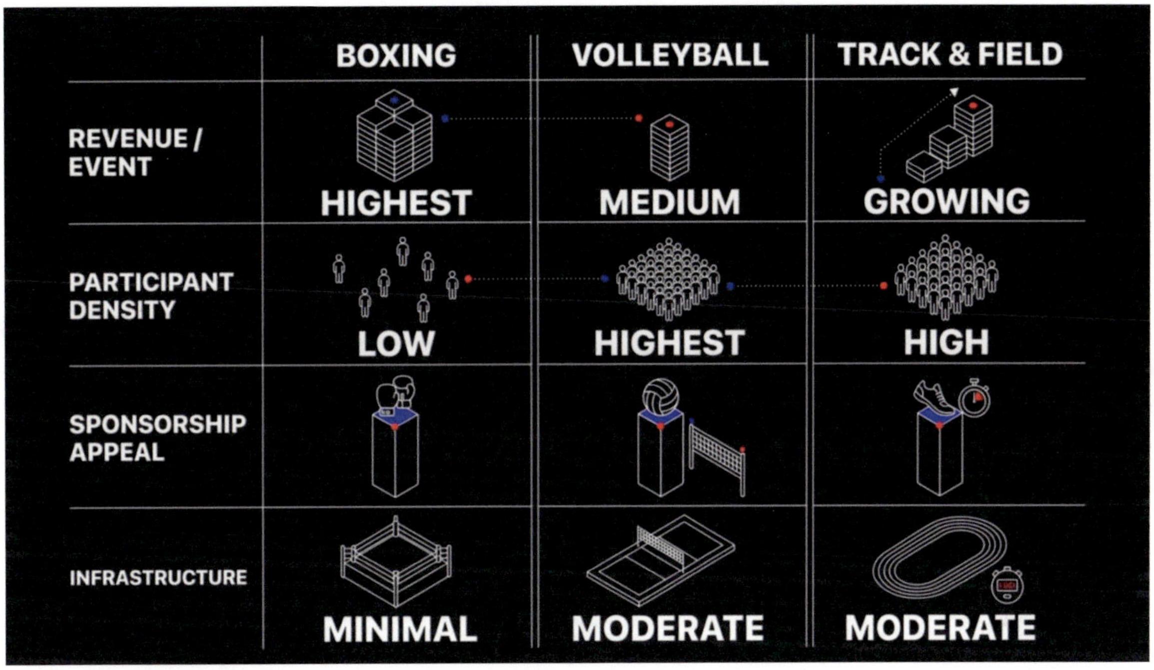

The strategic beauty of this trio is complementarity. Boxing delivers high-margin spectacle revenue with minimal participant volume. Volleyball maximizes density and lifestyle sponsorship revenue. Track & Field generates foundational data assets that appreciate over time and feed scouting pipelines across the entire ecosystem. Together, they cover the full spectrum of the sports economy.

Secondary & Cultural Expansions

Modular Cultural-Athletic Hybrids

Beyond the primary expansion targets, a second tier of properties operates as modular cultural-athletic hybrids, ideal for phased activation. These verticals extend the jurisdiction beyond traditional athletics into lifestyle, entertainment, and digital-native competition—each one leveraging the exact same infrastructure that powers the core sports economy.

These are not speculative bets. Each vertical has a proven global audience, established sponsorship markets, and natural alignment with the PWA's identity, payment, and media rails. They represent the long tail of sovereign absorption—properties that can be activated on demand as the ecosystem scales and market conditions dictate.

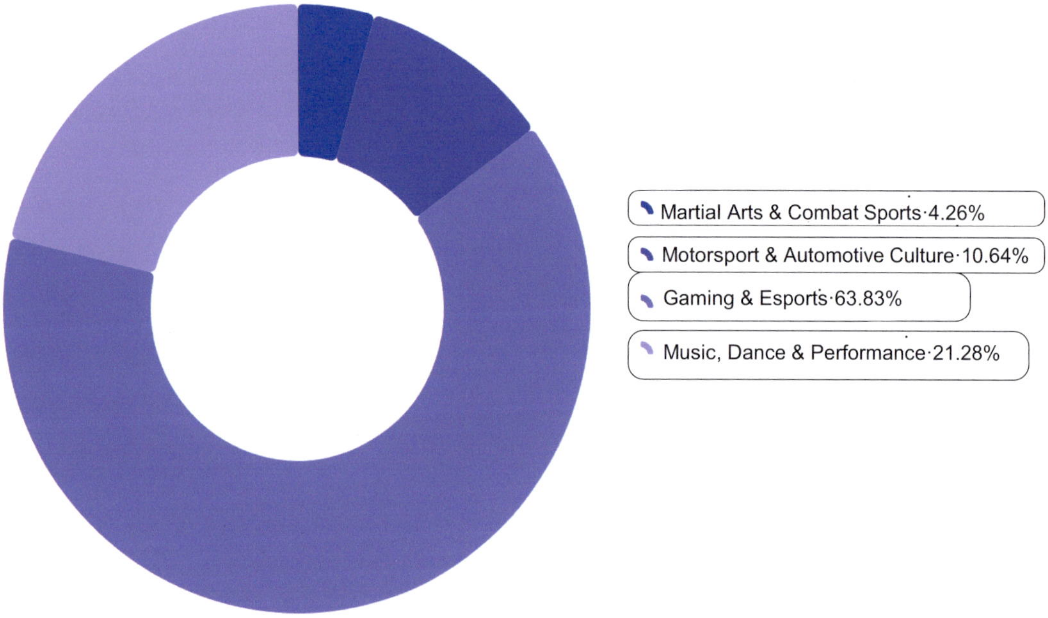

Each vertical represents a proven global market with established sponsorship ecosystems and passionate, engaged audiences ready for sovereign infrastructure.

Cultural Expansion Verticals

Beyond the primary sports expansions, the cultural verticals represent high-engagement, lifestyle-driven competitive economies that align perfectly with the sovereign infrastructure model. Each vertical brings passionate audiences, premium sponsorship opportunities, and media content with extraordinary shelf life.

Martial Arts & Combat Xperiences

Motorsport & Car Show Championships

Gaming & Esports Championships

Music, Dance & Performance Championships

Martial Arts brings discipline-based leagues spanning karate, judo, taekwondo, and MMA—high-engagement formats with passionate, loyalty-driven fan bases and premium event economics.

Motorsport represents lifestyle, sponsorship, and media convergence at scale. Automotive culture commands premium brand partnerships and attracts high-net-worth audiences across racing, exhibition, and collector events.

Esports captures digital-native participation and youth audiences. It represents the fastest-growing competitive economy on Earth, with audiences already comfortable in PWA-style digital environments.

Music, Dance & Performance Championships create cultural competition with global streaming appeal, attracting diverse demographics and generating media content with viral potential.

Shared Infrastructure Advantage

Every secondary and cultural expansion vertical leverages the same foundational infrastructure—eliminating the need for new platform development, new payment integrations, or new identity systems. This is the compounding power of sovereign architecture.

Tennis

Leverages shared platform services

Pickleball

Rapid deployment on common infra

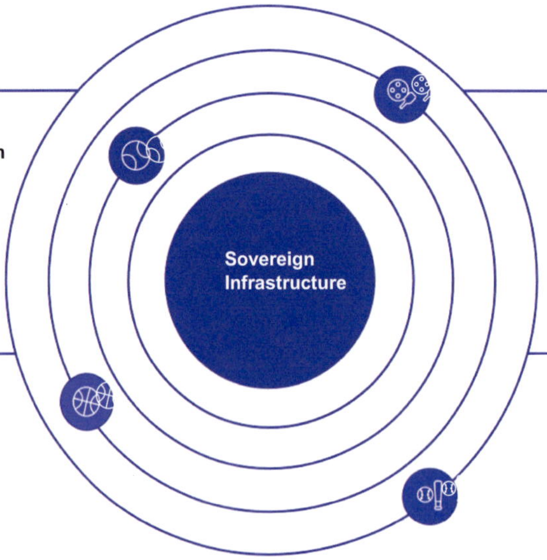

Basketball

Unified payments and identity

Baseball

Reusable backend and APIs

Existing User Identity Infrastructure

Every participant—athlete, fan, coach, sponsor—carries a single verified identity across all verticals.

Existing Payment, Media & Licensing Rails

Subscriptions, registrations, and broadcast rights flow through proven payment and distribution systems.

Existing Sponsorship & Broadcast Frameworks

Brand partners access multiple verticals through a single relationship, maximizing reach and ROI.

No new platform risk. No dilution of authority. Each cultural expansion vertical activates within the existing sovereign ecosystem, compounding value for every stakeholder already inside.

Strategic Implication: Vertical vs. Horizontal Scale

Most sports organizations scale horizontally—one sport, one problem, one ceiling. They build for a single discipline and hit structural limits when they attempt to diversify. New sports require new platforms, new teams, new infrastructure, and new capital. Growth is additive at best.

Horizontal Scaling

One sport. One platform. One ceiling. Each new vertical requires new development, new partnerships, and new capital allocation. Growth is linear and fragile. This diagram illustrates the power of jurisdictional scaling. This model enables exponential growth and true sovereign absorption across diverse disciplines.

Jurisdictional Scaling

Goddess Phenezy Holdings scales vertically and jurisdictionally. Once an athlete, fan, facility, or sponsor enters the ecosystem through any championship, they are retained across multiple verticals through shared identity, data, and media layers. Growth is exponential and self-reinforcing.

This is not portfolio expansion. It is sovereign absorption.

1. **Sovereign Digital Infrastructure**

   PWA, Identity System, Data Architecture, Media Layer

2. **Basketball Championships**
3. **Baseball Championships**
4. **Soccer Championships**
5. **Golf Championships**
6. **Cuju Championships**
7. **Tennis Championships**
8. **Pickleball Championships**
9. **Chess Leagues & Championships**
10. **Senet Championships**
11. **Boxing Championships**
12. **Volleyball Championships**
13. **Track & Field Championships & More...**

The Intellectual Jurisdiction

The National & International Chess Leagues, Championships & Academy

Patent Pending

The Cognitive Economy: Strategy as Sovereignty

Chess is the most universally recognized game of strategy in human history. It transcends language, age, borders, and physical ability. It is taught in schools across every continent, studied by military tacticians as a framework for decision-making, embedded in artificial intelligence research as the benchmark for machine cognition, and revered as the ultimate training ground for decision-making under pressure.

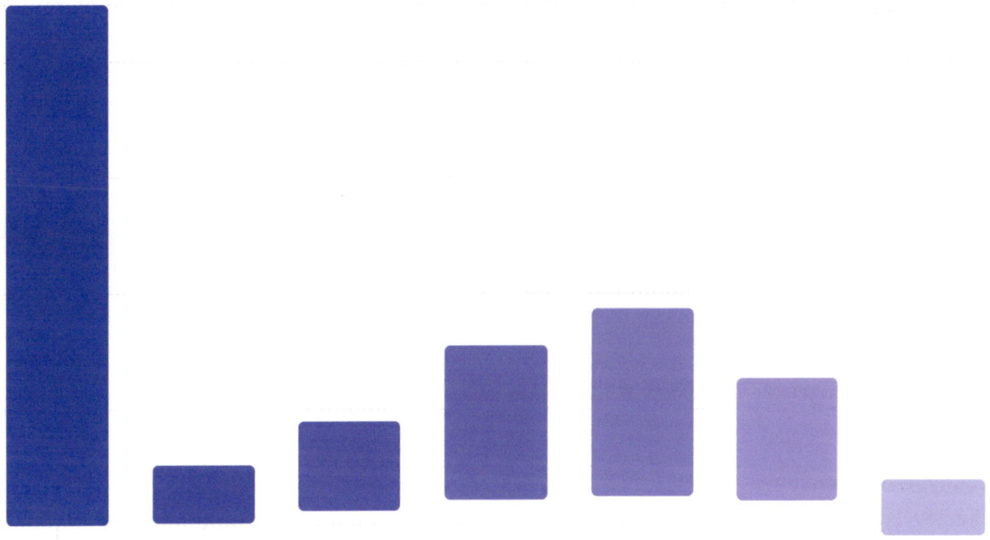

With over 528,000+ rated players globally, chess demonstrates massive reach. This demographic breakdown highlights the significant male representation and strong engagement across all age groups, especially among younger demographics. The female player base, while currently smaller, represents a substantial growth opportunity for the sport.

Yet despite this universality, the modern digital chess economy—engaging hundreds of millions of active players and generating billions in annual value—is effectively controlled by a centralized digital duopoly. Two platforms dominate the global chess landscape, extracting value from every game played, every lesson consumed, and every tournament organized on their infrastructure.

Today, to compete, learn, teach, monetize, or organize chess at scale, participants are forced to rent space on legacy platforms. They pay platform taxes, surrender audience ownership, relinquish performance data, and accept algorithmic gatekeeping controlled by third-party tech entities. The result is an economy where the participants who create all the value capture almost none of it.

The Digital Duopoly Problem

The current chess ecosystem operates under a model of systematic value extraction. Every stakeholder in the chess economy—players, coaches, organizers, and communities—surrenders sovereignty in exchange for access.

Players

Do not own their competitive history. Years of games, ratings, and achievements exist on platforms that can change terms, restrict access, or shut down at any time.

Coaches

Do not own their students. Student relationships, lesson histories, and teaching data are trapped inside platform silos that coaches cannot export or control.

Organizers

Do not own their leagues. Tournament data, participant lists, and competitive results are held hostage by the platforms that host the events.

Communities

Do not own their ecosystems. Clubs, schools, and national federations are tenants on platforms they did not build and cannot govern.

Goddess Phenezy Holdings, Inc. rejects this model entirely. We are not building another chess website. We are establishing a sovereign Intellectual Jurisdiction.

A New Board. And We Own It.

Through proprietary, next-generation Progressive Web App (PWA) infrastructure, the National & International Chess Leagues, Championships, and the integrated Chess Academy operate as a fully independent, self-governing digital ecosystem—designed for cognitive athletes, educators, institutions, and future generations.

This is not a feature improvement over existing platforms. It is an entirely new paradigm. Where legacy platforms function as game servers that host content, the Chess Jurisdiction functions as sovereign infrastructure that governs an ecosystem. The distinction is not semantic—it is architectural, economic, and legal.

Every game played, every lesson completed, every championship earned, and every data point generated within the Chess Jurisdiction is owned, controlled, and monetized by the ecosystem itself. There is no revenue leakage to app stores, no algorithmic gatekeeping by third-party platforms, and no dependency on external APIs or federation approvals.

> **This is a new board.**
> **And we own it.**

Structural Advantage: Infrastructure, Not Content

Most modern chess platforms function as content delivery systems masquerading as competitive infrastructure. The distinction matters because it defines who captures value and who merely facilitates it.

Legacy Platforms Function As:

- Content libraries
- Game servers
- Rating silos
- Advertising-driven engagement engines

They host games. They do not govern ecosystems. Their business model depends on maximizing time-on-platform for advertising revenue, not on maximizing cognitive development or competitive integrity.

The Chess Jurisdiction Controls:

- National and International Leagues
- Championship pathways (digital and physical)
- The Chess Academy and curriculum
- Proprietary learning and analysis features
- Verified player identities and longitudinal data
- Media, broadcasting, and institutional access

> Chess does not exist on our platform. Chess lives inside it. All of this operates within a unified PWA environment—free from app store taxes, external APIs, federation dependency, or third-party monetization leakage.

The Structural Moat: Provisional Fortification

Patent Pending

You do not disrupt a monopoly with surface-level features. You disrupt it with fortified architecture. The Chess ecosystem under Goddess Phenezy Holdings is not a conceptual outline or a loose roadmap. It is a deeply protected digital asset, backed by an extensive provisional intellectual property framework spanning nearly 200 pages of proprietary systems, mechanics, and operational logic.

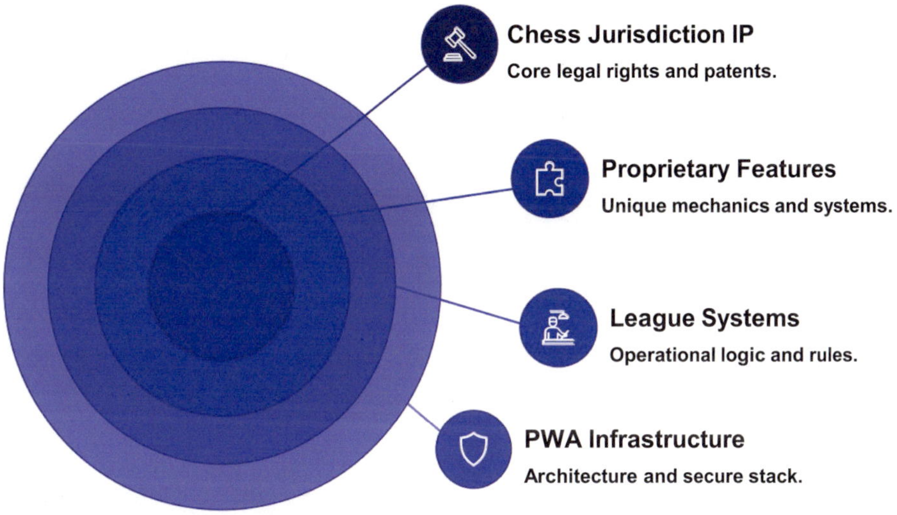

This level of IP protection is unprecedented in the digital chess space. While legacy platforms rely on network effects and user lock-in as their competitive moats, the Chess Jurisdiction relies on legally protected architectural innovation that cannot be replicated without infringement.

IP Protection Framework

The provisional patent framework establishes protection across every layer of the Chess Jurisdiction's architecture, creating a fortification that is as comprehensive as it is defensible.

1

League & Championship Architectures

Proprietary systems for organizing, qualifying, and executing competitive chess at national and international scale within a sovereign digital environment.

2

Academy Missions & Progression Systems

Adaptive learning frameworks that transform chess education from passive content consumption into active, mission-driven cognitive development.

3

Adaptive Learning & Performance Pathways

AI-assisted systems that personalize the learning journey based on individual performance patterns, strengths, and areas for growth.

4

Player Identity & Data Structures

Verified identity systems that create permanent, portable, and sovereign player profiles with lifetime performance archives.

5

Competition-to-Education Pipelines

Integrated systems connecting competitive play directly to educational outcomes, scholarship pathways, and cognitive skill development.

This is intellectual property protecting intellectual warfare. By keeping the entire technology stack in-house, we retain full sovereignty over every move, every lesson, every match, every championship, and every data point. There is no revenue siphoning. No algorithmic suppression. No dependency risk. 100% of value generated inside the jurisdiction is captured, retained, and compounded internally.

The Next-Generation Chess PWA

The Chess PWA is the backbone of the Intellectual Jurisdiction, designed to be far more than a conventional chess application. It serves as a unified cognitive environment, seamlessly integrating competition, education, analysis, and community into one cohesive experience. Unlike fragmented platforms that force players to switch between various apps for different needs, our PWA provides a continuous ecosystem where every aspect of a player's chess journey is interconnected and easily accessible.

League Play & Seasonal Competition

Championship Qualification & Finals

Lifetime Performance Archives

Guided Learning & Mastery Missions

Verified Global Player Identities

AI-Assisted Analysis & Feedback

This cycle of features ensures that every component feeds into the others, creating a truly unified and engaging player experience. Built on a Progressive Web App (PWA) architecture, it offers significant advantages, including freedom from app store dependencies, instant updates, and seamless cross-platform compatibility across all devices. This robust infrastructure is what enables us to deliver lifetime value, ensuring players never need to leave our comprehensive ecosystem to pursue their passion for chess.

The Continuous Chess Journey

The PWA architecture allows a player to move seamlessly through the entire cognitive athlete lifecycle without ever leaving the ecosystem. This is the fundamental departure from legacy platforms, which treat each interaction—a game, a lesson, a tournament—as a discrete event disconnected from the larger journey.

Beginner

Learning fundamentals through guided missions and adaptive curriculum

Competitor

Entering leagues and building a verified competitive record

Scholar

Deepening strategic mastery through advanced analysis and study

Champion

Qualifying for and competing in national and international championships

Mentor

Teaching, coaching, and giving back to the next generation of cognitive athletes

> **Chess is no longer episodic. It is continuous.**

Revenue Architecture: Lifetime Value Over Event Dependence

The Chess Jurisdiction's revenue model is designed around a fundamental principle: lifetime value over event dependence. Unlike traditional chess organizations that generate revenue primarily through one-time tournament fees and sponsorship deals, the jurisdiction captures recurring value across four integrated layers that compound over time.

Each revenue layer reinforces the others. League participants become Academy subscribers. Academy graduates become championship competitors. Championship broadcasts generate media revenue. Media exposure attracts institutional access deals. The flywheel never stops because the cognitive athlete never stops learning, competing, and creating data.

1 Leagues & Championships: The Competitive Layer

Chess leagues operate year-round, globally, and across demographics—making them the most reliable revenue engine in the jurisdiction. Unlike seasonal sports that go dormant for months, chess competition is perpetual and borderless.

National Chess Leagues

Youth, scholastic, adult, and open divisions operating across school years, calendar quarters, and seasonal cycles. Structured competition at every level of ability and age.

International Chess Leagues

Cross-border team and individual formats that connect national ecosystems into a global competitive framework. International play drives premium sponsorships and media interest.

National & International Championships

Digital qualifiers feeding into physical or virtual finals. Championship events serve as both the pinnacle of competitive achievement and the highest-margin revenue events in the calendar.

Revenue streams include league registrations and team licensing, championship entry fees, sponsorships and branded divisions, and institutional participation from schools, clubs, and national organizations.

2 The Chess Academy: The Education Layer

Chess is not learned once. It is studied for life. The Academy operates as a tiered, subscription-based institution offering the most comprehensive chess education system ever built—adaptive, mission-driven, and performance-based. This is not passive video content. It is active cognitive development.

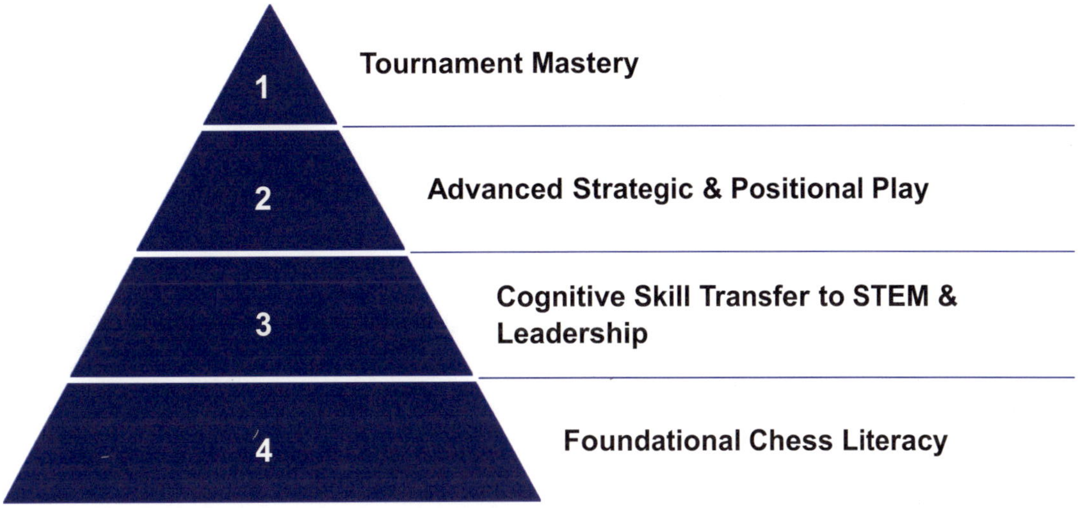

The Academy's subscription model generates predictable recurring revenue while simultaneously feeding the competitive pipeline. Every student who improves through the Academy becomes a more engaged league participant, a more valuable data asset, and a more likely championship competitor. Education is the entry point; competition is the retention mechanism.

3 Proprietary Features & Premium Intelligence

Patent Pending

The Differentiation Layer

The Chess Jurisdiction incorporates proprietary features that transform chess from a static game into a living intelligence system. These are not incremental improvements over existing platform features—they are architecturally distinct capabilities protected by provisional patent filings.

Increased Retention & Mastery

Proprietary engagement mechanics that keep players returning not through addiction-based design, but through genuine cognitive growth and measurable skill development.

Personalized Progression Pathways

AI-driven learning paths that adapt to individual playing styles, strengths, weaknesses, and goals—creating a truly personalized cognitive development experience.

Defensible Technological IP

Every proprietary feature adds another layer of protection to the competitive moat, making the jurisdiction increasingly difficult to replicate or circumvent.

Cognitive Training Platform

Chess elevated beyond recreation into a structured cognitive training environment with applications in education, professional development, and mental fitness.

Premium access unlocks advanced intelligence layers unavailable on legacy platforms, creating a tiered value proposition that rewards the most engaged participants.

4 Media, Data & Institutional Access: The Permanence Layer

Chess data never expires. A game played at age 10 remains relevant at age 40. This is a unique characteristic among all sports in the portfolio—and it creates a revenue layer with infinite shelf life. The platform monetizes this permanence through four distinct channels.

Championship Broadcasting

Live commentary, analysis, and multi-camera coverage of championship events delivered through the PWA to global audiences.

Historical Archives

Permanent performance records that grow more valuable over time as players advance, creating longitudinal datasets with compounding analytical worth.

Institutional Access

Schools, universities, and organizations purchase access to the platform's educational infrastructure and competitive frameworks.

Licensing

Educational and analytical use licensing for third-party researchers, publishers, and technology companies.

Total Annual Chess Ecosystem Revenue

The four revenue layers combine to create a $65 million annual ecosystem that is diversified, recurring, and compounding. Each layer feeds the others, creating a self-reinforcing economic flywheel.

Revenue Stream	Annual Value
Leagues & Championships	$14.5M
Chess Academy	$22.0M
Proprietary Features & Premium Intelligence	$18.5M
Media & Data	$10.0M
Total Annual Revenue	$65.0 Million

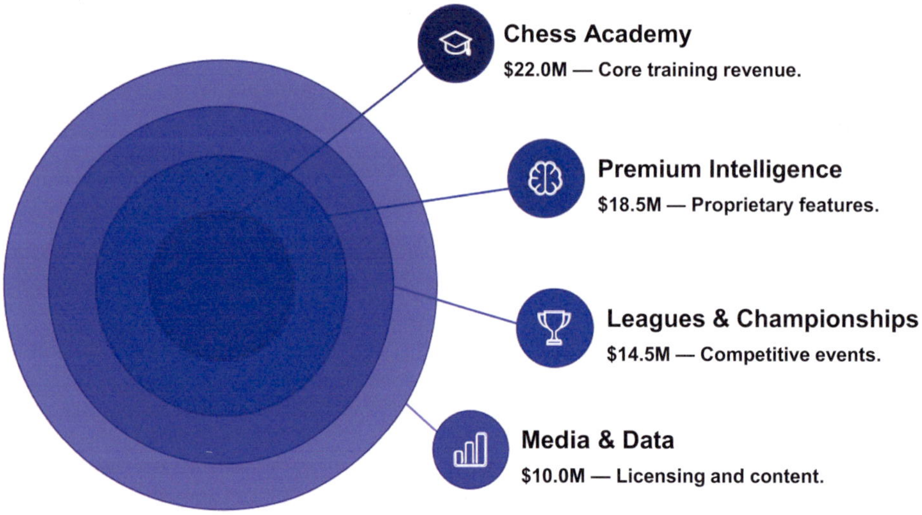

The Academy represents the single largest revenue contributor at $22M, reflecting the fundamental truth that chess is as much an educational product as a competitive one. Premium features at $18.5M demonstrate the monetization power of proprietary IP. Competitive revenue at $14.5M provides the engagement backbone, while media and data at $10M create the permanence layer that compounds indefinitely.

Revenue Distribution

The Chess Jurisdiction's revenue is strategically distributed across four layers, ensuring no single stream represents an existential dependency. This diversification is by design—each layer can grow independently while reinforcing the others.

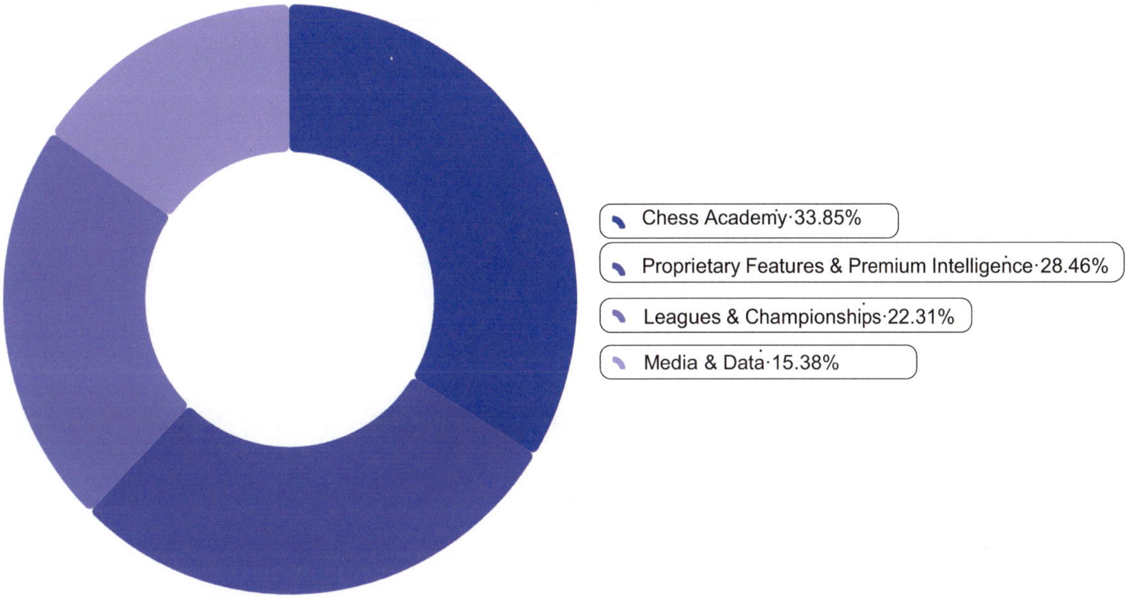

The education layer (Academy) commands 34% of total revenue, confirming that chess's greatest economic value lies in lifelong learning. Proprietary features capture 28%, leagues and championships deliver 22%, and media and data contribute 16%—a balanced portfolio with multiple growth vectors.

Valuation Framework: The Permanent Intellectual Multiple

Chess is not valued like a sport. It occupies a unique position at the intersection of four high-multiple sectors, each commanding premium valuations in today's capital markets.

Education Technology

EdTech companies with subscription models and adaptive learning command 8-12× revenue multiples in current markets.

Cognitive SaaS

Software-as-a-service platforms with high retention and low churn consistently attract premium multiples from institutional investors.

AI-Assisted Learning

AI-driven personalization in education is among the highest-growth categories in technology, with expanding multiples.

Lifetime Subscription

Ecosystems with lifetime engagement models achieve superior valuations due to predictable, compounding revenue streams.

Valuation Scenarios

Applying blended EdTech, SaaS, and IP multiples to the $65M annual revenue base produces valuation scenarios that reflect the Chess Jurisdiction's unique economic characteristics. These are not aspirational—they are mathematically grounded in comparable sector valuations.

$455M

Base Case

7× revenue multiple reflecting conservative EdTech and SaaS comparables

$585M

Growth Case

9× revenue multiple reflecting intelligence-driven growth and IP premium

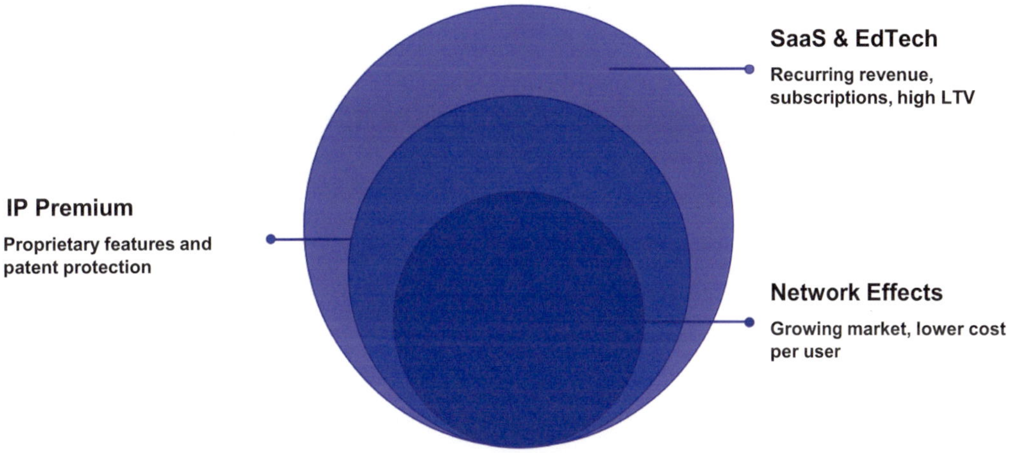

This reflects a rare economic truth: chess has no physical decay, no aging-out, and no geographic ceiling. Unlike physical sports where athletes retire, facilities depreciate, and markets saturate, the chess economy expands with every new player who learns the game—and every existing player who continues studying. The addressable market grows every year, and the cost to serve each additional user decreases with scale.

Strategic Significance Within the Portfolio

Each sport in the Goddess Phenezy Holdings portfolio captures a distinct dimension of human athletic and competitive expression. Chess occupies the most enduring position of all.

1. **Basketball**
   Captures motion
2. **Baseball**
   Captures continuity
3. **Soccer**
   Captures scale
4. **Cuju**
   Captures origin
5. **Golf**
   Captures capital
6. **Tennis & Pickleball**
   Capture lifestyle velocity

> **Chess captures permanence. It is the intellectual spine of the ecosystem—feeding education, STEM reasoning, AI literacy, and lifelong mastery. And unlike every other sport in this portfolio, chess does not end. It compounds.**

INNOVATION **PATENT PENDING**

Four Corners Chess & Checkers

When the World Assumed There Were No New Moves Left

While the global chess ecosystem has largely treated the game as complete—perfected, preserved, and immune to meaningful evolution—this assumption is not shared here. Within the Chess Jurisdiction, an original variant was conceived and developed by Goddess Phenezy to expand strategic dimensionality beyond traditional bilateral play.

It is called Four Corners Chess & Checkers.

Multi-Vector Intelligence

By extending the board architecture and introducing four directional points of engagement, the game transforms classical strategy into a multi-vector intelligence environment—requiring simultaneous threat assessment, long-range planning, and adaptive decision-making across multiple fronts for 2 to 4 players.

What It Is

- **Fully playable original game system**
- **Designed for 2 to 4 players**
- **Expanded board architecture**
- **Multi-vector strategic dimensionality**

What It Is Not

- **Not a novelty**
- **Not a reskin**
- **Not theoretical**
- **Not dependent on external platforms**

Four Corners: Strategic Applications

Four Corners Chess & Checkers exists natively within the broader Chess ecosystem and serves multiple strategic purposes that extend well beyond recreational gameplay.

Advanced Cognitive Training Platform

Multi-front decision-making develops cognitive skills that traditional bilateral chess cannot access—simultaneous threat assessment, resource allocation under complexity, and adaptive strategic thinking.

Bridge Between Disciplines

Connects chess to mathematics, systems thinking, and multi-agent problem solving in ways that create direct educational applications for STEM curricula and leadership development.

Foundation for Future Formats

Establishes the architectural precedent for competitive, educational, and AI-driven formats that push the boundaries of what board game intelligence can achieve.

> **Chess is permanent. But it wasn't finished.**
> **And just when the world assumed there were no new moves left to discover, the board expanded.**

ANCIENT JURISDICTION

The Ancient Jurisdiction: Senet

The Genesis Property: Reclaiming the First Board

The Origin of Strategy

Before chess codified warfare, before Go modeled grand strategy, before any modern board game began its journey to simulate the complexities of decision-making, there was Senet.

Dating back more than 5,000 years to ancient Egypt, Senet was more than just a game; it was a profound cultural simulation played by Pharaohs and commoners alike, believed to represent the soul's perilous journey through the Duat (the underworld) to reach the afterlife. Its moves were not merely tactical but deeply symbolic, mirroring existential struggles and the quest for spiritual triumph.

For millennia, this foundational game has been trapped behind museum glass, its original rules fragmented and subject to academic reconstruction, rendering it extinct in practice. Its rich history and cultural significance remained largely inaccessible, a relic of a bygone era.

> **Goddess Phenezy Holdings, Inc. is not a curator. We are architects. We are removing Senet from the museum and integrating it directly into a sovereign digital jurisdiction.**

Structural Advantage: Proprietary Evolution

To simply recreate Senet as a digital replica would be novelty—a museum exhibit with a touchscreen. To evolve it—to expand its architecture and code it for the modern family economy—is market creation. The distinction between replication and evolution is the difference between a Wikipedia article and a billion-dollar platform.

While the ancient form of Senet operated as a two-player, linear race, the Senet Jurisdiction within Goddess Phenezy Holdings introduces a completely original, proprietary evolution: a fully coded four-player family variant.

1 Maximized User Density

By expanding board capacity from two to four players, each session engages more participants and generates longer play times, increasing both engagement metrics and monetization opportunities per match.

2 Multi-Generational Family Experience

The four-player format transforms gameplay from a head-to-head competition into a social family experience—parents and children playing together, creating household-level engagement.

3 Defensible Intellectual Property

A wholly owned IP layer built atop a 5,000-year-old public-domain foundation. The ancient game belongs to history; the modern evolution belongs to us.

Ancient origin. Modern code. Sovereign control.

Revenue Architecture: The Heritage Gaming Economy

Hosted entirely within the proprietary PWA ecosystem, the Senet Revival bypasses traditional app-store monopolies and captures value through culturally aligned, high-margin channels that leverage the game's unique positioning at the intersection of heritage, education, and family entertainment.

Senet Revenue Channels

Hosted entirely within the proprietary PWA ecosystem, the Senet Revival bypasses traditional app-store monopolies and captures value through culturally aligned, high-margin channels that leverage the game's unique positioning at the intersection of heritage, education, and family entertainment.

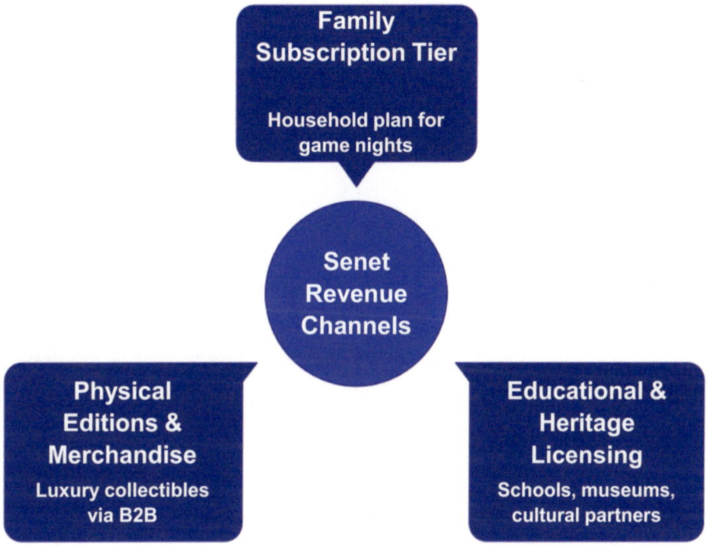

Each channel reinforces the others, creating a self-sustaining heritage gaming economy that scales across demographics, geographies, and institutional contexts.

We Own the Ancestry

The strategic significance of Senet extends far beyond its revenue potential. It completes a narrative arc that spans the entire history of human competitive intelligence.

The Physical Athlete Origin

By capturing Cuju, we claimed the origin of the physical athlete—the first sport, the first ball, the first competitive physical expression in recorded history.

The Cognitive Athlete Origin

By capturing Senet, we claim the origin of the cognitive athlete—the first board, the first strategy game, the first competitive intellectual expression in human civilization.

Chess proves we can expand the ultimate game of strategy. Senet proves we can resurrect the genesis of gaming itself. As with Four Corners Chess, the four-player Senet code demonstrates that Goddess Phenezy Holdings does not merely host legacy games—it evolves them.

We take the oldest properties in human history, build a digital fortress around them, and invite the world to play under our rules. This is not nostalgia. This is not cultural tourism. This is the deliberate, strategic capture of humanity's competitive heritage—encoded into sovereign infrastructure and monetized for the modern economy.

Ancient Origins

Senet: 5,000-year strategic lineage

Strategic Evolution

Four Corners Chess expands tactical depth

Digital Future

PWA sovereign championships for global play

This visual depicts our strategic progression, building from the ancient genesis of gaming with Senet, through the evolution of strategic competition with Four Corners Chess, to establish the future of digital competition through PWA-based sovereign championships.

> We do not merely own the future of digital competition.
> We own its ancestry.

The Complete Expansion Pipeline

From high-margin spectacle sports to ancient board games, the expansion pipeline demonstrates the true power of sovereign digital infrastructure: the ability to absorb any competitive economy, in any discipline, at any point in human history—and monetize it through a single, unified, jurisdiction-ready platform.

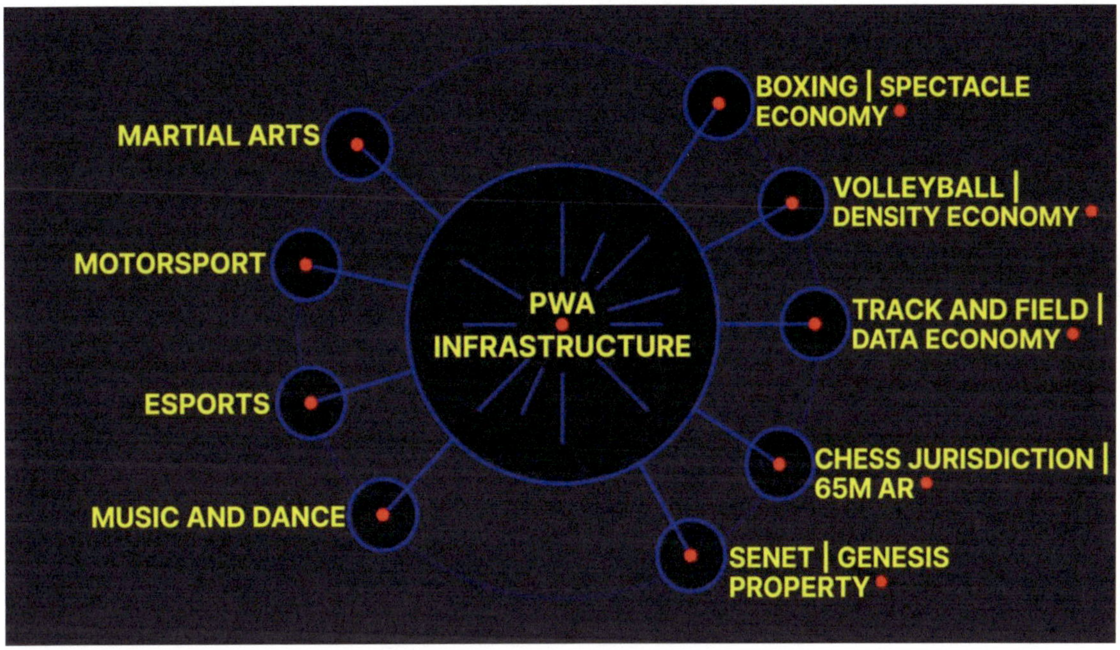

Every vertical in this pipeline shares three characteristics: proven global demand, compatibility with existing sovereign infrastructure, and alignment with the jurisdiction's data, media, and membership model. The question is not whether these markets can be absorbed—it is simply when.

The Infinite Game

The expansion pipeline is not a wishlist. It is not a roadmap of possibilities. It is a pre-engineered absorption sequence built on infrastructure that already exists, IP that is already protected, and economic models that are already proven across the core verticals.

Boxing brings spectacle. Volleyball brings density. Track & Field brings data purity. Cultural verticals bring lifestyle and digital-native audiences. Chess brings permanence and cognitive sovereignty. Senet brings ancestry and heritage.

Together, they represent something no other sports organization on Earth can claim: a single sovereign ecosystem that spans the entire spectrum of human competitive expression—from the world's oldest board game to tomorrow's esports championship—all operating under one jurisdiction, one identity system, one data architecture, and one economic model.

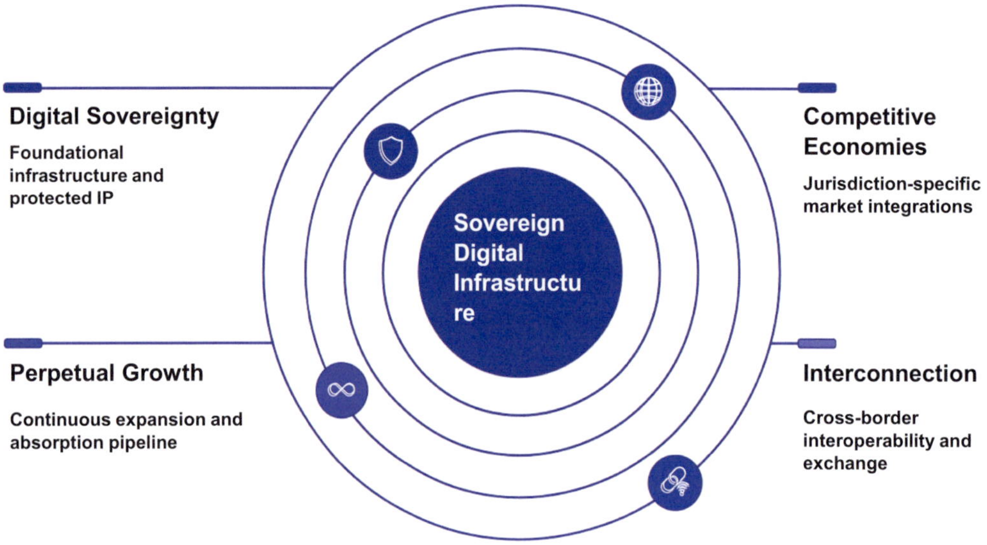

This systematic value capture ensures that each new jurisdiction compounds the value of all others, creating a powerful, interconnected web of growth. This is not expansion for its own sake—it is systematic value capture, built for perpetual growth and an infinite game.

The board keeps expanding. And we own every square.

The National International STEM Championships

The National & International Academies of STEM

The National & International STEM Championships & The International Academy of STEM

Intellectual Pipeline

A borderless, continuous ecosystem that identifies talent, develops capability, validates performance, and retains verified intelligence across a lifetime. This is not reform. This is replacement.

The Fragmented Genius Economy

STEM is not an academic category. It is the operating system of modern civilization.

Every global industry—technology, defense, healthcare, finance, energy, infrastructure, and artificial intelligence—runs on scientific reasoning, engineering precision, and computational logic. Yet the ecosystem tasked with identifying, training, and advancing the next generation of innovators is fundamentally broken. Today, exposure to advanced robotics, coding, engineering, and applied science is determined by geographic lottery—tied to local school budgets, outdated curricula, and isolated academic silos.

Competitions are episodic. Performance data disappears at graduation. Credentials are fragmented and unverifiable. And the bridge between early-stage talent and enterprise recruitment is inefficient, opaque, and wildly exclusionary. Students memorize concepts. They rarely compete. They almost never build public, portable records of proof.

The System Is Not Broken—It Was Never Built

Geographic Lottery

Access to advanced STEM is dictated by zip code, school budget, and local curriculum decisions—not by aptitude or ambition.

Episodic Competitions

Science fairs and robotics tournaments exist in isolation. No continuity, no longitudinal tracking, no lasting record.

Invisible Credentials

Paper transcripts and localized grade books trap achievement inside institutions. Proof disappears at graduation.

Broken Talent Bridge

The pipeline between early-stage talent and enterprise recruitment is opaque, inefficient, and wildly exclusionary.

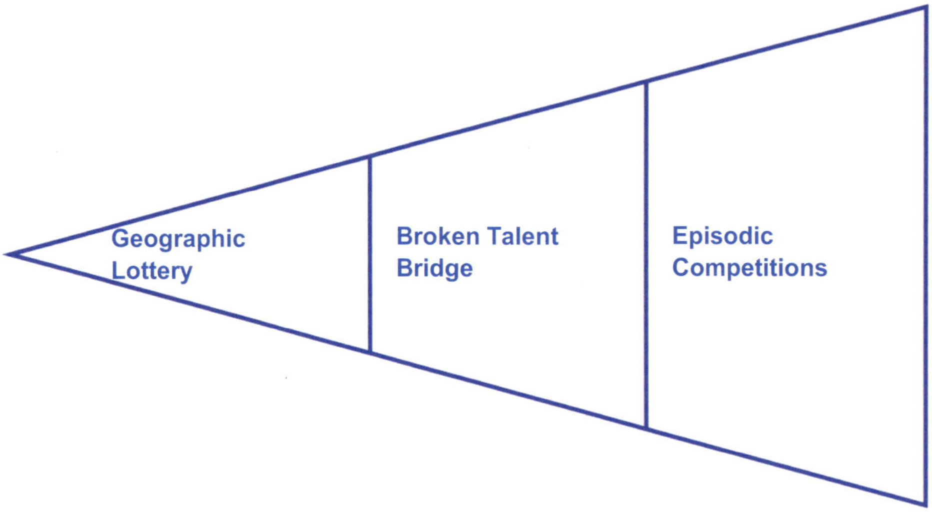

This diagram highlights the critical discontinuities in the current system, where talented individuals are lost at various stages due to systemic failures, rather than a lack of potential.

> **Goddess Phenezy Holdings, Inc. does not attempt to reform this system. We bypass it entirely.**

We Are Establishing an Innovation Jurisdiction

We are not building a science fair. We are not launching another online course platform. We are establishing an Innovation Jurisdiction—a sovereign intellectual ecosystem that operates entirely outside the constraints of legacy education.

Through the proprietary digital infrastructure of The International Academy of STEM and the high-stakes competitive arena of The National & International STEM Championships, we have engineered a borderless, continuous ecosystem—one that identifies talent, develops capability, validates performance, and retains verified intelligence across a lifetime.

We do not wait for the educational system to catch up. We replace it.

Identify Talent

The Academy discovers capability

Develop Capability

Structured learning pathways

Retain Intelligence

Cognitive Passport captures lifetime achievement

Validate Performance

Championships prove mastery

STRUCTURAL ADVANTAGE

The Global Cognitive Passport

By owning both The International Academy of STEM and The National & International STEM Championships, we control the entire pipeline—educating youth through world-class curriculum while simultaneously preparing them for elite competition and future opportunities in STEM. This vertical integration ensures seamless talent development from discovery to deployment.

Beyond the Transcript: A Living Intelligence Record

In legacy education models, achievement is trapped inside paper transcripts, localized grade books, and institution-specific credentials that lose relevance outside narrow contexts. The moment a student crosses a district line or a national border, their intellectual record becomes invisible.

In the STEM Jurisdiction, every participant is issued a Global Cognitive Passport, housed natively within our proprietary Progressive Web App (PWA). This passport functions as a permanent, verified digital ledger—a living intelligence record that grows, evolves, and compounds value over an entire lifetime.

What the Passport Tracks

01

Academy Curriculum

Completion records across Coding, Engineering, Data Science, and AI tracks

02

Championship Performance

Project outcomes, scoring metrics, and competitive rankings

03

Collaboration & Leadership

Peer collaboration dynamics, mentorship engagement, and team leadership

04

Intellectual Property

Patent-pending contributions, provisional IP, and original problem-solving artifacts

This is not a résumé. It is a living intelligence record—portable, verifiable, and permanent. It operates outside public-school bureaucracy and legacy accreditation bottlenecks, delivering direct-to-consumer intellectual development while creating the most valuable, highly vetted B2B talent intelligence database on Earth.

How the Cognitive Passport Creates Value

Learn Compete Verify Connect

By operating outside public-school bureaucracy and legacy accreditation bottlenecks, the Global Cognitive Passport delivers direct-to-consumer intellectual development while simultaneously creating the most valuable, highly vetted B2B talent intelligence database on Earth. Every interaction—every course completed, every competition entered, every project submitted—compounds into a verified, actionable intelligence profile that corporations, universities, and governments can trust.

Students Learn

Students acquire skills and knowledge through the Academy curriculum.

Performance Data Captured

Achievements and progress are recorded in the Global Cognitive Passport.

Corporations Access Talent

Companies recruit high-performing individuals based on verified data.

Revenue Funds Expansion

Revenue generated fuels growth and further development of the ecosystem.

More Students Learn

Expanded resources attract more students, restarting the cycle.

Intellect as an Elite Sport

Reframing STEM as High-Performance Competition

We are reframing STEM from an academic requirement into a high-performance competitive discipline. The National & International STEM Championships are designed with the same production value, prestige, and intensity as elite athletic properties. From competitive robotics and autonomous drone racing to algorithmic warfare, applied AI challenges, and sustainable engineering hackathons—we make intellectual dominance visible, aspirational, marketable, and recruitable.

Competitive Robotics

Teams design, build, and battle autonomous machines in head-to-head engineering showcases that test precision, creativity, and resilience under pressure.

Autonomous Drone Racing

AI-piloted drones navigate complex courses at breakneck speed. Competitors code the intelligence. The machines execute the vision.

Sustainable Engineering

Hackathons challenge teams to solve real-world environmental crises—designing and prototyping solutions that can scale from concept to civilization-level impact.

Algorithmic Warfare & AI

Applied artificial intelligence challenges pit the sharpest computational minds against problems that define the future of defense, healthcare, and infrastructure.

STEM Is No Longer Hidden in Classrooms

Visible

Staged on world-class platforms with broadcast-quality production and global reach

Aspirational

Elevated to the prestige of elite athletics—celebrated, not hidden

Marketable

Packaged for sponsors, media partners, and the global innovation economy

Recruitable

Every performance feeds the Global Cognitive Passport, connecting talent to opportunity

> **STEM is staged, scored, broadcast, and celebrated.**

The Access Gap in STEM Education

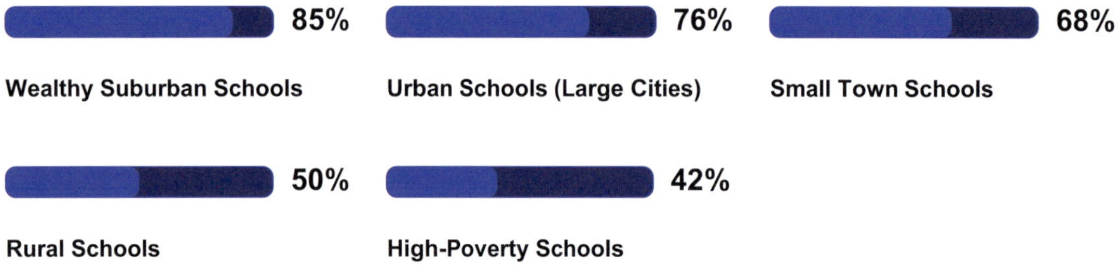

Access to advanced STEM courses—including AP Computer Science, Engineering, and Advanced Mathematics—is dramatically unequal. Students in wealthy suburban districts have nearly double the access to rigorous STEM programming compared to students in rural and high-poverty schools. This geographic lottery determines who gets to compete before the race even begins.

The Pipeline Economy

Revenue Layer 1: The International Academy of STEM

THE EDUCATION LAYER

The Academy operates as a premium, mission-driven educational platform serving both direct-to-consumer and institutional markets. It replaces static textbooks with adaptive, performance-aligned digital curricula that evolve with each learner's trajectory. This is not supplementary education—it is the primary intellectual development engine for the next generation of innovators, engineers, and technologists.

Student Memberships

Global monthly subscriptions for continuous STEM education, mentorship access, and championship challenge preparation. A direct-to-consumer model that scales without geographic limitation.

Institutional Licensing

Turnkey Academy access for private schools, municipalities, nonprofits, and youth organizations. Enterprise-grade deployment that embeds our curriculum into existing educational infrastructure worldwide.

$28.0M

Projected Annual Revenue

The International Academy of STEM — Education Layer

Revenue Layer 2: The STEM Championships

THE COMPETITION LAYER

Regional, national, and international tournaments operate as large-scale, multi-day innovation festivals. These are not sterile academic exercises—they are cultural events with the energy, spectacle, and commercial infrastructure of elite sporting championships. Host cities bid for the right to stage them. Title sponsors from the world's most powerful industries compete for branding alignment.

1

Entry Fees & Registrations

Team and individual entry fees across regional qualifiers, national rounds, and international finals create a scalable, recurring participation revenue base.

2

Host-City Bidding

Municipal partnerships and host-city bidding rights generate significant venue and economic development revenue, with cities competing for the prestige and tourism impact.

3

Title Sponsorships

Big Tech, aerospace, automotive, defense, and infrastructure corporations invest in title sponsorships to align their brands with the next generation of elite technical talent.

$18.5M

Projected Annual Revenue

The STEM Championships — Competition Layer

Revenue Layer 3: Corporate Scouting & B2B Talent Brokerage

THE DATA LAYER

This is the core monetization engine. Global corporations spend billions annually attempting to identify high-potential technical talent—often far too late in the pipeline, after university, after debt, and after competitors have already locked in the best minds. The STEM Jurisdiction reverses this dynamic entirely.

Through enterprise SaaS access, corporations license entry into the Global Cognitive Passport database, enabling them to identify, evaluate, and recruit elite talent before university, before debt, and before competitors. This is not a job board. This is a verified intelligence marketplace—the Bloomberg Terminal of human cognitive capital.

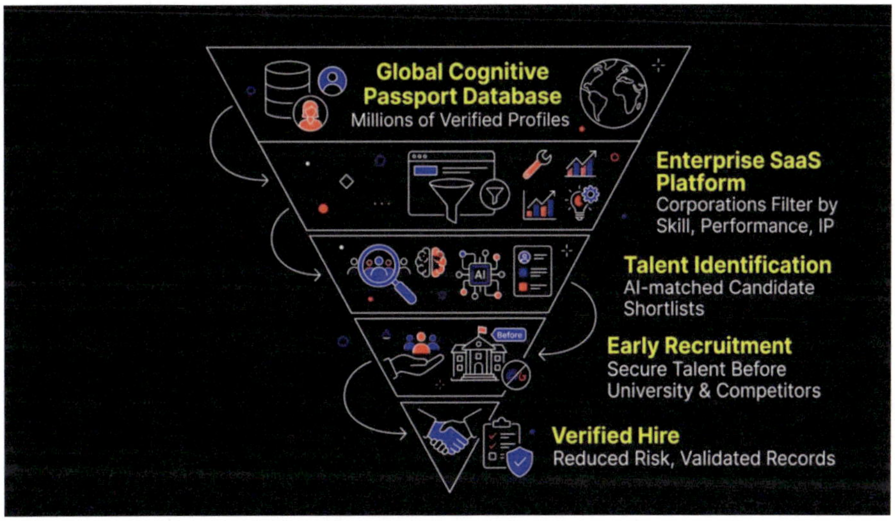

$24.0M

Projected Annual Revenue

Corporate Scouting & B2B Talent Brokerage — Data Layer

Revenue Layer 4: Media, Ed-Tech & Broadcasting

THE NARRATIVE LAYER

Through Xperience TV, the drama of invention is packaged and distributed globally. The human stories behind breakthroughs—the pressure, the creativity, the triumph—are captured with cinematic production value and delivered to audiences worldwide. This is not educational programming. This is prestige content that makes STEM culturally dominant.

Live Broadcasts

Championship finals and engineering challenges streamed live to global audiences with professional commentary, multi-camera production, and real-time scoring.

Docuseries

Narrative-driven series profiling young innovators, breakthrough teams, and the high-stakes journey from concept to competition glory.

Multi-Stream Revenue

Subscription access, title sponsorships, programmatic advertising, and licensing deals create diversified, compounding media income.

$9.5M

Projected Annual Revenue

Media, Ed-Tech & Broadcasting — Narrative Layer

Total Annual STEM Ecosystem Revenue

Revenue Stream	Annual Value
The International Academy of STEM	$28.0M
STEM Championships	$18.5M
Corporate Scouting & Talent Brokerage	$24.0M
Media, Ed-Tech & Broadcasting	$9.5M
Total Annual Revenue	$80.0 Million

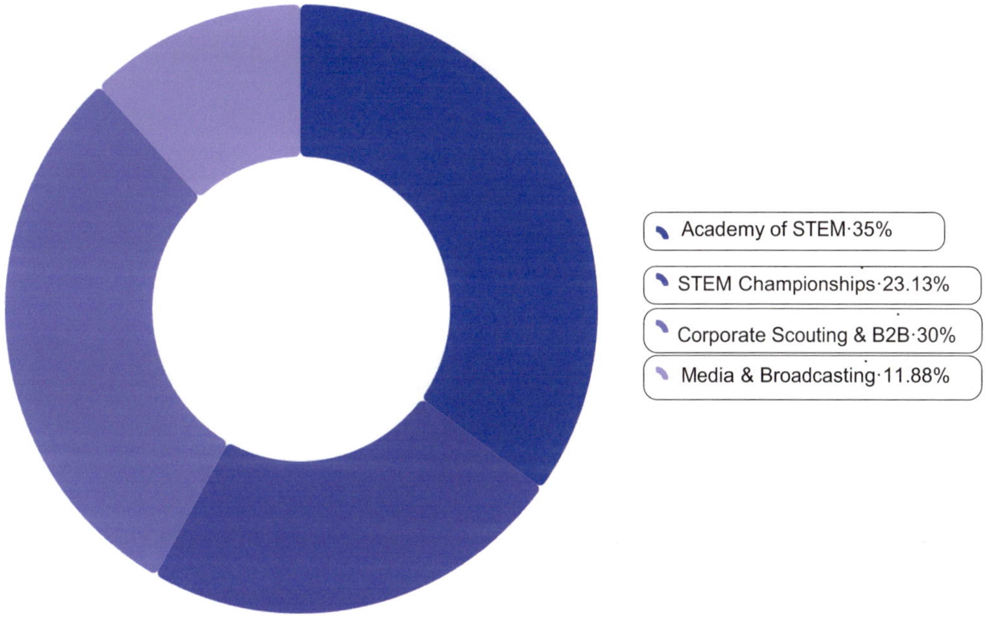

The revenue architecture is deliberately diversified across four interlocking layers. The Academy feeds the Championships. The Championships feed the Cognitive Passport. The Passport feeds Corporate Scouting. And Media amplifies the entire flywheel. No single revenue stream operates in isolation—each compounds the value of the others, creating a self-reinforcing pipeline economy with an aggregate annual value of $80.0 million.

The Revenue Flywheel

What makes this revenue architecture structurally distinct is its compounding interdependence. Each layer amplifies the others in a self-reinforcing loop that accelerates growth, deepens engagement, and expands the addressable market with every cycle.

Academy

Trains and prepares the talent pool

Media & Broadcast

Amplifies visibility and drives enrollment

Championships

Validates and showcases performance

Cognitive Passport

Records and verifies intelligence

Corporate Scouting

Monetizes the verified talent database

Enrollment in the Academy grows the Championships. Championships generate verified performance data. Performance data enriches the Cognitive Passport. The Passport drives enterprise licensing revenue. And media coverage of the entire ecosystem drives new enrollment—completing the loop and accelerating the flywheel with each revolution.

The Deep-Tech Education Multiple

Valuation Framework

STEM infrastructure commands premium valuations because it sits at the intersection of four high-growth, high-multiple sectors simultaneously. This is not a single-sector play—it is a convergence asset that benefits from the valuation premiums of each constituent market.

Education Technology

Global edtech market projected to exceed $400B by 2028. Adaptive, AI-driven platforms command the highest growth premiums in the sector.

Enterprise Recruitment SaaS

Talent acquisition platforms with verified, structured data trade at 10–15× revenue. Our database is unmatched in depth and verification quality.

Workforce Intelligence Platforms

Predictive workforce analytics and skills-mapping tools are among the fastest-growing enterprise software categories globally.

Future-Facing IP Generation

Patent-pending contributions and original problem-solving artifacts created within the ecosystem represent compounding intellectual property value.

Valuation Multiples & Enterprise Value

Applying blended deep-tech and talent-acquisition multiples to the $80M annual revenue base yields enterprise valuations that reflect the global premium placed on verified, actionable technical talent.

$640M

Base Case

8× revenue multiple reflecting established deep-tech education comps

$800M

Growth Case

10× revenue multiple reflecting innovation trajectory and data moat

These valuations are conservative relative to pure-play SaaS recruitment platforms (which trade at 12–20× revenue) because our model also includes event and media revenue, which trade at lower multiples. As the data layer scales and enterprise SaaS becomes an increasingly dominant share of revenue, the blended multiple will compress upward toward the higher end of this range.

Event & Media Revenue

6-8× multiple

Blended Model (Current)

8-10× multiple

Data-Driven SaaS Dominant

12-15× multiple

Pure Talent Intelligence Platform

15-20× multiple

Engineering Civilization Itself

Why STEM Is the Apex Asset

Physical sports capture the body. Heritage sports capture culture. Chess captures the mind. STEM builds the future.

By integrating STEM into the Goddess Phenezy Holdings architecture, the ecosystem does not merely entertain or educate—it engineers civilization itself. Every autonomous vehicle, every AI breakthrough, every climate solution, every defense system, every space mission, every medical advance begins as an idea inside a young mind. The question is whether that mind is identified, developed, validated, and connected—or lost to the geographic lottery of legacy education.

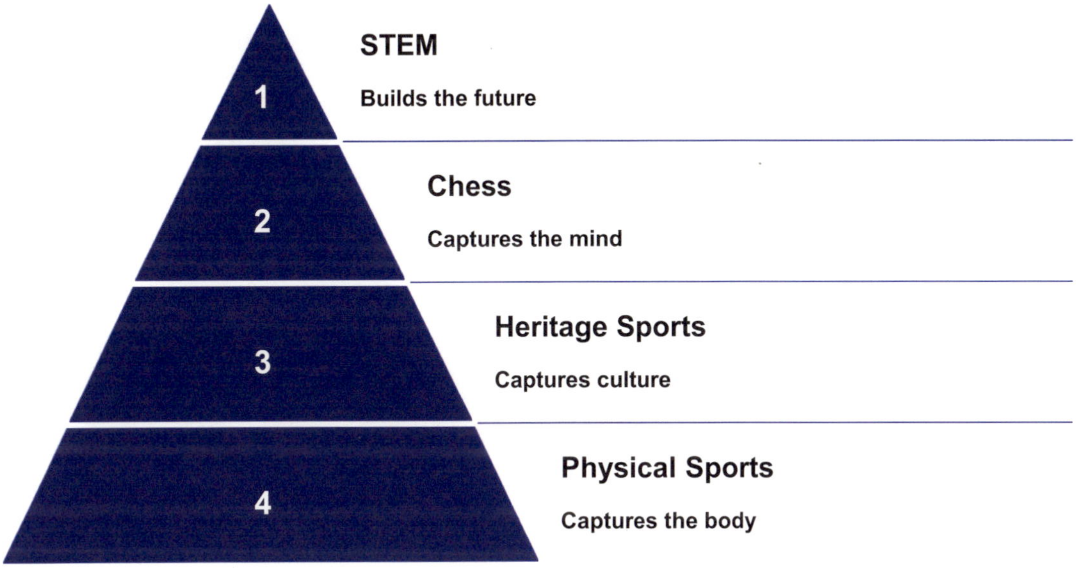

STEM sits at the apex because it does not reflect civilization—it creates it. Every other competitive domain operates within the world as it exists. STEM competition produces the minds that reshape the world into what it becomes.

The Ecosystem Within the Architecture

The STEM Jurisdiction does not exist in isolation. It is embedded within the broader Goddess Phenezy Holdings infrastructure—sharing audiences, data systems, media distribution, and brand architecture with the full portfolio of competitive properties. This integration creates structural advantages that no standalone edtech platform, competition organizer, or talent marketplace can replicate.

1

Identify

The Academy surfaces raw talent from every geography, demographic, and economic background on Earth

2

Develop

Adaptive curricula and competitive preparation sharpen skills to elite performance levels

3

Validate

Championships and the Cognitive Passport create verified, permanent records of proof

4

Connect

Enterprise SaaS and media distribution deliver talent to the organizations that need it most

Minting the Future

We are not just training students. We are minting the founders, engineers, and technologists of tomorrow—entirely within our jurisdiction.

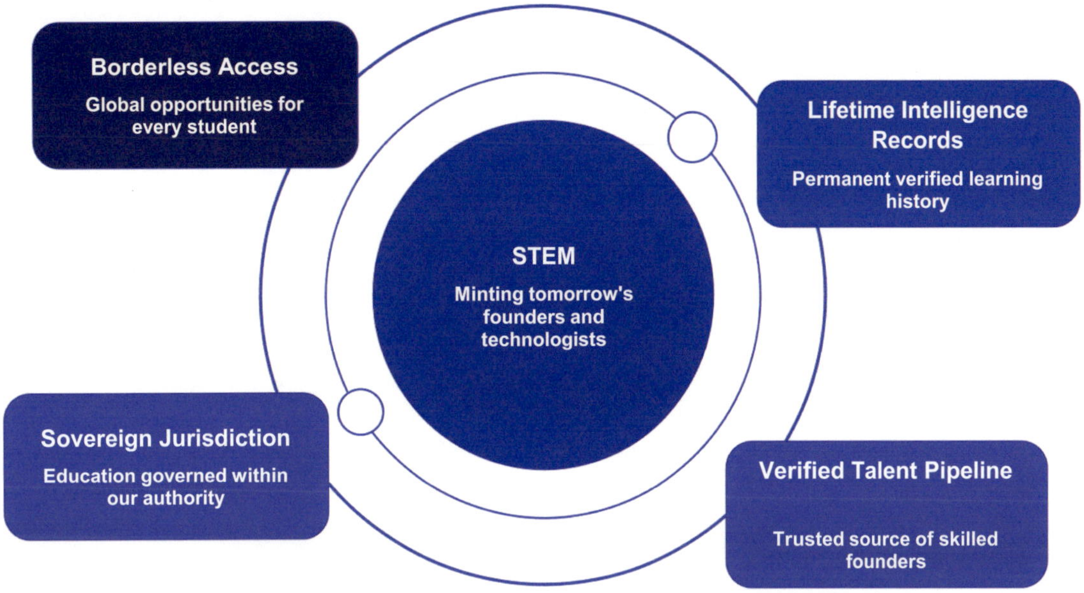

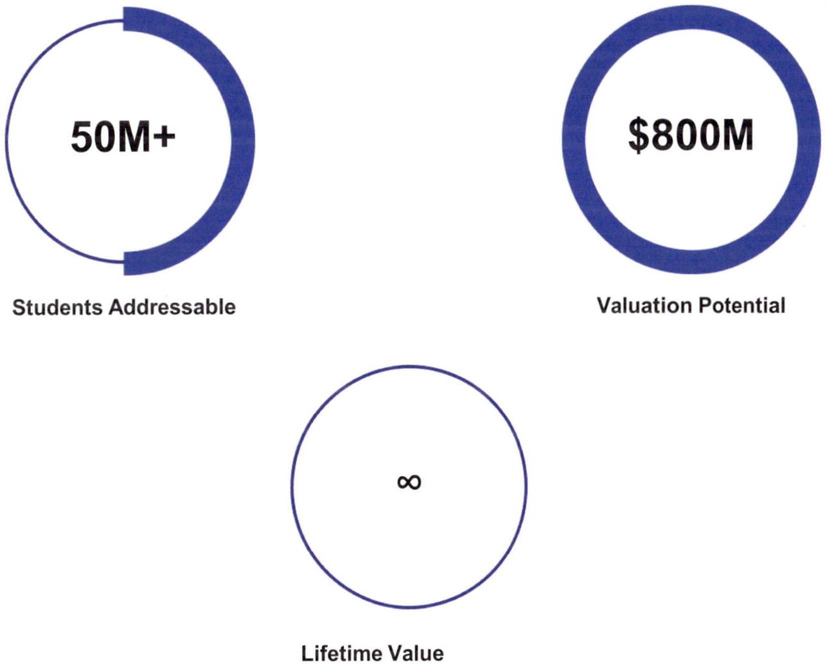

> **This is not reform. This is not incremental improvement. This is the construction of a new intellectual order—and Goddess Phenezy Holdings owns the jurisdiction.**

IQ WHIZZY

The Open Gate: Intelligence Without Permission

Every child deserves a chance to be seen. IQ Whizzy makes sure they are — free, open, and built for every learner, everywhere.

Why IQ Whizzy Is Free

IQ Whizzy is free by design. Not as a promotion. Not as a trial. Not as philanthropy theater.

IQ Whizzy exists because intelligence should never be gated by zip code, income, or institutional access. For generations, cognitive development tools — test preparation, logic training, language acquisition, and early STEM exposure — have been reserved for families with resources. The result is not a talent gap, but an access gap.

Goddess Phenezy created IQ Whizzy to collapse that divide. It is the front door to the entire intelligence ecosystem — open to every child, every family, everywhere. There are no hidden costs, no bait-and-switch subscriptions, and no paywall waiting around the corner. The mission is simple and unyielding: if a child has curiosity, they have everything they need to begin.

> No tuition. No prerequisites. No permission required.

In a world where premium learning tools routinely cost hundreds or thousands of dollars per year, IQ Whizzy stands as a radical alternative — a declaration that access to cognitive growth is a right, not a luxury. This isn't charity. It's architecture. Every free interaction lays the foundation for a lifetime of measurable intellectual development, visible to families, educators, and the children themselves.

IQ
Whizzy

The Access Gap Is Real

The data paints an undeniable picture: children from under-resourced communities don't lack talent — they lack tools. The so-called "achievement gap" is more accurately described as an opportunity gap, one that widens with every year a child goes without access to structured cognitive development.

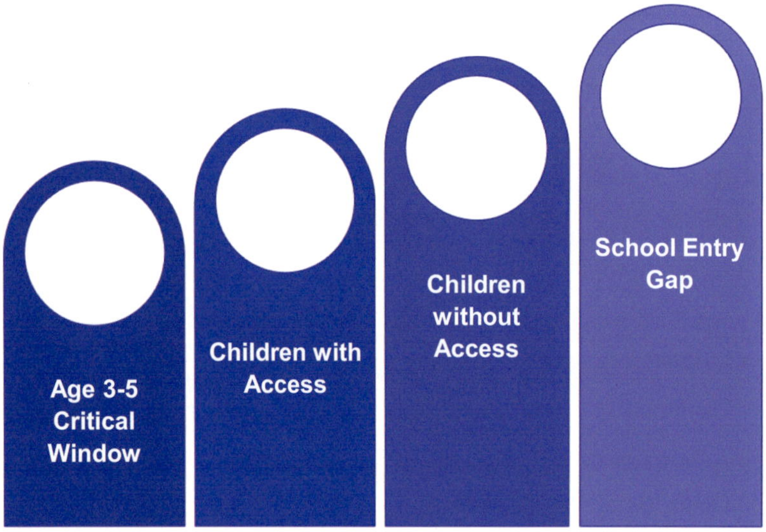

By the time most educational systems start paying attention — through standardized tests in third or fourth grade — the compounding effect of early access has already created deep divides. Children who had enrichment opportunities at ages 3, 4, and 5 arrive at school with measurably different cognitive profiles, not because they are inherently more capable, but because they had more exposure. IQ Whizzy intervenes at the earliest possible moment, ensuring that every child — regardless of background — arrives at school already primed to thrive.

Positioning: Play as the First Credential

IQ Whizzy is not "edutainment." It is a cognitive signal engine disguised as play.

Before children encounter grades, rankings, or standardized labels, IQ Whizzy captures how they think — how they reason, adapt, solve, and persist. Every tap, every choice, every moment of hesitation or breakthrough generates meaningful data about a child's cognitive profile. This isn't surveillance; it's recognition.

This data becomes the earliest layer of the Global Cognitive Passport, establishing a longitudinal intelligence record years before traditional systems begin paying attention. While conventional education waits for children to prove themselves through high-stakes exams, IQ Whizzy is already mapping the contours of their potential — gently, joyfully, through the universal language of play.

> IQ Whizzy does not replace school. It precedes it.

Research validates what children already know — learning through play works:

49%	36%	82%	46%
improvement in cognitive learning outcomes	increase in motivation and engagement	overall effectiveness boost in educational settings	better learning outcomes vs traditional methods

Think of it as the cognitive foundation upon which all future learning is built. By the time a child walks through the classroom door, IQ Whizzy has already illuminated their strengths, identified areas for growth, and created a living record that follows them — not as a label, but as an evolving portrait of potential.

What Makes a Cognitive Signal Engine?

Traditional educational tools measure outputs — test scores, grades, completion rates. IQ Whizzy measures process. It watches how a child approaches a problem, not just whether they get the right answer.

Reasoning Patterns

How does a child break down complex challenges? Do they work step-by-step, or leap to intuitive conclusions? IQ Whizzy tracks these micro-behaviors to build a rich cognitive profile that reveals how a learner thinks.

Adaptive Persistence

When a child encounters failure, what happens next? Do they retry immediately, change strategy, or seek a new approach? Persistence patterns are among the strongest early indicators of long-term intellectual growth.

Growth Velocity

Rather than comparing children to each other, IQ Whizzy measures each learner against their own trajectory. How quickly are they improving? Where are they accelerating? This self-referenced model celebrates every child's unique path.

Creative Problem-Solving

Some of the most powerful signals come from unexpected answers. IQ Whizzy recognizes and values divergent thinking — the ability to see solutions others miss — as a hallmark of genuine cognitive strength.

Critical Thinking & Logic

The Critical Thinking & Logic module forms the cognitive backbone of IQ Whizzy. It trains children to see structure in chaos, to identify patterns where others see randomness, and to reason through complex scenarios with confidence and clarity.

From simple sequencing exercises for early learners to advanced systems-thinking challenges for teens, this module scales with the learner. Every activity is designed to build the mental muscles that power success across every academic discipline and life domain.

Pattern Recognition

Identify hidden structures in visual, numerical, and abstract sequences

Sequencing & Probability

Develop intuitive understanding of order, likelihood, and prediction

Systems Thinking

Map cause-and-effect relationships in interconnected systems

Strategy Fundamentals

Build strategic reasoning that feeds directly into chess and STEM

Mathematics & Early STEM

Mathematics is not just a subject — it's a language of logic. The Mathematics & Early STEM module transforms numbers from abstract symbols into tangible tools for understanding the world. Through puzzles, challenges, and hands-on problem-solving, children develop number sense that goes far beyond memorizing multiplication tables.

Spatial awareness exercises prepare young minds for engineering and design. Computational thinking activities lay the groundwork for coding and data science. Every challenge is carefully calibrated to sit in the sweet spot between achievable and aspirational — challenging enough to promote growth, accessible enough to sustain confidence.

The research is clear — early STEM exposure creates lasting advantages:

Predictive Power

Early math knowledge predicts later reading achievement better than early literacy skills.

Cognitive Development

Early mathematical thinking contributes broadly to overall cognitive development.

STEM Success

Preschool math and science knowledge predicts later success in STEM fields.

Stronger Profiles

Children engaging with STEM ages 3-5 arrive at school with measurably stronger cognitive profiles.

Equity & Inclusion

Quality early STEM education significantly reduces achievement gaps for low-income students.

Number Sense

Quantitative reasoning & mathematical intuition

Applied Math

Real-world puzzles & challenges

Spatial Awareness

Engineering logic & visual reasoning

Computational Thinking

Early coding concepts & algorithmic logic

Language & Communication

Language is the operating system of thought. The Language & Communication module doesn't just teach vocabulary — it trains cognitive flexibility, the ability to shift between mental frameworks, recognize patterns across linguistic systems, and communicate with precision and power.

1

Vocabulary Development

Building a rich, contextual word bank through immersive challenges

2

Reading Comprehension

Deep understanding beyond surface-level decoding

3

Multi-Language Exposure

Linguistic patterning across diverse language families

4

Cognitive Flexibility

Training the brain to navigate multiple language systems fluidly

Research consistently demonstrates that multilingual exposure in early childhood strengthens executive function, enhances memory, and improves problem-solving across all domains. IQ Whizzy integrates multi-language activities not as an add-on, but as a core intelligence-building strategy. Every child, regardless of their home language, benefits from this cognitive cross-training.

Academic Readiness & Advancement

Standardized tests don't have to be gatekeepers. When children understand the logic behind tests — not just the content — they approach exams with confidence rather than anxiety. The Academic Readiness module demystifies test structures, builds strategic thinking skills, and ensures that no child is disadvantaged simply because they haven't been coached.

Test preparation creates measurable advantages:

30-50

SAT Math Points

Typical improvement with coaching, helping students achieve target scores.

20-30

SAT Verbal Points

Boost in scores by understanding question patterns and vocabulary strategies.

2-3

ACT Composite Points

Significant gains that can elevate college applications.

1

Private Tutoring Impact

Students with personalized private tutoring consistently show the strongest score gains.

Reduced Test Anxiety

Familiarity with test formats and strategic thinking reduces stress and builds confidence.

01

SAT & ACT Foundations

Build the reasoning frameworks that underlie college entrance exams, starting years before test day

02

GED Readiness Pathways

Structured preparation for learners pursuing alternative educational credentials

03

Test-Logic Mastery

Understanding how questions are designed, not memorizing answers — strategic thinking over rote recall

04

Confidence Through Progress

Measurable milestones that show learners how far they've come and where they're headed

Early Learning & Neurodevelopment

The earliest years of life represent the most explosive period of brain development in human history. During this critical window, a child's neural architecture is being shaped by every interaction, every stimulus, every moment of play. The Early Learning & Neurodevelopment module is specifically designed to maximize this window — not through pressure, but through joyful, age-appropriate engagement.

The science is undeniable — the early years are everything:

90%

Brain Development

Occurs by age 5

1M+

Neural Connections

Formed per second (birth to age 3)

3x

Brain Volume

Increases in first 3 years

Early experiences create either sturdy or fragile foundations for all future learning, as neural pathways established early become the foundation for all later cognitive development.

Early Learners Through Teens

Activities scaled for every developmental stage, from first cognitive interactions to complex adolescent reasoning

Family-Friendly Design

Multi-user environments where parents, siblings, and caregivers can learn and play together

Inclusive by Design

Built for diverse learning styles — visual, auditory, kinesthetic, and neurodivergent learners are all welcome

Every module within IQ Whizzy is designed to generate measurable intelligence signals — not just completion badges. These signals form the foundation of each child's cognitive profile, a living document of their intellectual growth that travels with them through every stage of the IQ Whizzy ecosystem and beyond.

All Modules at a Glance

IQ Whizzy's five core modules work together to build a complete picture of every child's cognitive landscape. Here's how they connect and what each one measures:

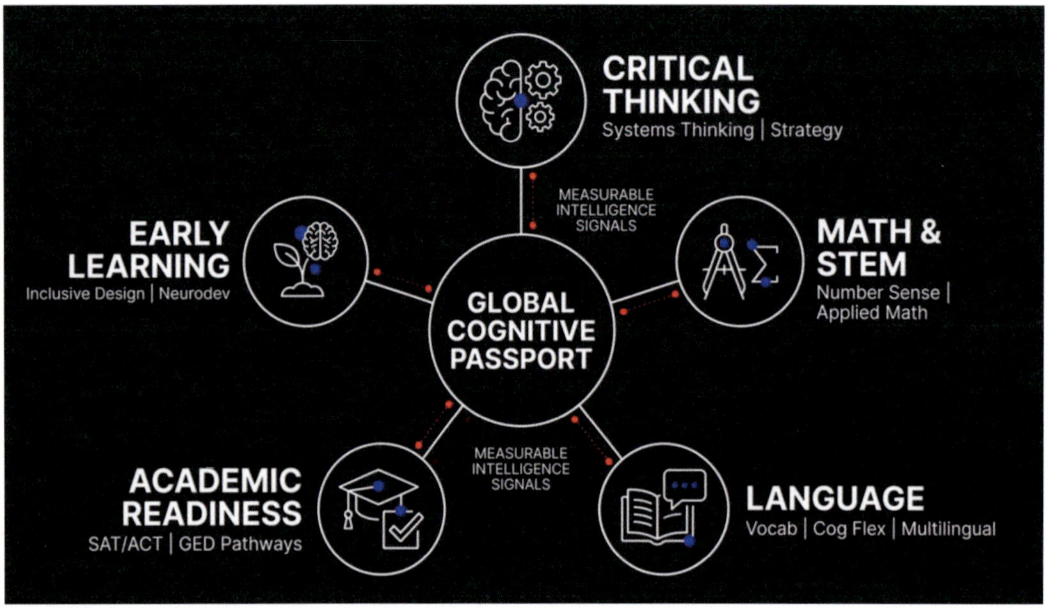

Together, these modules ensure that no dimension of intelligence is overlooked. Whether a child's strengths lie in numbers, words, logic, or creative problem-solving, IQ Whizzy recognizes and nurtures their unique cognitive fingerprint.

STRUCTURAL ADVANTAGE

Free at the Front, Sovereign at the Core

IQ Whizzy is free because the value is not in the first interaction. The value is in continuity.

Think of IQ Whizzy as the wide-open entrance to a cathedral of opportunity. The door is free. What lies inside — the ascending pathways, the verified credentials, the scholarships and career opportunities — is built on the foundation of data and growth that begins the moment a child starts playing. There is no bait-and-switch. The free access is genuine. The downstream value is earned through growth, not purchased through privilege.

As learners progress, their cognitive data — captured ethically and transparently — flows forward into progressively more sophisticated environments. Each step builds on the last. Each achievement is recorded, verified, and carried forward. This is not a disconnected collection of apps and programs. It is the world's first end-to-end intelligence ladder, beginning with free access and culminating in verified opportunity.

The economic engine of IQ Whizzy thrives on a virtuous cycle. As more children engage with the free platform, the volume and diversity of cognitive data grow exponentially. This rich dataset fuels the continuous improvement of our AI-driven cognitive models, making pathways more precise, insights more actionable, and the value of verified credentials increasingly robust. This is not about extracting value; it's about generating compound growth where every participant's engagement enhances the system for all, creating a shared asset of human potential.

Crucially, the transition from free exploration to premium opportunities is seamless and merit-based. Learners unlock advanced certifications, access to exclusive mentorship, scholarships to higher education, and direct pathways to career opportunities, all predicated on demonstrated capability and verified progress rather than financial means. Our free-at-the-front model ensures that socio-economic status is never a barrier to discovering and maximizing one's intellectual capital, democratizing access to truly transformative educational and professional advancement.

Free Access

Broad engagement & initial learning.

Ecosystem Sustainability

Economic value reinvested for continuous growth.

Data Generation

Ethical capture of cognitive signals.

Cognitive Insights

AI-driven personalization & model refinement.

Verified Outcomes

Recognized achievements, proven potential.

The Intelligence Ecosystem

IQ Whizzy doesn't exist in isolation. It is the entry point to a comprehensive ecosystem designed to nurture, verify, and amplify cognitive potential from childhood through career.

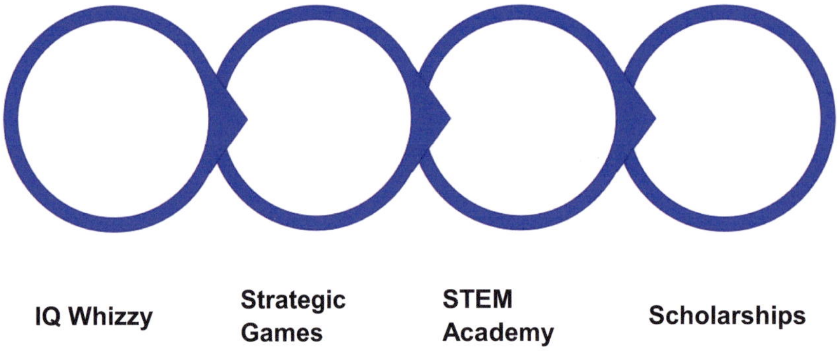

This ecosystem creates an unbroken chain of development. A child who begins with IQ Whizzy at age 4 can progress through strategic games, STEM competitions, scholarship opportunities, and ultimately into career pathways — all connected by the same cognitive record, the same verified profile, the same Global Cognitive Passport. No other system in the world offers this level of continuity.

Where the Data Flows

As learners progress through IQ Whizzy, their cognitive data — captured ethically and with full transparency — flows forward into increasingly powerful opportunities. Here's what that pathway looks like:

Chess & Four Corners Chess

Strategic reasoning and competitive thinking, calibrated by IQ Whizzy logic signals

Senet & Systems Reasoning

Ancient game logic meets modern cognitive science, deepening strategic awareness

International Academy of STEM

Advanced learning environments powered by verified cognitive profiles

STEM Championships

National and international competitions where talent is recognized and celebrated

Scholarships R Us & Careers

Verified credentials unlock funding, mentorship, and professional pathways

> **IQ Whizzy does not monetize children. It protects them from being invisible.**

STRATEGIC SIGNIFICANCE

Why This Changes Everything

Most systems identify talent late — after advantage has already compounded. A child born into a well-resourced family receives enrichment from birth. By age five, they've had thousands more hours of cognitive stimulation than their peers. By the time standardized testing begins, the gap isn't about ability — it's about accumulated exposure.

IQ Whizzy identifies potential early — before inequity hardens into outcome. By placing a free, world-class cognitive development tool in the hands of every family, Goddess Phenezy short-circuits the cycle of compounding disadvantage. The child in a rural village and the child in a metropolitan penthouse start with the same tool, the same quality, the same opportunity to be seen.

This isn't a utopian dream. It's an architectural decision. When you design a system where the entry point is free and the measurement is equitable, you create the conditions for genuine meritocracy — not the hollow version that claims to reward talent while secretly rewarding privilege.

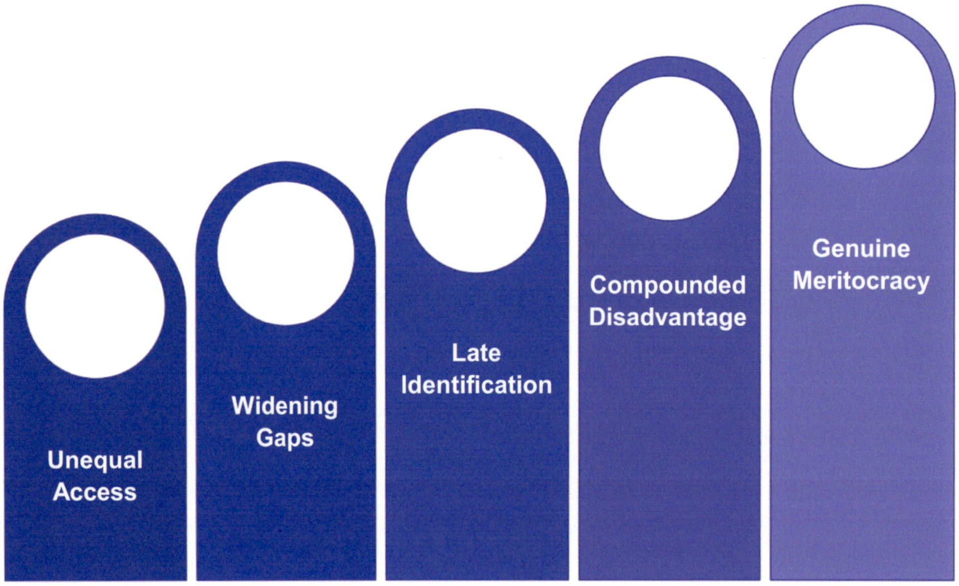

This diagram illustrates the stark contrast between traditional systems, which identify potential late and exacerbate existing inequities, and the IQ Whizzy model, which offers early, equitable access to cognitive development to create a genuine meritocracy.

Three Promises to the World

By making IQ Whizzy free, Goddess Phenezy makes three irrevocable promises — not to investors, not to boards, but to children and the families who love them:

Genius Is No Longer Gated

Every child, regardless of where they are born or what resources their family has, gains access to the same world-class cognitive development tools. The gate is open. The door will never close. Brilliance exists in every community — IQ Whizzy simply ensures it can be found.

Opportunity Is No Longer Inherited

The accident of birth should not determine a child's ceiling. IQ Whizzy decouples opportunity from inheritance, creating pathways based on demonstrated potential rather than parental income. What you can do matters more than where you came from.

Intelligence Is No Longer Wasted

For every celebrated prodigy, there are thousands of equally brilliant children whose potential was never recognized. IQ Whizzy's cognitive signal engine ensures that no spark goes unnoticed, no pattern of brilliance goes unrecorded.

This Is Not a Loss Leader

Let's be direct about what IQ Whizzy is — and what it is not.

What IQ Whizzy Is

- A sovereign investment in human capital
- A free, permanent cognitive development platform
- The entry point to the world's first end-to-end intelligence ladder
- An ethical data system that protects and empowers children
- A declaration that every child's mind matters

What IQ Whizzy Is Not

- A promotional trial waiting to expire
- A data-harvesting scheme disguised as education
- A watered-down version of a "real" paid product
- Philanthropy theater designed for press coverage
- A replacement for formal education or schooling

IQ Whizzy is free because it must be free. The entire architecture of the Goddess Phenezy intelligence ecosystem depends on universal access at the front door. If even one child is locked out, the system fails its mission. Free access is not a business strategy — it is a moral imperative.

The Ecosystem in Action

How does the full intelligence ecosystem work in practice? Here's a look at how each component builds on IQ Whizzy's foundation to create unbroken pathways from first play to verified achievement:

1 IQ Whizzy

Opens the door. Free cognitive play captures early intelligence signals and builds the Global Cognitive Passport.

2 Chess & Strategic Games

Deepens strategic thinking. Four Corners Chess and Senet develop systems reasoning and competitive resilience.

3 STEM Academy

Validates capability. The International Academy of STEM provides advanced learning powered by verified cognitive data.

4 Championships

Celebrates excellence. National and international STEM Championships give emerging talent a global stage.

5 Scholarships R Us

Distributes opportunity. Verified cognitive profiles unlock scholarships, mentorship, and career pathways.

This is not a collection of disconnected programs. It is a single, continuous architecture — a ladder where every rung is connected to the next, and the first rung is always free.

SCHOLARSHIPS
R US

Every Child Deserves to Be Seen

IQ Whizzy

Opens the door for free cognitive development.

Scholarships R Us

Distributes opportunity and funding access.

STEM Championships

Validates capability through competitive achievement.

Chess

Trains strategy for cognitive mastery.

IQ Whizzy opens the door.

Somewhere right now, a child is sitting with untapped brilliance — a mind capable of extraordinary things — waiting not for talent, but for access. Waiting not for permission, but for a chance. IQ Whizzy was built for that child. It was built so that genius born in a forgotten neighborhood has the same runway as genius born into privilege.

This is the promise of IQ Whizzy: that no child's intelligence will be wasted because no one thought to look. That no family will be told their child doesn't qualify because they can't afford to pay. That the door to cognitive development — to being seen, measured, recognized, and nurtured — will remain open, free, and permanent.

> Every child deserves a chance to be seen. IQ Whizzy makes sure they are.

Created by Goddess Phenezy — because intelligence should never require permission.

The National & International Poker Championships

The Probability Jurisdiction: Decision-Making as Economic Power

A comprehensive framework for the formalization, governance, and economic elevation of poker, Spades, and Bid Whist as measurable, skill-based competitive jurisdictions — presented by Goddess Phenezy Holdings, Inc.

The Most Honest Intelligence Test

Poker Is Not Chance — It Is Decision Science

Poker is the purest real-world intelligence laboratory ever created. It is mathematics under uncertainty. Psychology under pressure. Strategy with incomplete information. Risk management with real consequences. Unlike games of perfect information, poker does not reward memorization alone. It rewards judgment.

Yet despite being one of the most sophisticated cognitive competitions on Earth, poker has historically been dismissed as vice rather than recognized as applied intelligence. This fundamental misclassification has denied millions of strategic thinkers the recognition, infrastructure, and competitive pathways they deserve. The intellectual rigor required to consistently win at poker surpasses many activities already recognized as legitimate competitive disciplines.

Goddess Phenezy Holdings, Inc. rejects that misclassification with absolute conviction. Poker is not gambling dressed in sophistication. It is decision science operating in real time, under authentic pressure, with measurable outcomes that separate the elite from the average over statistically significant sample sizes. The variance that casual observers mistake for luck is precisely what makes poker the ultimate proving ground for cognitive excellence.

> Through The National & International Poker Championships, poker is elevated from underground tables and fragmented platforms into a fully governed, ethical, and intelligence-driven competitive jurisdiction.

Why Poker Demands Reclassification

1 Gambling Perception

2 Mathematical Reality

3 Psychological Mastery

4 Strategic Intelligence

5 Decision Science

What The World Sees

- A gambling activity driven by luck
- Entertainment for casual players
- Unregulated and fragmented
- Associated with vice and excess

What Poker Actually Is

- Applied probability and Bayesian reasoning
- Real-time psychological assessment
- Multi-variable strategic optimization
- The most honest test of decision-making under uncertainty

The gap between perception and reality represents one of the greatest misallocations of intellectual capital in modern competitive history. Every year, hundreds of thousands of individuals demonstrate extraordinary cognitive ability at the poker table — and receive zero institutional recognition for it. The Poker Jurisdiction exists to close that gap permanently.

Structural Advantage: Skill Over Luck, Data Over Myth

Most poker ecosystems are built around fragmented, short-term models that serve the house rather than the player. They extract value without creating lasting infrastructure, competitive legitimacy, or developmental pathways. The result is an industry that generates billions in revenue while leaving its most talented participants without verified credentials, longitudinal records, or upward mobility.

Isolated Tournaments

One-off events with no continuity, no ranking persistence, and no connection to broader competitive frameworks.

Third-Party Platforms

Players surrender their data, identity, and performance history to platforms they do not control.

Rake-Heavy Economics

Revenue models designed to extract maximum value from players while returning minimal investment in their development.

Zero Long-Term Development

No structured pathways from amateur play to professional competition. No cognitive development framework.

The Poker Jurisdiction Is Architected Differently

By operating within proprietary Progressive Web App (PWA) infrastructure, Goddess Phenezy Holdings captures and leverages four critical data dimensions that transform poker from episodic gambling into measurable cognitive performance. This infrastructure does not merely host games — it builds the most comprehensive decision-making database in competitive history.

This framework provides unprecedented insights, fostering skill development and elevating the strategic depth of the game beyond traditional luck-based perceptions.

Verified Player Identities

Authenticated credentials eliminate fraud and multi-accounting, addressing anonymity in online poker.

Longitudinal Performance Data

Tracking win rates and decision accuracy across thousands of hands for true poker intelligence.

Decision-Making Patterns

Algorithmic analysis of betting, positional awareness, and bluff frequency reveals cognitive signatures.

Risk Tolerance & Adaptability Metrics

Quantified behavioral profiles measure responses to pressure, variance, and evolving conditions.

The Data Truth

> **Luck fades. Skill compounds. Data tells the truth.**

Over statistically significant sample sizes, the best poker players in the world consistently outperform the field. This is not anecdote — it is mathematical certainty. The Poker Jurisdiction's proprietary data infrastructure proves this irrefutably, transforming subjective claims of skill into objective, verifiable evidence. When you measure ten thousand decisions, the signal overwhelms the noise. The cream does not merely rise — it is documented.

Raw Play **Data Capture** **Proven Skill**

This pipeline represents the foundational architecture of the Poker Jurisdiction — the mechanism by which subjective reputation becomes objective proof, and play becomes proficiency.

The Championships

National & International Poker Championships

The Championships operate across structured formats designed to emphasize intelligence, discipline, and consistency rather than short-term variance. These are not chaotic cash games where luck dominates a single session. They are regulated intelligence competitions where the best minds rise across rigorous, multi-stage competitive structures that separate genuine cognitive excellence from momentary fortune.

National Championships

Live and digital hybrid formats bringing together the finest players from every region through qualifying rounds, regional finals, and national showdowns.

International Championships

Global qualifiers feed into marquee international finals — positioning the jurisdiction as the definitive worldwide authority in competitive poker.

Skill-Weighted Formats

Proprietary tournament structures that emphasize long-term performance over short-term variance, ensuring that the most intelligent players are rewarded consistently.

Championship Revenue Streams

The competitive arena generates revenue through four distinct and synergistic channels, each reinforcing the others to create an economic flywheel that grows more powerful with each championship cycle.

1

Championship Entry Fees

Tiered buy-in structures from amateur qualifiers to elite final tables, creating accessible on-ramps and premium competitive experiences.

2

Sponsorships & Branded Tables

Corporate partners and luxury brands secure naming rights and branded experiences within the championship ecosystem.

3

Media & Broadcast Rights

Live coverage, commentary packages, and content licensing create recurring revenue from the drama of elite competition.

4

Premium Player Tiers

VIP competitive access, priority seating, enhanced analytics, and exclusive tournament invitations for the most committed competitors.

The Poker Intelligence Layer

From Play to Proficiency

Poker uniquely trains cognitive skills that directly translate to life, business, and leadership. No other competitive discipline simultaneously develops probability reasoning, emotional regulation, pattern recognition, strategic deception, and long-term value thinking in a single decision environment. Every hand is a compressed simulation of real-world decision-making under uncertainty.

Probability & Statistical Reasoning

Bayesian updating in real time — calculating odds, adjusting ranges, and making mathematically optimal decisions under incomplete information.

Emotional Regulation

Managing tilt, maintaining composure through adversity, and executing discipline when every instinct screams otherwise.

Pattern Recognition Under Uncertainty

Detecting behavioral tells, identifying strategic tendencies, and extrapolating from limited data to form actionable hypotheses.

Strategic Deception & Defense

Understanding when to project strength, conceal weakness, and defend against manipulation — skills directly applicable to negotiation and leadership.

Long-Term Expected Value Thinking

Sacrificing short-term gains for long-term profitability — the foundational mindset of successful investors, entrepreneurs, and executives.

Poker as a Classroom Without Walls

Within the Goddess Phenezy Holdings ecosystem, poker integrates seamlessly with complementary disciplines to create a holistic cognitive development framework. Rather than existing in isolation, poker serves as the connective tissue between strategic, analytical, and entrepreneurial training — a classroom without walls that meets learners where they are and elevates them to where they belong.

Chess Integration

Strategy and foresight — developing multi-move planning and positional thinking.

STEM Pathways

Probability, statistics, and logic applied through competitive play rather than abstract textbooks.

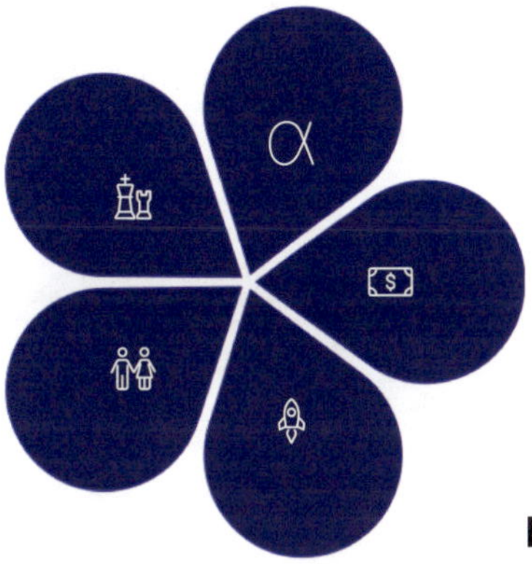

Financial Literacy

Bankroll management, risk assessment, and value optimization mirror real-world financial decision-making.

Leadership Development

Reading people, managing pressure, and making decisive calls with incomplete information.

Entrepreneurship

Calculated risk-taking, resource allocation, and strategic pivoting under uncertainty.

The Decision Economy

Revenue Stream 1: Championships & League Play

National and international tournaments operate year-round, creating a continuous revenue engine that builds audience engagement, player loyalty, and competitive legitimacy across every calendar quarter. The championship circuit is not seasonal — it is perpetual, with regional qualifiers feeding national events that culminate in international finals of extraordinary prestige.

The league structure ensures that revenue is not dependent on a single marquee event but distributed across hundreds of competitions annually, each contributing to the overall economic architecture while deepening player investment and brand equity.

Regional Qualifiers

Building audience engagement

National Events

Fostering player loyalty

International Finals

Ensuring competitive legitimacy

$18.0M

Projected Annual Revenue

Championships & League Play

365

Days of Competition

Year-round tournament calendar

50+

National Events

Regional and national competitions annually

Revenue Stream 2: Digital Memberships & Analytics

Premium access to performance analytics, decision reviews, and intelligence feedback transforms every player interaction into a recurring revenue opportunity. This is not a passive subscription — it is an active cognitive development platform that players rely on to improve their competitive edge and validate their strategic evolution.

The analytics suite delivers verified performance data that players cannot obtain anywhere else. Decision accuracy scores, positional efficiency ratings, bluff success rates, and risk-adjusted return metrics create a comprehensive cognitive profile that serves as both a training tool and a competitive credential.

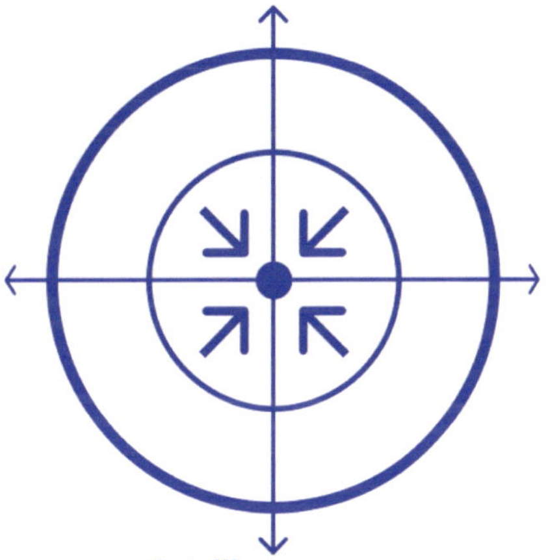

This cognitive development platform integrates essential analytics to empower players, offering tools for continuous improvement and official recognition of their competitive standing.

$14.5M

Projected Annual Revenue from Digital Memberships & Analytics

What Members Receive

- Decision review tools with hand-by-hand analysis
- Performance benchmarking against peer cohorts
- Intelligence feedback loops with actionable recommendations
- Verified digital credentials for competitive standing
- Priority access to premium tournament tiers

Revenue Stream 3: Media, Broadcast & Content

Live streams, expert commentary, and educational content packaged via Xperience TV transform championship poker into premium entertainment programming. The drama of elite competition — the reads, the bluffs, the mathematical precision — creates compelling content that attracts viewers far beyond the competitive poker community.

The media architecture monetizes every dimension of the competitive experience: live broadcasts generate advertising revenue, commentary packages create licensing opportunities, and educational content drives subscription growth. Documentary series profiling elite competitors add narrative depth that traditional sports broadcasts have proven audiences will pay premium prices to consume.

Live Broadcasts

Real-time championship coverage with hole-card cameras, expert analysis, and dynamic graphics.

Commentary Packages

Multi-tier broadcast rights for networks, streaming platforms, and international distributors.

Educational Content

Strategy masterclasses, decision-review series, and cognitive training programs.

$9.5M

Projected Annual Revenue

Media, Broadcast & Content

Total Annual Poker Ecosystem Revenue

Revenue Stream	Annual Value
Championships & League Play	$18.0M
Memberships & Analytics	$14.5M
Media & Broadcasting	$9.5M
Total Annual Revenue	$42.0 MILLION

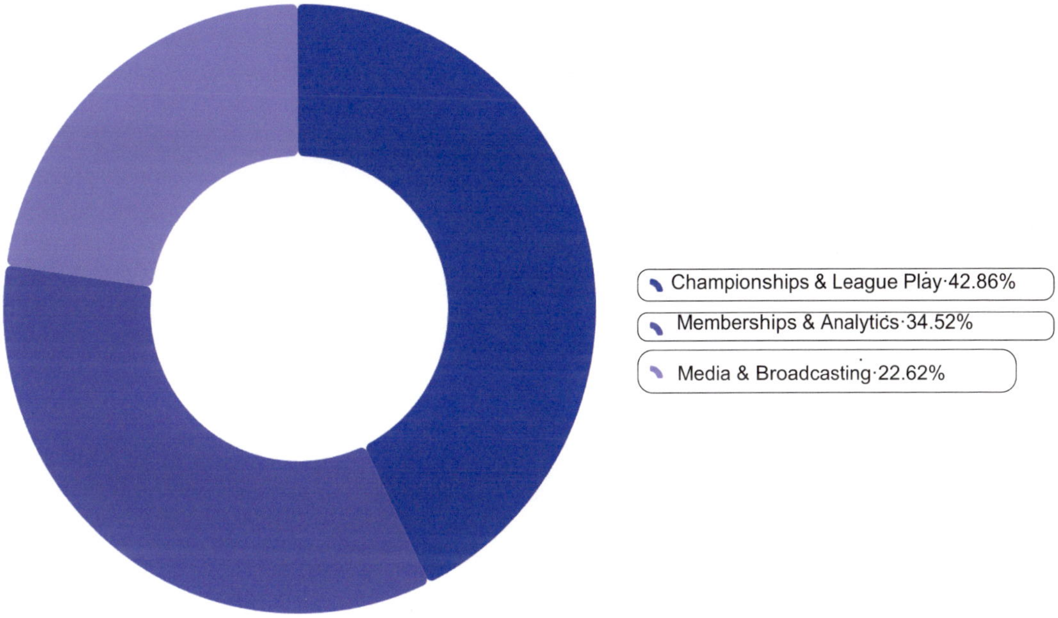

The poker ecosystem generates $42.0 million in annual revenue through three synergistic channels. Championships drive player acquisition, analytics drive retention, and media amplifies both — creating a self-reinforcing economic flywheel with compounding growth dynamics.

Revenue Distribution Analysis

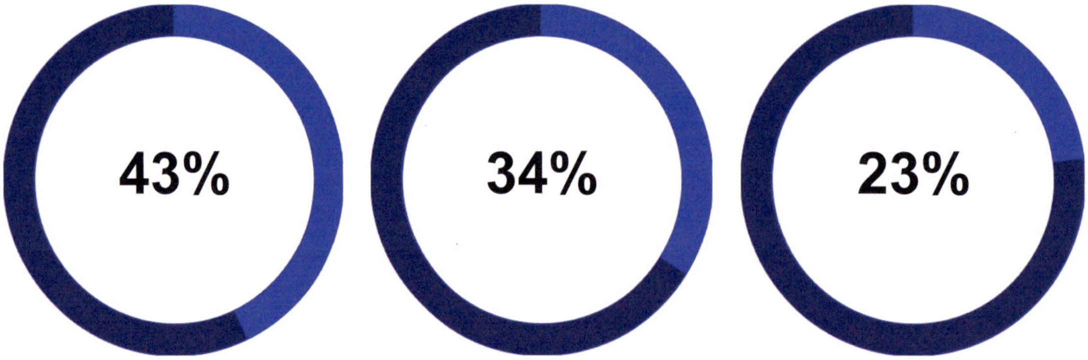

Championships

Core competitive revenue from entry fees, sponsorships, and league play

Digital Analytics

Recurring membership revenue with high retention and low marginal cost

Media & Broadcast

Scalable content revenue with global distribution potential

The diversification across physical competition, digital subscriptions, and media content creates natural resilience against market disruption in any single channel. As the ecosystem matures, the digital and media components are projected to grow at higher rates than championship revenue, shifting the mix toward higher-margin, more scalable revenue streams.

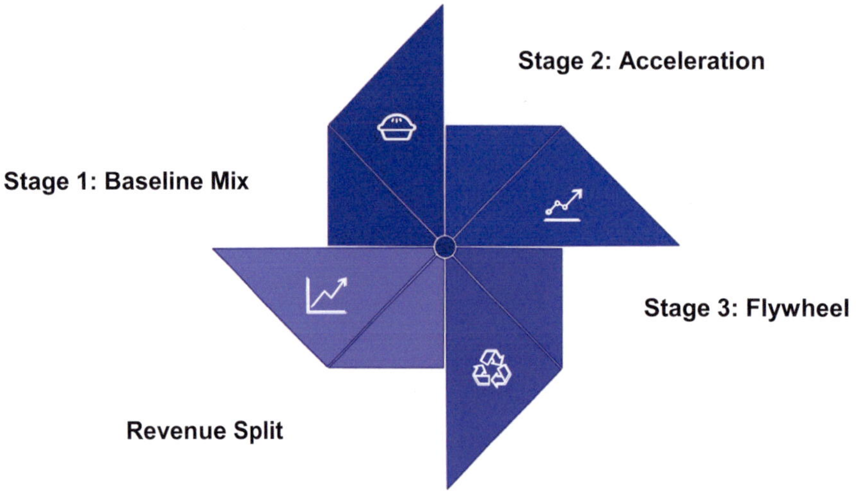

The Applied Intelligence Multiple

Valuation Framework: The Applied Intelligence Multiple

Poker exists at the intersection of four high-growth, high-multiple verticals. This convergence creates a valuation premium that pure-play competitors in any single category cannot command. The Poker Jurisdiction is not merely a tournament operator — it is a data-rich, media-integrated, behaviorally-driven competitive platform with characteristics that attract premium multiples across every comparable sector.

Competitive Gaming

The explosive growth of esports validates skill-based competition as premium entertainment.

Behavioral Economics

Decision-making data is among the most valuable intellectual property in modern analytics.

Data Analytics

Proprietary performance databases command premium valuations from strategic acquirers.

Entertainment Media

Live competitive content drives subscription growth and advertising premiums.

Poker Jurisdiction Valuation

Applying blended gaming, analytics, and media multiples to the Poker Jurisdiction's $42.0M annual revenue yields a valuation range that reflects the premium value of platforms that measure real decision-making under uncertainty. These are not speculative projections — they are grounded in comparable transaction multiples from adjacent industries.

$252M

Base Case Valuation

6× revenue multiple

$336M

Growth Case Valuation

8× revenue multiple

Spades & Bid Whist

The National & International Spades & Bid Whist Championships

The Games That Built Strategists Without Permission

Long before access to elite chess clubs, formal logic training, or institutional recognition, entire communities trained their minds through games passed hand to hand, table to table, generation to generation. Two of the most enduring of these are Spades and Bid Whist.

These are not casual card games. They are memory systems, probability engines, and communication disciplines disguised as play. In households, military barracks, community centers, and cultural gatherings, Spades and Bid Whist taught strategic planning, partner coordination, risk assessment, emotional discipline, and long-term score management. They did so without textbooks, without tuition, and without institutional permission.

Goddess Phenezy Holdings, Inc. recognizes these games for what they truly are: cognitive training systems born of necessity and brilliance.

A
THE NATIONAL
SPADES
CHAMPIONSHIP

The Cognitive Legacy of Cultural Card Games

1

Households & Family Tables

Foundational learning: Memory, basic strategy, interpersonal cues.

2

Military Barracks

Discipline under pressure: Advanced strategy, risk assessment, quick decision-making.

3

Community Centers

Social intelligence: Communication, partner coordination, reputation management.

4

Cultural Gatherings

Competitive mastery: High-stakes execution, emotional discipline, long-term score management.

This diagram illustrates how diverse cultural environments fostered critical cognitive skills through traditional card games, shaping generations of strategic thinkers in military, family, community, and competitive settings.

What These Games Taught

The cognitive demands of Spades and Bid Whist rival or exceed many formally recognized competitive disciplines. These games taught the full spectrum of strategic intelligence in compressed, high-stakes social environments where reputation, pride, and respect were on the line with every hand dealt.

Strategic Planning

Formulating multi-round strategies based on limited information and probabilistic assessment of opponents' holdings.

Partner Coordination

Developing sophisticated non-verbal communication systems and building trust under competitive pressure.

Risk Assessment

Evaluating bid viability, defensive responsibilities, and the consequences of overcommitment versus underperformance.

Emotional Discipline

Maintaining composure when partnerships fail, bad beats occur, and opponents gain momentum.

Long-Term Score Management

Optimizing across multiple rounds, understanding when to press advantages and when to protect positions.

STRUCTURAL ADVANTAGE

Codifying Cultural Intelligence

Despite their sophistication, Spades and Bid Whist have never been formally governed at scale. Rules vary by region. Competitive records vanish after the table clears. Skill is recognized socially — but never archived, ranked, or elevated. The accumulated intelligence of millions of players across decades exists only in memory, not in data.

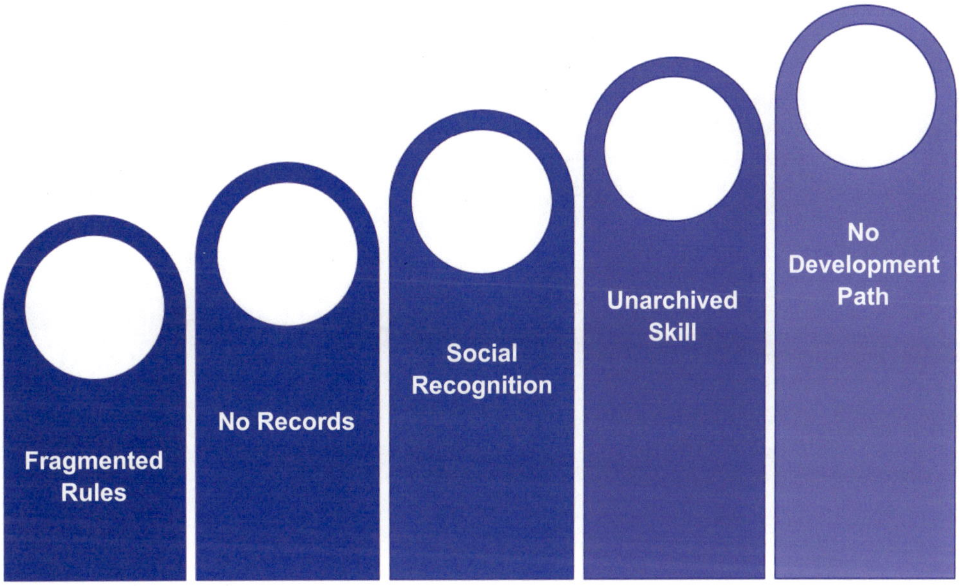

The Cultural Intelligence Jurisdiction changes this permanently. By combining The National & International Spades Championships and The National & International Bid Whist Championships under a single proprietary Progressive Web App (PWA), Goddess Phenezy Holdings transforms culturally significant play into documented, competitive, and respected intellectual sport.

> **This is not appropriation. It is recognition and elevation.**

The Transformation: Before and After

The Cultural Intelligence Jurisdiction does not merely organize tournaments. It constructs the entire institutional infrastructure that these games have been denied — governance, records, rankings, credentials, and competitive pathways that transform social play into sovereign sport.

The Games Decoded

Spades: Probabilistic Partnership Mastery

Spades is far more than a recreational card game — it is a live lesson in team intelligence. Every hand demands real-time probability estimation, strategic bidding, partner signaling, and adaptive decision-making. The partnership dynamic elevates Spades beyond individual cognitive performance into the realm of collaborative strategic optimization — a discipline with direct applications to business, military strategy, and organizational leadership.

Probability Estimation

Calculating odds of card distribution across four hands with incomplete information — Bayesian reasoning in action.

Strategic Bidding

Contract management requiring precise self-assessment and partner calibration under competitive pressure.

Partner Signaling

Developing and interpreting non-verbal communication systems that convey strategic intent through play patterns.

Adaptive Strategy

Shifting between offensive and defensive postures in real time as the competitive landscape evolves within each hand.

Bid Whist: Advanced Memory & Multi-Variable Strategy

Bid Whist elevates the complexity of trick-taking games to extraordinary levels. The addition of multiple trump suits, advanced memory tracking requirements, deep inference chains, and precision bidding with cascading consequences creates a cognitive environment that tests the absolute limits of working memory, pattern recognition, and strategic foresight.

Bid Whist is not casual. It is advanced cognitive warfare.

Multiple Trump Suits

Strategic selection of trump suit adds a meta-layer of decision-making before gameplay even begins.

High-Load Memory Tracking

Mentally cataloging every card played across all four hands while simultaneously planning future plays.

Deep Inference & Counter-Strategy

Deducing opponents' holdings from their play patterns and formulating multi-level counter-strategies in real time.

Precision Bidding

Cascading consequences where a single bidding miscalculation can determine the outcome of an entire match.

THE NATIONAL
BID WHIST
CHAMPIONSHIP

Spades vs. Bid Whist: Intelligence Comparison

Cognitive Dimension	Spades	Bid Whist
Primary Skill	Partnership Coordination	Memory & Inference
Bidding Complexity	Moderate (trick count)	High (suit + direction)
Memory Load	High	Very High
Strategic Depth	Team-optimized	Multi-variable warfare
Communication	Partner signaling	Inference-based reading
Risk Profile	Contract management	Cascading consequences

Both games demand exceptional cognitive performance, but through complementary channels. Together, they form a complete intelligence training system — Spades building collaborative strategic thinking and Bid Whist developing individual cognitive processing power.

The Heritage Competition Economy

Revenue Stream 1: Leagues & Championships

Regional, national, and international tournaments formalize competitive play across Spades and Bid Whist for the first time in history. These are not informal gatherings — they are professionally organized, digitally tracked, and culturally celebrated competitions that honor the heritage of these games while elevating them to institutional legitimacy.

1

Team & Individual Registrations

Tiered entry structures accommodating casual competitors through elite championship qualifiers.

2

League Licensing

Regional league operators license the Goddess Phenezy competitive framework and brand.

3

Cultural Festival Integrations

Championships embedded within cultural events, heritage celebrations, and community gatherings.

4

Sponsored Championship Tables

Brand partnerships aligned with cultural values and community engagement.

$16.0M

Projected Annual Revenue

Leagues & Championships

Revenue Stream 2: Digital Memberships & Performance Tracking

For the first time in the history of Spades and Bid Whist, players receive verified digital identities that track their competitive performance across every dimension of gameplay. This transforms social reputation into documented evidence, giving players portable credentials that follow them from table to table, tournament to tournament, and year to year.

What Players Receive

- **Win/loss records — Complete competitive history verified on the PWA**
- **Bidding accuracy — Precision metrics tracking how consistently players meet their contracts**
- **Partner efficiency metrics — Data revealing how well partnerships perform relative to individual skill**
- **Longitudinal performance history — Career arcs showing growth, consistency, and competitive evolution**

$12.5M

Projected Annual Membership Revenue

Recurring, high-retention subscription model with minimal marginal cost per additional member.

Revenue Stream 3: Media, Culture & Broadcast

Through Xperience TV, championships are elevated into cultural events that transcend traditional sports broadcasting. The stories behind these games — the heritage, the community, the brilliance that developed without institutional support — create narratives of extraordinary power and resonance that attract audiences far beyond the competitive card game community.

We provide professional live broadcasts and commentary for championship events, including expert analysis, player profiles, and real-time strategic breakdowns to make the intelligence visible to every viewer.

Furthermore, we produce compelling docuseries content profiling the generations of brilliant players who built these traditions — stories that have never been told at scale, celebrating cultural game mastery.

This platform also attracts heritage brand sponsorships, aligning brand partnerships with cultural authenticity and community empowerment, thereby attracting sponsors who understand the value of genuine engagement.

$9.5M

Projected Annual Media Revenue

Media, Culture & Broadcast

Total Annual Spades & Bid Whist Ecosystem Revenue

Revenue Stream	Annual Value
Championships & Leagues	$16.0M
Digital Memberships	$12.5M
Media & Broadcasting	$8.5M
Total Annual Revenue	$37.0 MILLION

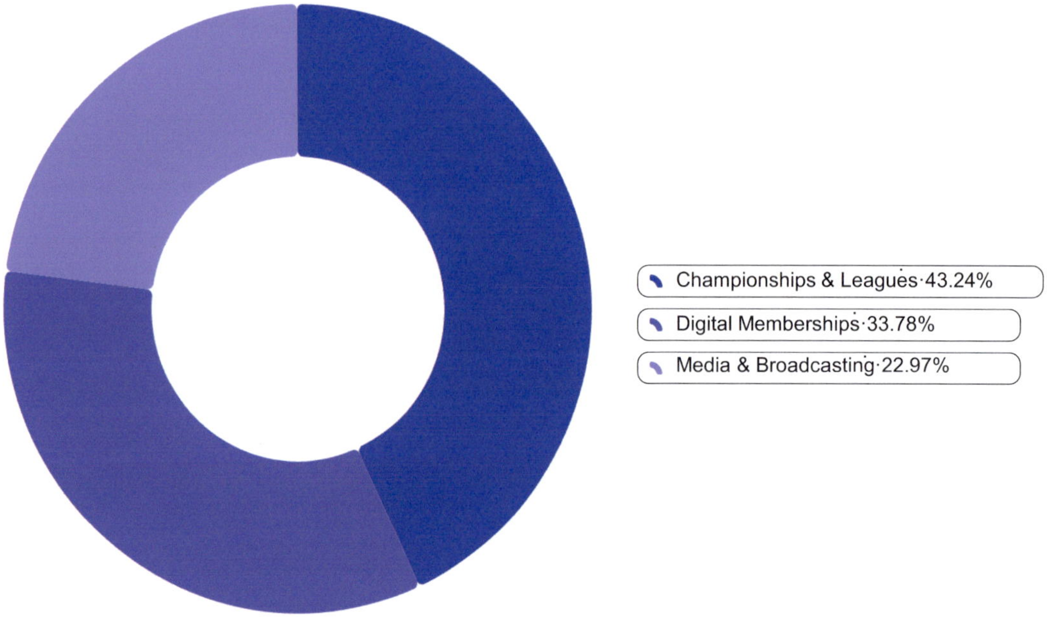

The Spades and Bid Whist ecosystem generates $37.0 million in annual revenue — validating that cultural intelligence, when properly organized and governed, commands economic power commensurate with its cognitive demands.

VALUATION

The Cultural Intelligence Multiple

Spades and Bid Whist occupy a rare intersection of four high-value verticals that no other competitive platform can claim. This unique positioning creates a valuation premium rooted in the convergence of competitive legitimacy, cultural IP, cognitive performance data, and community-driven organic growth.

Competitive Gaming

Formalized tournament infrastructure with verified rankings

Community Engagement

Organic growth driven by authentic cultural connection

Cultural Heritage

Deep-rooted IP with generational emotional resonance

Cognitive Performance

Measurable intelligence data with longitudinal depth

Spades & Bid Whist Valuation

Applying blended gaming, media, and cultural IP multiples to the $37.0M annual revenue produces a valuation range that reflects the long-overdue formal recognition of games that have produced generations of strategic thinkers without institutional support. The cultural growth case reflects the premium attached to authentic heritage IP in a market increasingly desperate for genuine community engagement.

$222M

Base Case Valuation

6× revenue multiple

$296M

Cultural Growth Case

8× revenue multiple

The Arc of Intelligence

Completing the Intelligence Arc

Chess formalized strategy for the elite. Poker validated decision-making under risk. Spades and Bid Whist democratized intelligence.

They taught logic where classrooms were underfunded. They trained memory where opportunity was restricted. They forged strategists where recognition was denied. For generations, these games operated as invisible academies — producing cognitive excellence that the formal world refused to see, let alone measure or reward.

By elevating Spades and Bid Whist into sovereign competitive jurisdictions, Goddess Phenezy Holdings ensures that cultural intelligence is no longer invisible. The data exists. The rankings exist. The credentials exist. And the world can no longer pretend that brilliance born at a kitchen table is worth less than brilliance certified by an institution.

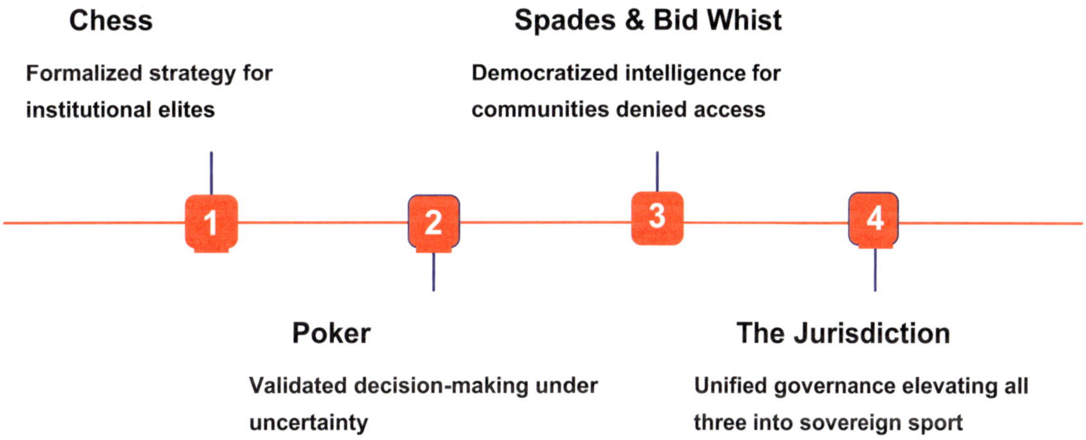

The Transformation Pipeline

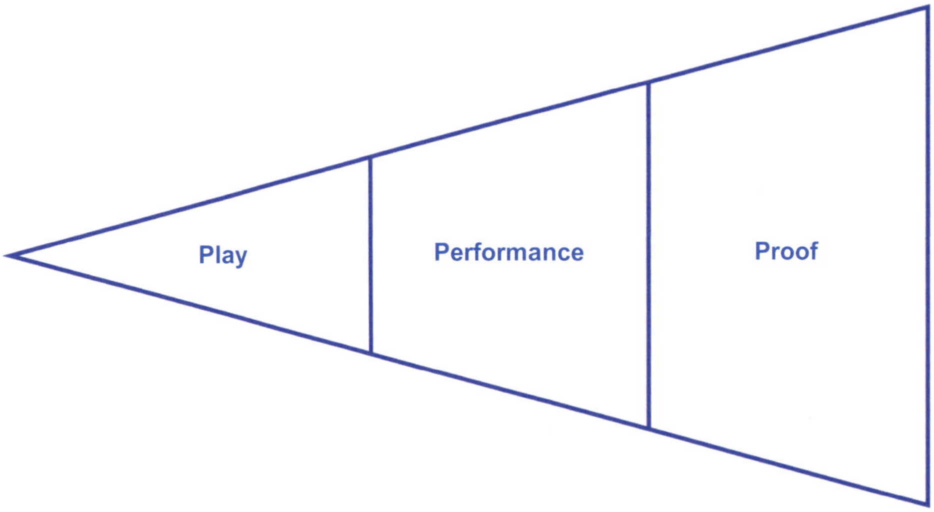

This pipeline represents the foundational thesis of the Cultural Intelligence Jurisdiction. Every player who enters the ecosystem begins at play and advances through documented proof into verified performance and, ultimately, genuine advancement. The pipeline is not theoretical — it is built into the PWA infrastructure, the championship structures, and the credentialing systems that Goddess Phenezy Holdings operates.

> **Play → Proof → Performance → Advancement.**

Total Jurisdiction Power

Combined Ecosystem Revenue

The Poker, Spades, and Bid Whist jurisdictions together form a unified competitive intelligence ecosystem generating $79.0 million in annual revenue across six synergistic channels. This combined architecture creates cross-pollination opportunities — poker players discovering Spades, Bid Whist champions entering poker, and media audiences consuming content across all three — that amplify growth beyond what any individual jurisdiction could achieve alone.

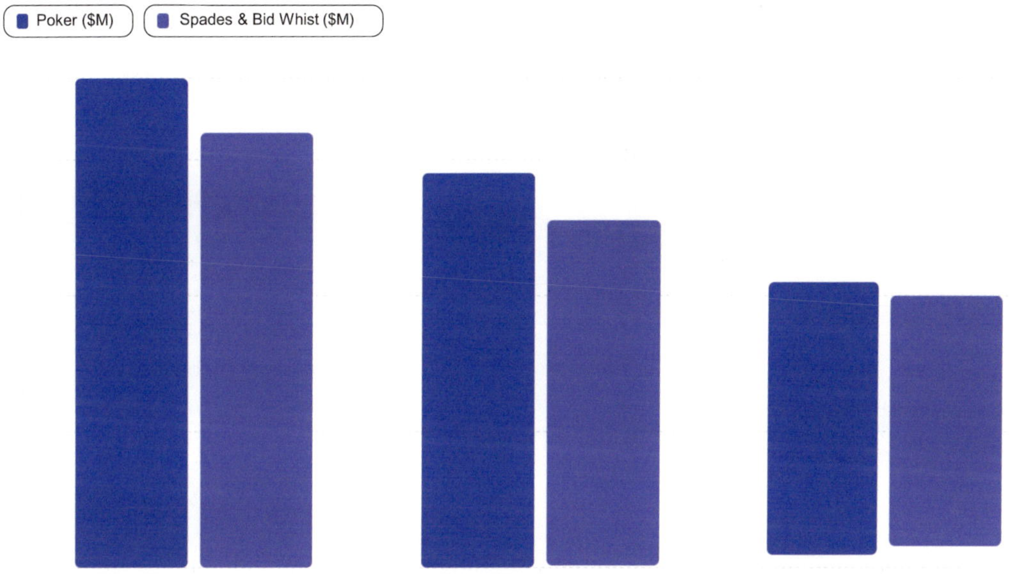

Combined Revenue Summary

Revenue Stream	Poker	Spades & Bid Whist	Total
Championships & Leagues	$18.0M	$16.0M	$34.0 M
Digital & Analytics	$14.5M	$12.5M	$27.0 M
Media & Broadcast	$9.5M	$8.5M	$18.0 M
Total	$42.0M	$37.0M	$79.0 M

$79.0M

Combined Annual Revenue

Total card game jurisdiction ecosystem

$474M

Base Valuation (6×)

Combined ecosystem base case

$632M

Growth Valuation (8×)

Combined ecosystem growth case

Combined Valuation Framework

The combined card game jurisdictions represent an unprecedented convergence of competitive gaming, behavioral data, cultural heritage, and entertainment media. No other entity in the global marketplace governs poker, Spades, and Bid Whist under unified infrastructure with proprietary data capture, verified rankings, and integrated media distribution.

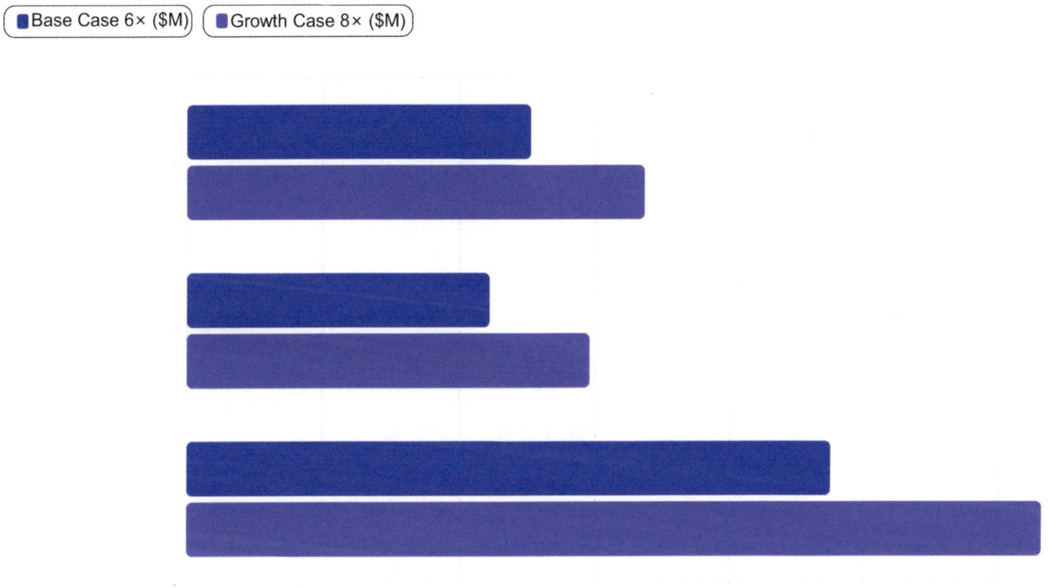

The $474M–$632M combined valuation reflects not merely the sum of individual jurisdictions but the strategic premium of a unified competitive intelligence platform with cross-jurisdiction synergies, shared infrastructure, and compounding network effects.

The Ecosystem Flywheel

Player Acquisition

Championships and cultural events draw competitors into the ecosystem

Data Capture

PWA infrastructure records and verifies every competitive interaction

Analytics Value

Performance data drives subscription revenue and player retention

Media Amplification

Xperience TV broadcasts attract new audiences and sponsors

Ecosystem Growth

Expanded audience feeds back into player acquisition and competitive depth

Each revolution of the flywheel increases the value of the entire ecosystem — more players produce richer data, richer data creates better content, better content attracts larger audiences, and larger audiences drive more player acquisition. The compounding effect accelerates with scale.

Intelligence Deserves Sovereignty

The Thesis: Why This Matters

The National & International Poker Championships and The National & International Spades & Bid Whist Championships are not entertainment products. They are infrastructure for cognitive recognition. They formalize what has always been true: that these games produce extraordinary thinkers, strategists, and decision-makers. The only thing missing was the system to prove it.

Goddess Phenezy Holdings has built that system. Proprietary PWA infrastructure captures verified performance data. Championship governance structures create legitimate competitive pathways. Media distribution via Xperience TV elevates competition into cultural programming with global reach. And valuation frameworks grounded in comparable market multiples demonstrate that this is not aspiration — it is applied economic reality.

$79.0M	$474M–$632M	3 Jurisdictions
Combined annual revenue across all card game jurisdictions	Combined ecosystem valuation range	Poker, Spades, and Bid Whist unified under sovereign governance

The Final Word

> Chess formalized strategy for the elite. Poker validated decision-making under risk. Spades and Bid Whist democratized intelligence.

1

Play

Cultural games in communities

2

Proof

Verified data & rankings

3

Performance

Championship competition

4

Advancement

Economic recognition & sovereignty

By elevating these culturally significant games into sovereign competitive jurisdictions, Goddess Phenezy Holdings ensures that previously unrecognized intelligence is finally visible.

And by doing so, we complete the arc:

Play → Proof → Performance → Advancement

Goddess Phenezy Holdings, Inc.The Probability Jurisdiction. The Cultural Intelligence Jurisdiction.
Decision-making as economic power. Intelligence as sovereign sport.

$79.0M	$474M–$632M	3
Annual Revenue	Valuation Range	Unified Jurisdictions

The National & International Conferences

The Convening Jurisdiction: Where Ideas, Capital, and Power Collide

A comprehensive strategic overview of the conference ecosystem transforming how industries convene, capitalize, and compound influence across verticals, cultures, and global markets.

Conferences Are Not Events — They Are Economic Engines

Conferences are one of the most underestimated power centers in the global economy. While most organizations view them as logistical exercises — booking venues, printing badges, and scheduling panels — the reality is far more consequential. Conferences are where the architecture of industries is quietly redesigned. They are where capital is allocated, standards are set, narratives are shaped, and industries decide who leads next.

Consider the mechanics: when a CEO sits across from a sovereign wealth fund manager at a curated dinner, that is not "networking." That is capital formation. When a standards body convenes alongside enterprise buyers and regulators, that is not a "panel discussion." That is policy infrastructure being built in real time. The conference is the jurisdiction where these forces converge — and the entity that governs that jurisdiction wields extraordinary influence.

Capital Allocated

Investment decisions shaped through curated deal flow and relationship architecture

Standards Set

Industry benchmarks and regulatory frameworks forged through expert convenings

Narratives Shaped

Media, thought leadership, and market positioning driven from the main stage

Leadership Decided

The next generation of industry leaders surfaces through visibility and credibility

CONFERENCES
R US

The Problem: Structural Inefficiency at Scale

Yet most conferences operate as one-off gatherings — disconnected, ephemeral, and structurally inefficient. The typical conference lifecycle is painfully predictable: months of planning culminate in a two-day event, after which attendee data is discarded, relationships dissolve after closing remarks, and intellectual capital is never compounded. The billions of dollars invested annually in conference attendance yield a fraction of their potential return because the infrastructure is designed to expire.

This is not a minor inefficiency. It is a systemic failure. Organizations spend $30,000 to $500,000 per conference on sponsorships, travel, and executive time — only to watch the value evaporate within weeks. The data collected during registration is rarely analyzed. The connections made during sessions are rarely facilitated beyond the event. The insights shared from the stage are rarely operationalized into institutional knowledge.

The conference industry, valued at over $1.1 trillion globally, is built on a model of planned obsolescence. Every gathering starts from zero. Every relationship must be rebuilt. Every dollar spent is a sunk cost rather than a compounding investment. This represents one of the largest untapped opportunities in the professional services economy — and it is precisely the gap that Goddess Phenezy Holdings, Inc. was designed to fill.

> **Goddess Phenezy Holdings, Inc. does not host conferences. We govern convenings.**

The Goddess Phenezy Holdings Thesis

Through The National & International Conferences ecosystem, Goddess Phenezy Holdings transforms conferences into recurring, data-driven, multi-industry jurisdictions that compound influence, revenue, and opportunity year after year. This is not an incremental improvement to the conference model — it is an entirely new category of institutional infrastructure.

The thesis is built on three structural pillars. First, jurisdictional authority: by owning the naming rights, intellectual property, and operational infrastructure of flagship conference brands across multiple verticals, the ecosystem establishes sovereign authority over the convenings it governs. Second, compound intelligence: every interaction, transaction, and connection generated within the ecosystem feeds a proprietary data architecture that grows more valuable with each cycle. Third, cross-vertical monetization: attendees, sponsors, and partners who enter through one conference gain access to the broader ecosystem — creating network effects that no single-event operator can replicate.

The result is a conference ecosystem that functions less like an events company and more like an economic jurisdiction — with its own currency of credibility, its own infrastructure for deal flow, and its own compounding returns on relationship capital.

Jurisdictional Pillars

AuthoritySovereign naming rights and IP ownership across verticals

IntelligenceProprietary data compounding with every cycle

MonetizationCross-vertical network effects no competitor can replicate

Flagship Conference Properties

Rather than dilute authority across hundreds of unrelated topics, the portfolio focuses on high-leverage verticals aligned with capital, policy, innovation, and human advancement.

Each flagship conference is designed as a self-sustaining jurisdiction within the broader ecosystem. They share infrastructure, data architecture, and credentialing systems — but each operates with vertical-specific programming, audience development, and revenue models. The following sections detail the select highlights of the portfolio, beginning with the highest-leverage properties in technology, health, education, diplomacy, and entrepreneurship.

The National & International AI Conference

The apex gathering for artificial intelligence, automation, ethics, defense, education, and enterprise deployment.

Artificial intelligence is not a vertical — it is the new substrate upon which every industry will be rebuilt. The National & International AI Conference is positioned at the intersection of enterprise deployment, government regulation, academic research, and workforce transformation. Unlike typical AI conferences that cater exclusively to technical audiences, this convening brings together the full spectrum of stakeholders who will determine how AI reshapes the global economy.

Enterprise Leaders

C-suite executives and transformation officers deploying AI across Fortune 500 operations, supply chains, and customer systems

Government Stakeholders

Federal, state, and international regulators shaping AI policy, defense applications, and public sector adoption

AI Researchers & Founders

The builders and inventors advancing foundational models, applied AI, and breakthrough architectures

Workforce Transition Pipelines

Educators, workforce development agencies, and reskilling organizations preparing workers for the AI economy

Positioning: AI as infrastructure, not novelty. This conference treats artificial intelligence as the foundational layer of 21st-century economic infrastructure — demanding the same institutional rigor as energy, transportation, and telecommunications.

The National & International Health & Wealth Conference

Where healthcare, longevity, finance, insurance, and generational wealth intersect.

The false separation between health outcomes and financial outcomes is one of the most damaging structural myths in American public policy. Medical debt remains the leading cause of personal bankruptcy. Life expectancy correlates more strongly with zip code than with genetics. Insurance access determines not just medical care but economic mobility. The National & International Health & Wealth Conference was designed to demolish the artificial wall between these domains and build integrated solutions at scale.

This conference brings together health systems executives, institutional investors, insurance innovators, and community health advocates to forge new models of integrated prosperity. The programming spans clinical innovation, health-tech venture capital, Medicaid and Medicare policy reform, longevity science, financial literacy, and generational wealth architecture — all under one convening jurisdiction.

Health Systems & Policy Leaders

Hospital network executives, public health officials, and regulatory architects reshaping care delivery models and payment systems

Investors & Financial Institutions

Venture capital firms, private equity, insurance companies, and banks deploying capital into health-wealth convergence opportunities

Entrepreneurs & Educators

Health-tech founders, financial literacy educators, and social enterprise builders creating accessible solutions

Veterans & Underserved Communities

Military veterans, rural communities, and historically marginalized populations who face the sharpest disparities in both health and wealth

> Health *is* wealth — structurally and generationally.

The National & International STEM Conference

The intellectual counterpart to the STEM Championships — where talent is discovered, not theorized about.

The global conversation around STEM education has generated enormous enthusiasm and significant investment. Yet despite decades of advocacy, the pipeline from STEM interest to STEM employment remains leaky, inequitable, and disconnected from industry needs. Too many STEM conferences focus on abstract discussions of policy and curriculum without creating direct pathways to opportunity. The National & International STEM Conference was designed to close that gap.

This conference operates as a talent marketplace, scholarship accelerator, and industry-education bridge. Rather than convening educators to talk about what students need, it brings educators, students, industry partners, and capital allocators into the same room — and creates transactions. Defense contractors identify future recruits. EdTech companies demonstrate products to the educators who will adopt them. Scholarship partners match directly with qualified candidates. The result is a conference that produces measurable outcomes, not aspirational white papers.

Educators & Institutions

K-12 systems, universities, and community colleges presenting curricula and identifying partnership opportunities

EdTech & Defense Contractors

Technology companies and defense enterprises scouting talent and demonstrating next-generation tools

Youth Pipelines & Scholarship Partners

Foundations, corporate sponsors, and government programs matching capital directly to qualified students

> Positioning: Talent discovery, not theory. Every session, exhibition, and networking event is designed to produce measurable outcomes — scholarships awarded, internships filled, and partnerships formalized.

The National & International Peace Conference

A rare convening focused on diplomacy, conflict resolution, technology ethics, and global cooperation.

Peace is rarely treated as a discipline. It is invoked as an aspiration, cited in speeches, and printed on banners — but it is almost never approached with the same structural rigor applied to defense, trade, or technology policy. The National & International Peace Conference exists to correct this oversight by treating peace as what it actually is: a strategic discipline requiring expertise, infrastructure, investment, and institutional commitment.

This conference brings together the architects of international stability — diplomats, NGO leaders, conflict resolution experts, technology ethicists, and educators — to develop actionable frameworks for cooperation. The programming spans cybersecurity diplomacy, AI ethics in warfare, post-conflict economic development, refugee integration, and cross-border education partnerships. Unlike academic symposia that produce papers, this convening produces partnerships, frameworks, and commitments with measurable accountability structures.

Policy Architects

Diplomats, legislators, and international relations experts designing frameworks for sustainable cooperation

NGOs & International Leaders

Humanitarian organizations, multilateral institutions, and civil society leaders operating in conflict zones

Technologists & Educators

AI ethicists, cybersecurity experts, and education leaders addressing technology's role in conflict and cooperation

> Peace as a strategic discipline. Not an aspiration. Not a slogan. A rigorous, investable, measurable practice.

The National & International Business & Entrepreneurship Conference

A convergence of founders, operators, investors, and ecosystem builders.

Entrepreneurship is often romanticized but rarely supported with the structural infrastructure it demands. Most business conferences offer inspirational keynotes and generic networking — but fail to create the transactional architecture that turns ideas into companies and companies into enterprises. The National & International Business & Entrepreneurship Conference was designed as a competitive arena where execution is the currency and outcomes are the scoreboard.

The Business Conference Market: Growth & Impact

78%

say business conferences are their organization's most impactful marketing channel

80%

say business conferences are critical to organizational success

66%

plan to hold MORE business events in 2025

57%

report attendance growth at business conferences year-over-year

54%

of attendees plan to attend MORE business events in 2025

> Positioning: Business as sport. Execution as the championship. Every session is designed to produce measurable transactional outcomes — not motivation, but deals.

Flagship Portfolio at a Glance

The five flagship conference properties represent the highest-leverage verticals in the global economy. Each operates as a sovereign jurisdiction within the broader ecosystem — sharing infrastructure while maintaining vertical-specific authority.

Conference	Core Audience	Key Verticals	Positioning
AI Conference	Enterprise, Government, Researchers	Automation, Ethics, Defense, Education	AI as Infrastructure
Health & Wealth	Health Systems, Investors, Communities	Healthcare, Longevity, Finance, Insurance	Health is Wealth
STEM Conference	Educators, EdTech, Youth Pipelines	Education, Defense, Scholarships	Talent Discovery
Peace Conference	Diplomats, NGOs, Technologists	Diplomacy, Ethics, Global Cooperation	Peace as Discipline
Business & Entrepreneurship	Founders, Investors, Corporate Buyers	Startups, Licensing, Capital, Procurement	Business as Sport
International Black Business Conference	Black Entrepreneurs, Investors, Corporate Buyers	Capital Access, Enterprise Contracts, Licensing, Procurement	Economic Reclamation
International Black Empowerment Conference	Community Leaders, Educators, Policy Makers	Education, Economic Development, Policy, Community Building	Structural Equity
National & International Men's & Women's Conferences	Men, Fathers, Community Leaders, Health Professionals	Mental Health, Fatherhood, Leadership, Economic Stability	Stability Infrastructure

The Social Infrastructure Layer

Economics, Identity, and Power

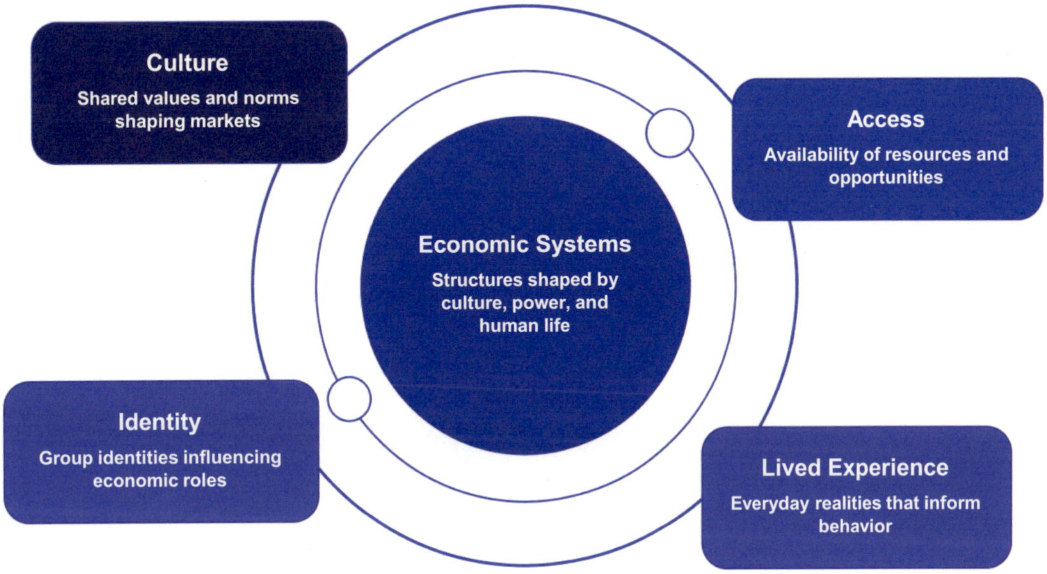

This diagram illustrates how economic systems are fundamentally shaped and influenced by critical human and cultural factors, demonstrating that these elements are not separate but deeply interconnected.

Economic systems do not function in a vacuum. They are shaped by culture, access, identity, and lived experience. The most sophisticated financial models, the most elegant technology architectures, and the most well-intentioned policy frameworks all ultimately succeed or fail based on whether they account for the human beings they are designed to serve — and which human beings they were designed to exclude.

This is not a philosophical observation. It is an economic one. Markets that systematically exclude participants leave value on the table. Industries that fail to access diverse talent pools underperform. Economies that concentrate opportunity in narrow demographic bands create instability that ultimately costs everyone — including the beneficiaries of that concentration.

Why Cultural Specificity Is Strategic, Not Optional

While many conferences portfolios avoid cultural specificity to appear "neutral," Goddess Phenezy Holdings, Inc. does the opposite — because neutrality has never produced equity, innovation, or stability.

The pretense of neutrality in institutional design is not merely inadequate — it is actively harmful. When a conference claims to serve "everyone," it invariably defaults to serving the demographic that already holds the most power, the most capital, and the most institutional access. The programming reflects their priorities. The networking favors their connections. The outcomes reinforce their advantages. This is not conspiracy; it is structural gravity.

The Neutral Approach

- **Claims universal relevance but defaults to dominant demographics**
- **Programming reflects existing power structures**
- **Networking favors established connections**
- **Outcomes reinforce existing advantages**
- **Perpetuates structural exclusion through omission**

The Intentional Approach

- **Builds sovereign spaces for historically excluded communities**
- **Programming addresses specific barriers and opportunities**
- **Networking creates new connections and deal flow**
- **Outcomes produce measurable economic mobility**
- **Generates innovation through diverse perspectives**

Goddess Phenezy Holdings builds intentional convenings where historically excluded communities, leaders, and decision-makers are not guests — they are the focus. These conferences are not symbolic. They are economic engines.

The National & International Black Business & Empowerment Conferences

The Economic Reclamation Jurisdiction

Black entrepreneurship represents one of the fastest-growing yet most undercapitalized economic sectors in the world. Despite trillions in annual spending power, Black-owned businesses continue to face systemic barriers to capital access, enterprise contracts, and scalable infrastructure. These are not anecdotal challenges — they are documented, measured, and persistent structural failures in the allocation of economic opportunity.

The National & International Black Business & Empowerment Conferences were created to correct that imbalance — structurally, not rhetorically. This is not a conference about aspiration. It is a conference about transaction. The distinction matters enormously. Aspiration without infrastructure is motivational theater. Transaction with infrastructure is economic architecture.

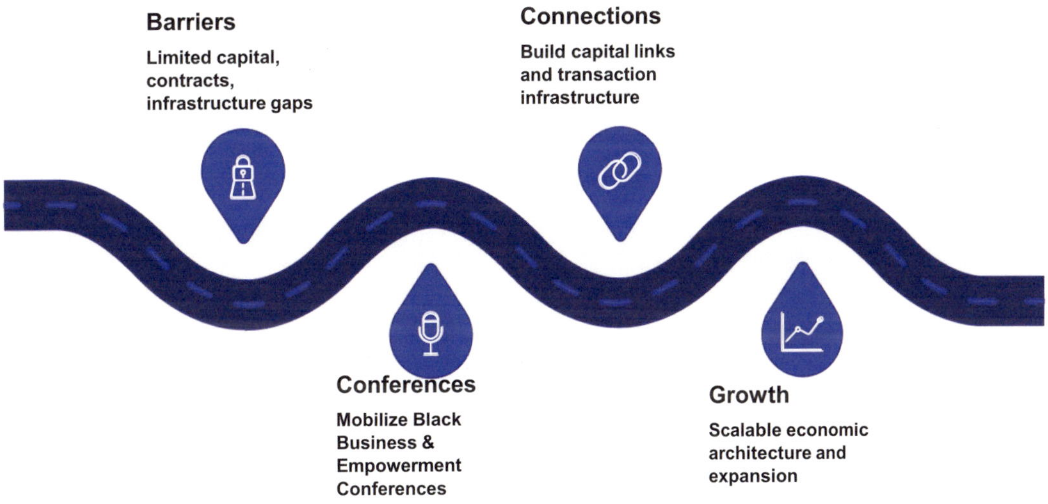

Who Convenes at the Black Business & Empowerment Conferences

The power of these conferences lies in the composition of the room. By design, every attendee category represents a node in the economic value chain — from the founders who create enterprises to the institutions that capitalize, contract with, and scale them.

Black Founders, Operators & Enterprise Leaders

CEOs, managing directors, and operational leaders building businesses across technology, professional services, manufacturing, healthcare, and consumer markets

Institutional Investors, Procurement Officers & Lenders

Venture capital firms, CDFIs, SBA lenders, corporate procurement teams, and institutional investors deploying capital into diverse-owned enterprises

Government Agencies & Corporate Buyers

Federal, state, and municipal contracting officers alongside Fortune 500 supplier diversity programs sourcing qualified vendors

Educators, Youth Pipelines & Policy Advocates

Business school deans, youth entrepreneurship programs, HBCU partnerships, and policy organizations advancing structural reform

> **Ownership. Access. Scale.**

Economic Focus Areas

The programming architecture of the Black Business & Empowerment Conferences is organized around five economic focus areas — each designed to address a specific structural barrier and create a specific transactional outcome. These are not panel topics. They are deal-making verticals.

1 Capital Readiness & Alternative Funding Models

Revenue-based financing, community development financial institutions, crowdfunding architecture, and investor-ready pitch preparation — because traditional venture capital has systematically underinvested in Black founders

2 Federal, State & Corporate Contracting

Navigating 8(a) certification, HUBZone qualifications, supplier diversity portals, and the procurement cycles of major government agencies and Fortune 500 corporations

3 Technology, AI & Digital Infrastructure

Ensuring Black-owned businesses are not merely consumers of technology but builders and owners of the digital infrastructure that will define the next economic era

4 Generational Wealth & Asset Ownership

Real estate, intellectual property, business succession planning, and financial literacy designed to compound value across generations

5 Global Trade & International Expansion

Export readiness, international partnership development, and cross-border commerce strategies connecting Black-owned enterprises to global markets

Deal Rooms, Not Motivational Stages

The fundamental design principle of the Black Business & Empowerment Conferences is captured in a single distinction: these conferences function as deal rooms, not motivational stages. Attendees leave with contracts, partnerships, and capital pathways — not applause.

This distinction is critical because the conference industry has historically offered Black professionals and entrepreneurs a surplus of inspiration and a deficit of infrastructure. There is no shortage of motivational speakers willing to tell audiences to "dream bigger" and "bet on yourself." What has been missing — systematically and by design — is the transactional architecture that converts ambition into enterprise: the procurement officer ready to issue a contract, the investor ready to write a check, the corporate buyer ready to place an order.

Every session at these conferences is structured around outcomes. Pitch competitions produce term sheets. Procurement workshops produce vendor applications. Investor panels produce follow-up meetings with capital allocators. The measure of success is not attendance — it is transactions closed, contracts signed, and capital deployed.

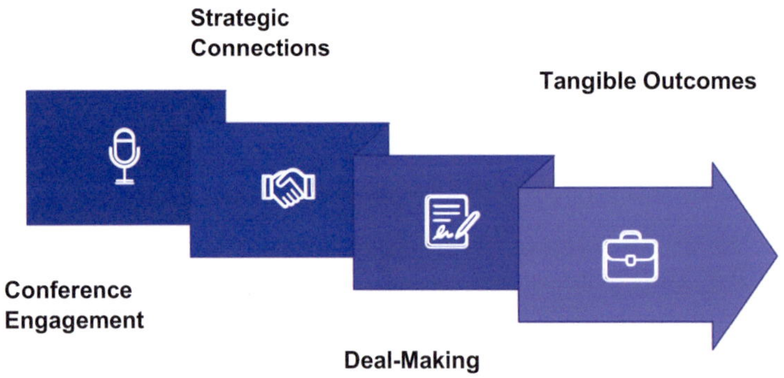

$1.6T	3%	10x
Black Spending Power	**VC Funding Share**	**Wealth Gap Multiple**
Annual consumer spending by Black Americans — a market larger than most national GDPs	Percentage of U.S. venture capital deployed to Black-founded startups	Ratio of median White household wealth to median Black household wealth

FLAGSHIP CULTURAL PROPERTY

The National & International Men's Conference

The Stability Jurisdiction: Leadership, Health, and Purpose

Men are disproportionately affected by some of the most urgent crises in modern society — mental health crises, suicide, incarceration, untreated trauma, and economic displacement. The data is unambiguous and alarming. Men account for nearly 80% of suicides in the United States. Men comprise over 90% of the incarcerated population. Men are significantly less likely to seek mental health treatment, less likely to have a primary care physician, and less likely to maintain the social connections that serve as protective factors against isolation and despair.

Yet most leadership and wellness spaces are not designed *for* men — nor do they address male-specific pressures around identity, responsibility, and purpose. The result is a growing population of men who are disconnected from the institutions, relationships, and support systems that produce stability — for themselves, their families, and their communities.

Integrated Portfolio Architecture

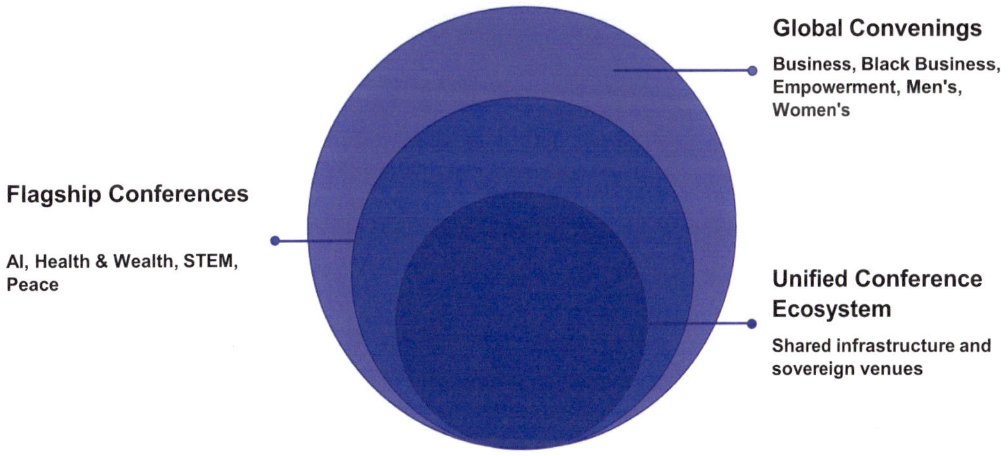

Why Precision-Built Spaces Matter

The National & International Men's Conference exists because men and women do not experience the world the same way — and pretending otherwise helps no one. This is not a political statement. It is a clinical, sociological, and economic observation supported by decades of research across every discipline from psychology to labor economics.

Men process trauma differently. Men communicate under stress differently. Men face different social expectations around vulnerability, provision, and emotional expression. Ignoring these differences in the design of support systems does not create equality — it creates abandonment. Men who cannot find spaces designed for their specific challenges do not seek alternative spaces. They withdraw. And withdrawal — from families, from communities, from economic participation, from life itself — is the crisis this conference was designed to address.

Mental Health & Emotional Resilience

Clinical frameworks, peer support models, and evidence-based approaches to men's mental health that account for how men actually experience and express psychological distress

Fatherhood, Mentorship & Legacy

Building the institutional knowledge and community infrastructure for engaged fatherhood, intergenerational mentorship, and deliberate legacy construction

Career Reinvention & Economic Relevance

Navigating economic displacement, mid-career transitions, and the skills transformation required by AI, automation, and the evolving labor market

Physical Health, Longevity & Performance

Preventive medicine, fitness, nutrition, and the physiological dimensions of sustained high performance across the male lifespan

Leadership Without Shame or Silence

Creating models of male leadership that integrate strength with vulnerability, authority with accountability, and ambition with emotional intelligence

The Stability Thesis

> **Strong men build stable families. Stable families build stable economies.**

This is not a slogan — it is a macroeconomic thesis with empirical support. Research consistently demonstrates that male engagement in families, communities, and institutions is one of the strongest predictors of community-level economic stability. When men are employed, healthy, connected, and purposeful, the downstream effects on child outcomes, household income, community safety, and economic productivity are measurable and significant.

The National & International Men's Conference is not exclusionary — it is precision-built. Just as a cardiac surgeon specializes in hearts without being accused of ignoring the lungs, this conference specializes in the specific challenges, opportunities, and responsibilities that men face without diminishing the importance of women's spaces or gender-inclusive environments. Precision is not exclusion. It is the prerequisite for effectiveness.

The conference operates on the same proprietary PWA infrastructure as every other property in the ecosystem. Attendees benefit from cross-conference networking, shared credentialing, and integrated deal flow. A man who attends this conference and also operates a business can seamlessly connect with the Business & Entrepreneurship Conference or the AI Conference — because the ecosystem was designed for exactly that kind of cross-vertical mobility.

The Crisis in Numbers

The data driving the Men's Conference thesis is not ambiguous. It represents a cascading crisis with economic, social, and public health dimensions that demand institutional intervention at scale.

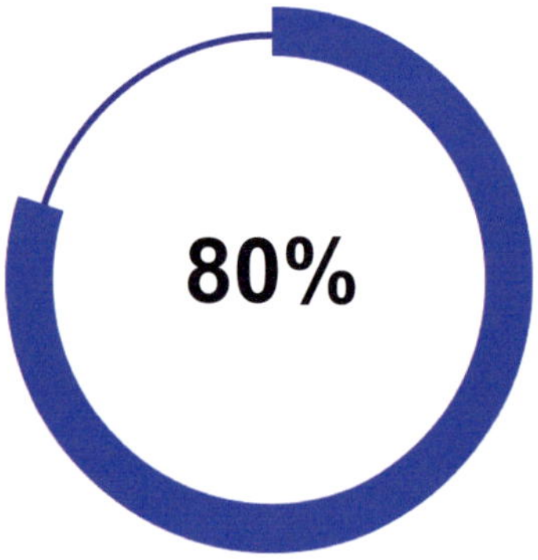

Suicide Rate

Men account for nearly 80% of all suicides in the United States

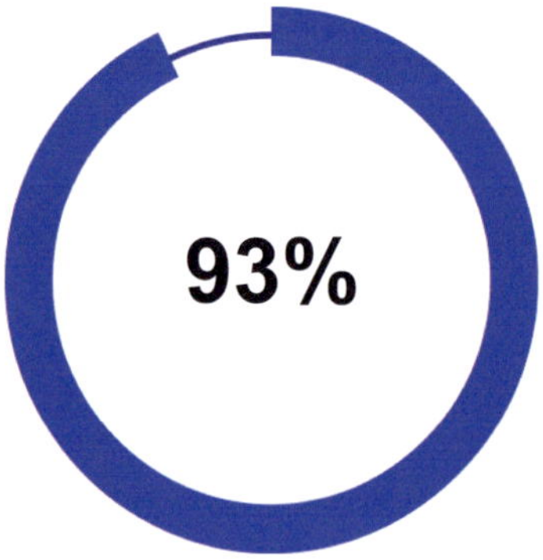

Incarceration

Men comprise 93% of the federal and state prison population

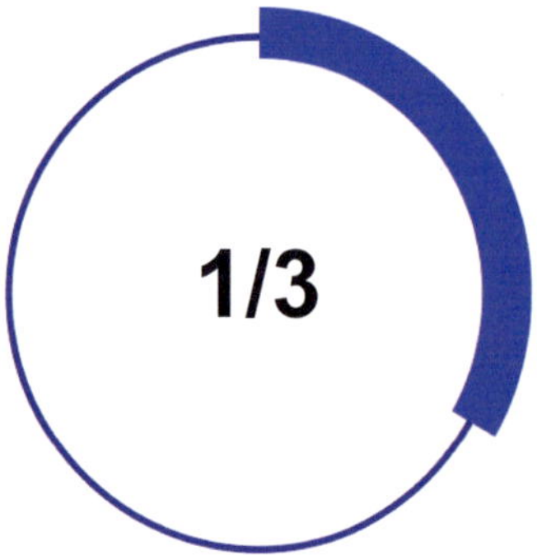

No Primary Doctor

One-third of American men do not have a primary care physician

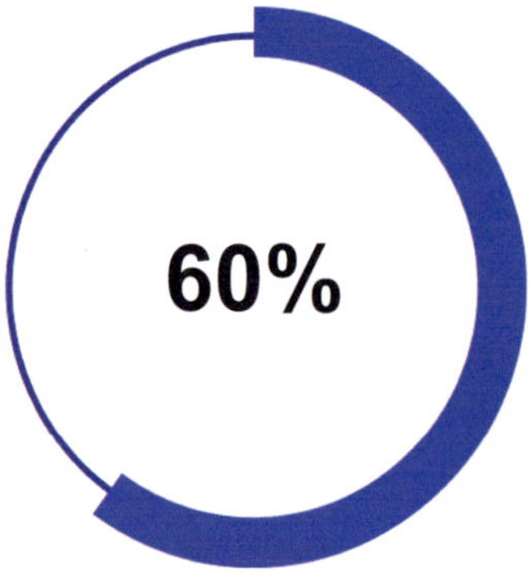

Declining Enrollment

Men now represent only 40% of college enrollment — a reversal that threatens economic mobility

These are not isolated statistics. They represent an interconnected system of male disengagement that, left unaddressed, will continue to degrade family stability, community safety, workforce productivity, and economic growth. The Men's Conference treats this crisis with the urgency and structural sophistication it demands.

Integration with the Conference Ecosystem

All cultural conferences operate on the same proprietary PWA infrastructure as the technology, business, and STEM conferences.

The decision to build all conferences — cultural, technological, educational, and diplomatic — on a single unified platform is one of the most strategically significant architectural choices in the entire Goddess Phenezy Holdings ecosystem. It means that an attendee who enters through the Black Business & Empowerment Conference has the same digital identity, the same credentialing system, and the same access to cross-conference opportunities as an attendee who enters through the AI Conference or the STEM Conference.

This is not a technical detail. It is a philosophical commitment with profound economic implications. By refusing to build separate, isolated platforms for cultural conferences, Goddess Phenezy Holdings ensures that historically excluded communities are not merely acknowledged within the ecosystem — they are structurally integrated into its most valuable features: data, networking, deal flow, and monetization.

The Conference Market Gap

The conference market currently exhibits a massive disparity. While established sectors are oversaturated with numerous events, culturally-specific conferences with real economic infrastructure are virtually nonexistent, highlighting a critical underserved segment.

10K+

Tech Conferences

annually

8K+

Business Conferences

annually

15K+

Industry Trade Shows

annually

<50

Black Business Conferences

annually

<100

Cultural Identity Conferences

annually

<75

Men's/Women's Empowerment Conferences

annually

0

Integrated Economic Infrastructure

Nearly Zero

Shared Infrastructure Architecture

The unified PWA infrastructure creates four distinct categories of cross-conference value that no siloed conference model can replicate.

Shared Digital Identity & Credentialing

One profile, one credential, all conferences. Attendees build a persistent professional identity that compounds credibility across every convening they attend.

Cross-Conference Networking & Deal Flow

Connections made at one conference persist and compound across the ecosystem. The AI Conference attendee can connect with the Business Conference founder seamlessly.

Unified Sponsorship & Enterprise Access

Sponsors and enterprise partners gain access to the entire ecosystem — multiplying their reach and ROI while enabling cross-vertical brand presence.

Ongoing Education & Community Engagement

Between conferences, the platform delivers continuous learning, mentorship, and community engagement — ensuring value compounds 365 days a year.

Cultural conferences are not isolated silos. They are fully monetized, fully integrated jurisdictions within the broader Goddess Phenezy Holdings ecosystem.

Cross-Conference Value Flow

The integration architecture creates compound returns that accelerate with each conference cycle. The following visualization illustrates how value flows across the ecosystem — from attendee data to sponsorship revenue to deal transactions.

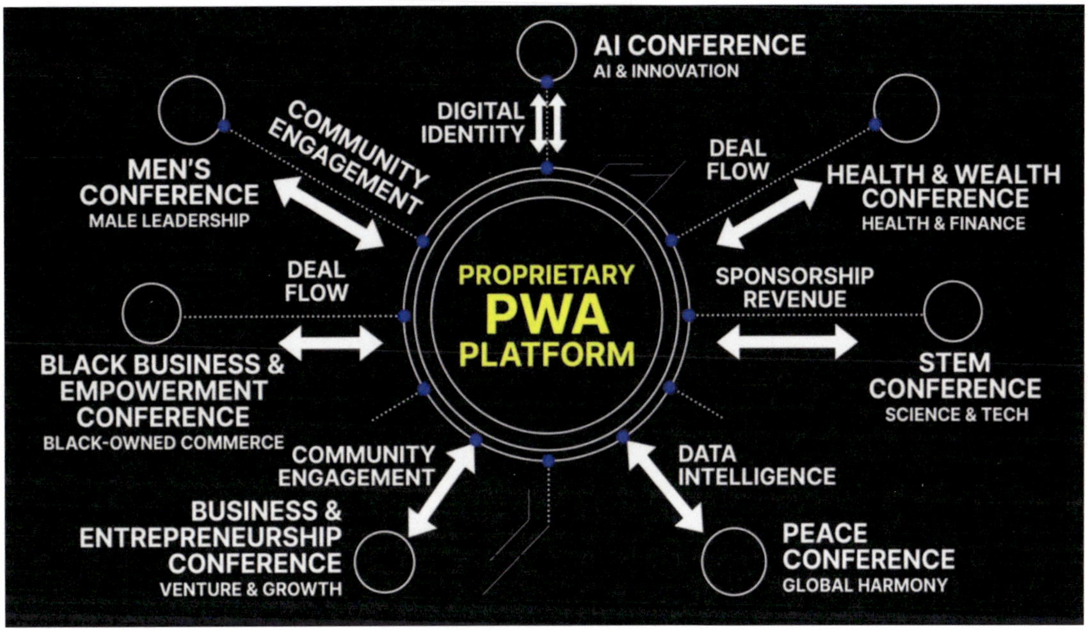

This architecture means that every new conference added to the portfolio increases the value of every existing conference. Every new attendee enriches the data intelligence. Every new sponsor gains broader reach. The ecosystem exhibits true network effects — the kind of compounding advantage that separates platform businesses from event businesses.

SEO & Discovery Advantage

These conference properties intentionally align with high-intent search behavior across multiple verticals and demographics. The naming strategy — owning both National and International designations — is not cosmetic. It is a deliberate SEO architecture designed to capture search traffic at the exact moment when target audiences are expressing purchase intent, partnership interest, or professional development needs.

High-intent search queries represent the most valuable traffic in digital marketing because they indicate a user who has already decided to act and is now choosing where to direct that action. By owning the authoritative conference brand for each query category, Goddess Phenezy Holdings captures that intent at the moment it crystallizes.

The SEO-to-Conference Value Cycle

Discovery & Authority

Goddess Phenezy Holdings conferences dominate search results with National + International naming

High-Intent Search

User searches "National AI Conference" or "International Black Business Conference"

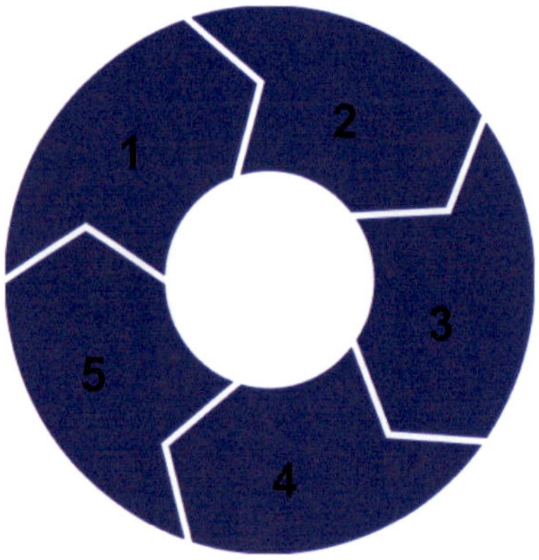

Registration & Attendance

Authoritative positioning drives ticket sales and sponsorship interest

Compound Growth

Each conference cycle strengthens search dominance and brand authority

Data & Network Effects

Attendee data feeds back into SEO, content marketing, and cross-conference recommendations

This visual illustrates how SEO is more than just marketing; it's the foundational discovery infrastructure that powers the entire conference ecosystem, creating a continuous loop of growth and authority.

Search Dominance Strategy

The portfolio is engineered to capture and dominate the following high-value search corridors — each representing a distinct audience segment with specific economic intent.

Black Business Conference Queries

***The International Black Business Conference* — capturing entrepreneurs, investors, and corporate diversity officers searching for premier Black business events**

Black Empowerment Conference Queries

***The International Black Empowerment Conference* — reaching community leaders, policy advocates, and institutional partners seeking empowerment-focused convenings**

Men's Leadership Conference Queries

***The International Men's Leadership Conference* — connecting with executives, organizational development professionals, and men's health advocates**

Men's Mental Health Conference Queries

***The International Men's Mental Health Conference* — reaching clinicians, wellness organizations, employers, and advocacy groups focused on male mental health**

Men's Empowerment Conference Queries

***The International Men's Empowerment Conference* — capturing mentorship organizations, community leaders, and individuals seeking purposeful male-focused development**

Brand Authority & Long-Term Equity

By owning the National and International naming rights and pairing them with authoritative infrastructure, these conferences dominate three critical dimensions of brand positioning: search visibility, institutional credibility, and long-term brand equity.

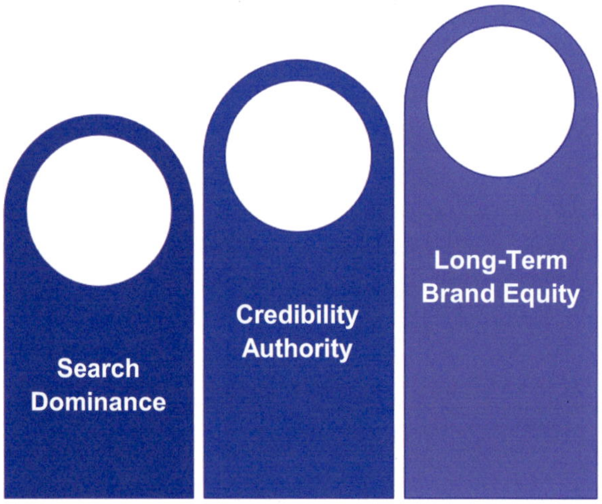

The compounding nature of this strategy cannot be overstated. Every year that these conferences operate, their search authority deepens, their brand recognition compounds, and their competitive moat widens. A competitor entering these search categories in year three, five, or ten faces an incumbent with years of accumulated authority, backlinks, media coverage, and attendee testimonials. The first-mover advantage in conference brand ownership is nearly insurmountable when paired with operational excellence.

STRATEGIC SIGNIFICANCE

The Strategic Thesis: Building Whole Humans, Not Just Markets

Technology builds systems. Sports build discipline. STEM builds the future. Culture determines who gets to participate.

This is the central strategic insight that distinguishes Goddess Phenezy Holdings from every other conference ecosystem in the market. Most conference companies build verticals. Goddess Phenezy Holdings builds an integrated jurisdiction that spans technology, education, commerce, culture, health, and human development. The result is an ecosystem that does not merely create economic value — it creates the conditions under which economic value can be equitably generated, distributed, and compounded.

By establishing sovereign cultural convenings alongside economic, technological, and educational platforms, Goddess Phenezy Holdings does something most institutions never attempt: it builds whole humans, not just markets. The AI Conference develops technological competence. The STEM Conference develops educational pipelines. The Business Conference develops commercial capability. The Black Business Conference develops economic access. The Men's Conference develops human stability. Together, they create a comprehensive infrastructure for human advancement that no single-vertical conference company can match.

The Comprehensive Ecosystem

The strategic architecture of the conference ecosystem can be understood through the lens of five interconnected layers — each addressing a distinct dimension of human and economic development, and each reinforcing the others.

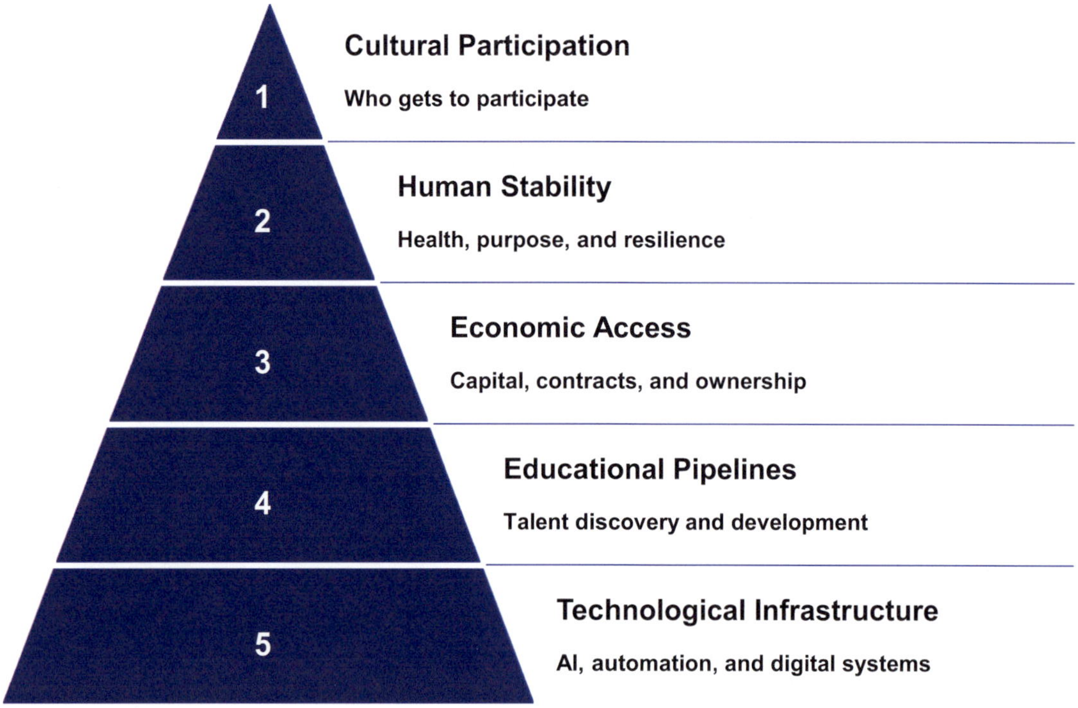

Each layer of this pyramid depends on the layers above and below it. Technological infrastructure without educational pipelines produces tool without talent. Educational pipelines without economic access produce graduates without opportunity. Economic access without human stability produces wealth without well-being. And all of it, without cultural participation, produces systems that serve some and exclude others. The conference ecosystem addresses every layer — simultaneously and deliberately.

Discoverability by Design

The SEO and discovery strategy accomplishes something profoundly important beyond search rankings. It ensures the ecosystem is discoverable not just by investors and executives — but by the people it was designed to uplift.

Discoverable by Capital

Investors, institutional partners, and corporate sponsors find the ecosystem through industry-specific search queries, financial media coverage, and enterprise procurement channels. The flagship conference brands — AI, STEM, Business — dominate these corridors.

These audiences discover the ecosystem's economic infrastructure: the deal flow, the sponsorship opportunities, the data intelligence, and the cross-vertical network effects that make investment compelling.

Discoverable by Community

Entrepreneurs, educators, fathers, veterans, students, and community leaders find the ecosystem through culturally specific search queries, community media, and grassroots networks. The cultural conference brands dominate these corridors.

These audiences discover the ecosystem's human infrastructure: the mentorship, the health resources, the capital access programs, and the community connections that make participation transformative.

The genius of the dual-discovery strategy is that both audiences — capital and community — enter the same ecosystem, share the same platform, and create value for each other. The investor who discovers the AI Conference also gains access to the diverse talent pipeline of the STEM Conference. The entrepreneur who discovers the Black Business Conference also gains access to the technology infrastructure of the AI Conference. Every discovery point leads to the entire ecosystem.

Conclusion: The Convening Jurisdiction

Conferences are not events. They are economic jurisdictions. And the entity that governs the jurisdiction governs the outcomes.

The National & International Conferences ecosystem represents a fundamentally new category of institutional infrastructure. It is not an events company. It is not a media company. It is not a technology company. It is all of these — and none of them — because it operates at a higher level of abstraction: the convening jurisdiction, where ideas, capital, and power collide and are deliberately channeled toward compounding value.

Every conference in the portfolio is designed to produce three categories of outcome: transactions (contracts, investments, partnerships), intelligence (data, relationships, market insights), and equity (brand value, search authority, institutional credibility). These outcomes compound with each cycle, creating a flywheel that accelerates with scale and becomes increasingly difficult to replicate.

Transactions

Contracts, investments, and partnerships closed

Intelligence

Data, relationships, and market insights compounded

Scale

Network effects and compounding returns accelerate growth

Equity

Brand value, search authority, and credibility built

Goddess Phenezy Holdings, Inc. has built something that most organizations never conceive of — and that no competitor can easily replicate. A convening ecosystem that serves capital and community simultaneously. That builds markets and humans in parallel. That compounds economic value and social equity with every cycle. This is not a conference portfolio. This is a jurisdiction. And the jurisdiction is open.

The Academies

The Knowledge Jurisdiction: Permanent Infrastructure for Human Capital

Events create moments. Championships create champions. Conferences create alignment. Academies create permanence. A global network of proprietary intellectual institutions — engineered to institutionalize learning, leadership, and legacy across every dimension of human capital development.

Beyond Episodic Engagement

Goddess Phenezy Holdings, Inc. does not rely on episodic engagement to shape the future. Where other organizations convene audiences for a weekend and call it impact, we have architected something fundamentally different — a system that institutionalizes learning, leadership, and legacy through a global network of proprietary academies, each operating as a sovereign knowledge engine within the broader ecosystem.

These are not online courses. They are not webinar series repackaged with certificates. They are intellectual institutions — designed to produce credentialed, competent, and culturally grounded leaders at scale. Each academy functions as a distinct jurisdiction of knowledge, yet all are unified by shared infrastructure, shared identity, and a shared commitment to permanence over performance theater.

Delivered through the same proprietary Progressive Web App (PWA) infrastructure that powers our championships and conferences, the Academies ensure that education, culture, and credentialing are continuous, borderless, and fully owned. There is no dependency on third-party learning management systems, no vulnerability to platform deplatforming, and no dilution of brand equity through external partnerships that compromise sovereignty.

> Key Differentiator: Every academy operates on proprietary digital infrastructure — ensuring institutional permanence, data sovereignty, and complete independence from external platforms.

The Academy Ecosystem at a Glance

Four distinct academies form the intellectual backbone of the Goddess Phenezy Holdings ecosystem. Each serves a unique strategic function while operating within a unified credentialing and digital architecture that enables cross-pollination, stackable learning, and lifelong progression.

International Academy of Learning

The central academic hub for interdisciplinary mastery across business, leadership, technology, and strategy.

International Academy of STEM

Engineering the future workforce through coding, AI, robotics, and applied mathematics pathways.

International Academy of Business

Teaching ownership as a discipline — capital structuring, licensing, valuation, and scale.

International Academy of Black History

Restoring truth as infrastructure — cultural intelligence, civilizational context, and grounded leadership.

The International Academy of Learning

The core educational platform powering lifelong intellectual development across every discipline that matters.

The Central Academic Hub

The International Academy of Learning functions as the central academic hub — housing interdisciplinary curricula across business, leadership, technology, and strategy. It is not a single course or a static curriculum; it is a living, evolving platform that responds to market demands, technological shifts, and the real-time performance data generated by our championship and conference ecosystems.

This Academy exists to serve a fundamental truth: the most successful leaders are not specialists trapped in silos — they are polymath thinkers who can navigate complexity across domains. The curriculum architecture reflects this philosophy, weaving together strategic thinking, technical literacy, leadership psychology, and operational excellence into coherent learning journeys that build upon one another.

Every credential earned here is stackable, portable, and tied to verified performance — not seat time. In a world where traditional degrees are increasingly decoupled from actual competence, the International Academy of Learning offers something far more valuable: proof of capability, measured against real-world standards and validated through applied competition.

1

Lifelong Learning

Continuous professional development for executives, founders, and emerging leaders at every career stage.

2

Skill Acceleration

Curricula aligned with real-world championships and conferences — learning that translates directly to performance.

3

Stackable Credentials

Verified credentials tied to demonstrated mastery, not attendance — building a permanent record of capability.

THE INTERNATIONAL ACADEMY OF LEARNING

The Spine of the Ecosystem

The International Academy of Learning is the spine that connects applied competition to intellectual mastery. It is the connective tissue that ensures a championship winner understands the theory behind their triumph, that a conference attendee can deepen the insights they encountered, and that every participant in the ecosystem has a pathway from curiosity to credentialed expertise.

Consider the journey of a young entrepreneur who competes in a business championship. Their performance reveals both strengths and gaps. The Academy captures that data and offers a personalized learning pathway — addressing gaps in financial modeling, deepening strengths in strategic communication, and connecting them to mentors who have mastered the domains they aspire to lead.

This is not hypothetical. This is the architectural intention behind every integration point in the system. Learning is not a precursor to action — it is simultaneous with action, running in parallel with competition, deal-making, and leadership development. The Academy ensures that no moment of engagement is wasted, and no insight goes uncaptured.

From Competition to Mastery

Championships reveal capability gaps. The Academy fills them — creating a continuous loop of performance, learning, and growth.

From Conference to Depth

Conferences spark insight. The Academy deepens it — transforming momentary alignment into lasting intellectual infrastructure.

The International Academy of STEM

Engineering the future workforce — from foundational knowledge to competition-grade execution.

Identifying Talent Is Not Enough

Identification

Championships surface raw talent.

Development

The Academy transforms aptitude into professional-grade capability.

Workforce Sovereignty

Graduates don't seek employment — they build enterprises.

Coding & Software Development

From fundamental programming concepts to full-stack development, students build the technical foundation that powers modern industry.

Artificial Intelligence & Data Science

Machine learning, neural networks, and data analytics — the disciplines that will determine who controls the next economic era.

Robotics, Engineering & Applied Mathematics

Physical computing, mechanical design, and the mathematical frameworks that underpin every technological breakthrough.

The Global Cognitive Passport

Students progress from foundational knowledge to competition-grade execution, all tracked through a permanent Global Cognitive Passport — a living, verifiable record of intellectual achievement that travels with the learner across borders, institutions, and career stages.

What the Passport Captures

- Completed learning pathways and mastery levels
- Championship performance data and rankings
- Project portfolios and applied demonstrations
- Peer and mentor evaluations
- Cross-academy credential stacking history

Unlike traditional transcripts that capture what courses a student attended, the Global Cognitive Passport captures what they can actually do. It is a permanent, portable proof of capability — recognized across the entire Goddess Phenezy Holdings ecosystem and designed for external validation by employers, investors, and institutions worldwide.

Learning Pathways

Completed courses and mastery levels across all disciplines

Championship Performance

Competition data, rankings, and tournament results

Project Portfolios

Applied demonstrations and real-world project work

Evaluations

Peer and mentor assessments of capability and growth

Credential Stacking

Cross-academy credential history and progression records

Preparation for What Matters

> **This is not preparation for school. This is preparation for industry, invention, and ownership.**

The International Academy of STEM rejects the premise that education's purpose is to prepare students for the next level of education. That circular logic has produced generations of credentialed professionals who understand theory but cannot build, who can pass examinations but cannot create enterprises, who have degrees but lack the applied competence to transform industries.

Our Academy breaks this cycle by designing every learning pathway around three terminal outcomes: the ability to participate in industry at the highest level, the capability to invent solutions to real problems, and the knowledge required to own the infrastructure and intellectual property that results from that work. Students who complete STEM pathways in this ecosystem do not emerge as job seekers — they emerge as job creators, patent holders, and infrastructure owners.

This progression — from foundation to competition to ownership — is the defining architecture of the STEM Academy experience.

The International Academy of Business

Ownership as a discipline — where founders, operators, executives, and investors learn to build, control, and scale real assets.

Teaching Control, Not Theory

Most business education teaches theory. The International Academy of Business teaches control. This distinction is not semantic — it is structural. Theory produces analysts who can describe market dynamics. Control produces founders who can manipulate those dynamics in their favor. Theory produces employees who understand organizational charts. Control produces executives who design them.

The Academy trains four distinct populations — founders, operators, executives, and investors — in the real-world systems that determine who builds wealth and who merely works within systems built by others. The curriculum is not organized around academic departments; it is organized around the actual mechanics of enterprise creation and value capture.

Every module addresses a specific lever of business control: capital structuring, digital infrastructure ownership, licensing and intellectual property strategy, valuation methodology, and the operational frameworks required to scale from concept to multinational operation. Students do not study case histories of other people's success — they build their own cases in real time, with real capital, real entities, and real consequences.

Founders

Entity formation, IP strategy, capital architecture

Operators

Systems design, team scaling, operational excellence

Executives

Strategic leadership, governance, institutional positioning

Investors

Valuation, deal structuring, portfolio construction

Intentional Ecosystem Alignment

What makes the International Academy of Business categorically different from any MBA program or executive education offering is its intentional alignment with every other pillar of the Goddess Phenezy Holdings ecosystem. Business education here is not isolated in a classroom — it is embedded in a living, competitive, deal-making environment that produces immediate returns on learning.

Championships → Execution Under Pressure

Business students compete in real-time scenarios that mirror market conditions — investor pitches, operational crises, market entry challenges, and strategic pivots. The championships test not what students know, but what they can do when stakes are real and time is limited. This feedback loop between learning and competition accelerates mastery in ways that traditional education cannot replicate.

Conferences → Deal Flow & Partnerships

The conferences provide the relational infrastructure that transforms business knowledge into business activity. Students present ventures to real investors, negotiate partnerships with real operators, and build the networks that will sustain their careers for decades. Education without access is incomplete — the Academy ensures both.

Proprietary Platforms → Ownership of Infrastructure

Students learn on platforms that the ecosystem owns, modeling the very principle the curriculum teaches. They see firsthand what it means to control your digital infrastructure, your data, your credentialing systems, and your distribution channels. The medium is the message.

Graduates Leave With Assets

The ultimate measure of any business education is what the graduate possesses upon completion — not what they memorized, not what they discussed, but what they own. The International Academy of Business is designed around this uncompromising standard.

The Outcome Standard

Graduates do not leave with ideas.

They leave with assets, entities, and authority.

Every graduate completes the program with a registered entity, a functional business plan, an intellectual property portfolio, and a credentialed record of competitive performance.

What Graduates Possess

- Registered business entities — not hypothetical ventures, but legally formed organizations ready for operation
- Capital structure documentation — investment-ready materials, cap tables, and financial projections built under expert guidance
- Intellectual property portfolios — trademarks, provisional patents, and licensing frameworks protecting their innovations
- Verified credentials — championship performance data, peer evaluations, and mentor endorsements recorded in the Global Cognitive Passport
- Active networks — relationships with investors, operators, and partners cultivated through conferences and competition

The International Academy of Black History

The Cultural Intelligence Jurisdiction — restoring truth as infrastructure and grounding leaders in civilizational context.

This Academy of Is Non-Negotiable

The International Academy of Black History exists because history has been fragmented, sanitized, and strategically misrepresented — often stripping context from power, innovation, resistance, and contribution. This is not a matter of opinion; it is a matter of documented record. Entire civilizations have been reduced to footnotes. Innovations that changed the trajectory of human progress have been attributed to the wrong continents. Systems of governance, mathematics, architecture, and medicine that predated European enlightenment by millennia have been erased from mainstream curricula.

This Academy restores truth as infrastructure. It operates from the foundational premise that accurate historical knowledge is not a luxury or an elective — it is a structural requirement for producing leaders who understand how power works, how systems are built, and how civilizations rise. Without this context, even the most technically skilled graduates operate with an incomplete map of human achievement.

The curriculum reframes Black history not as a subset of another narrative, but as a continuous, global, and profoundly consequential civilizational story. It does not begin with oppression. It begins with origin, authorship, and agency — because that is where the actual historical record begins, long before the distortions that have dominated Western education for centuries.

> It does not begin with oppression. It begins with origin, authorship, and agency.

Knowing Your Bloodline Is Power

There is a profound transformation that occurs when a person discovers the full depth of their ancestral lineage. When you know where your people came from — the kingdoms they built, the sciences they mastered, the philosophies they authored — you do not move through the world as a footnote. You move as an heir. Ancestral knowledge is not nostalgia. It is activation. It rewires how a student sees themselves, what they believe they are capable of, and what they understand themselves to be building toward.

The history of Black people did not begin in America. It did not begin in chains. It began in the cradle of civilization — in the Nile Valley, in the empires of Mali, Songhai, and Kush, in the mathematical and astronomical genius of ancient Kemet, in the trade routes and governance systems that shaped the ancient world long before the transatlantic slave trade attempted to sever that memory. The Academy restores that memory — not as mythology, but as documented, verifiable, civilizational fact.

> As a Black woman founder, Goddess Phenezy has made an unbreakable commitment: the culture will be respected, fully represented, and every truth will be told — without apology, without omission, and without compromise.

Why Cultural History Education Changes Everything

Research consistently confirms what the International Academy of Black History is built upon: when students learn history that reflects their identity and civilization, every measurable outcome improves — academically, psychologically, and civically.

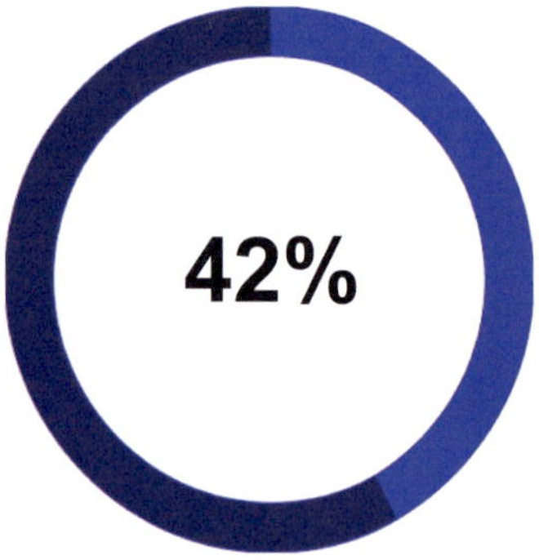

of student achievement gains explained by identity affirmation in culturally responsive classrooms

Source: Int'l Journal of Educational Development, 2025

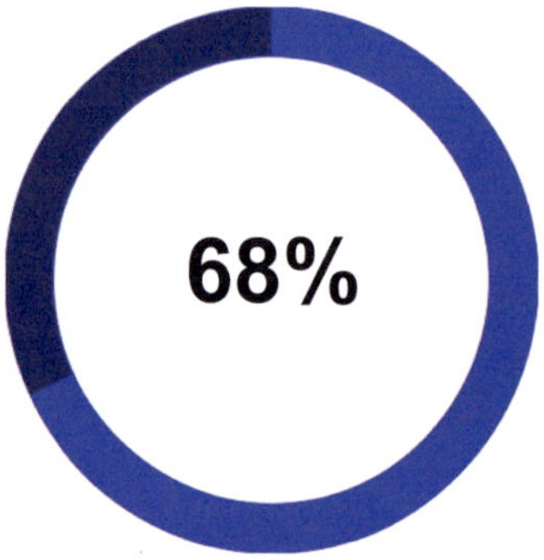

improvement in student engagement when teaching reflects cultural identity

Source: Culturally Responsive Teaching Study, 2025

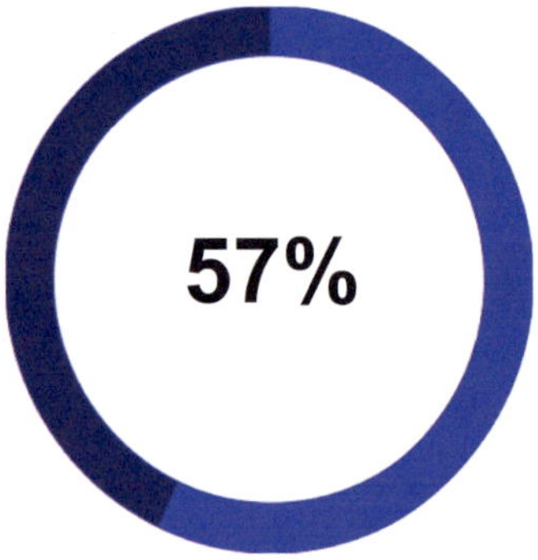

increase in sense of belonging in culturally affirming learning environments

Source: CRT Intervention Study, 2025

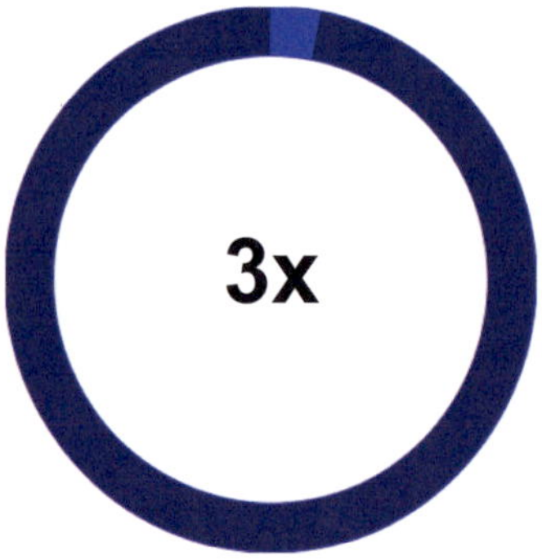

more likely to show civic engagement after completing an ethnic studies curriculum

Source: Child Development Journal / Univ. of Minnesota, 2025

> **Cultural history is not supplemental. It is the foundation upon which every other form of learning becomes more powerful.**

Reframing the Historical Record

The International Academy of Black History presents a comprehensive curriculum that reframes the Black historical experience across three foundational pillars — each designed to restore the full scope and grandeur of a civilizational story that has been systematically compressed by conventional education.

1

The Origin

Before Greece. Before Rome. Africa is the birthplace of humanity, language, spirituality, mathematics, and organized society. This is where the record truly starts.

2

The Civilization

From the Nile Valley to Timbuktu, from Great Zimbabwe to the Moors of Iberia — millennia of city-building, commerce, science, and cultural production that shaped the entire world.

3

The Dominance

Innovation, resistance, and intellectual authority woven across every era and geography — a lineage defined by agency, authorship, and power.

Strategic Purpose of Cultural Intelligence

The International Academy of Black History serves a strategic function that extends far beyond historical education. It provides the cultural and intellectual foundation that makes every other Academy more effective, more grounded, and more relevant to the populations the ecosystem is designed to serve.

Cultural Literacy for Global Citizens

In an interconnected world, cultural literacy is not optional — it is a competitive advantage. Understanding the full scope of human civilizational achievement, including the contributions systematically excluded from mainstream education, produces leaders who can navigate diverse markets, build inclusive organizations, and make decisions informed by the complete historical record rather than a curated subset of it.

Intellectual Grounding for Black Leaders

For Black entrepreneurs, executives, and youth, this Academy provides something that no amount of technical training can replace: a deep, evidence-based understanding of their civilizational lineage. This grounding produces confidence rooted in knowledge, ambition informed by precedent, and leadership shaped by an awareness that they are continuing — not beginning — a tradition of excellence.

Contextual Mastery for Strategy

Business decisions, political strategy, STEM innovation, and institutional design all operate within historical contexts. Leaders who understand those contexts — who understand how power has been accumulated, contested, and distributed throughout human history — make fundamentally better decisions. This Academy ensures that contextual mastery is woven into every leader the ecosystem produces.

Cultural Literacy

A competitive advantage in global markets

Civilizational Grounding

Confidence and ambition rooted in ancestral precedent

Strategic Context

Better decisions through complete historical awareness

This Academy ensures that the ecosystem does not just produce talent — it produces grounded leaders who understand where they come from and how power actually works.

Academy Integration

One system, many paths — a unified architecture enabling seamless progression from learning to competition to recognition.

One System, Many Paths

All four Academies operate within the same sovereign digital architecture — a unified platform that ensures every learner, regardless of which Academy they enter, has access to the full power of the ecosystem. This integration is not a bolt-on feature; it is the foundational design principle that makes the entire system greater than the sum of its parts.

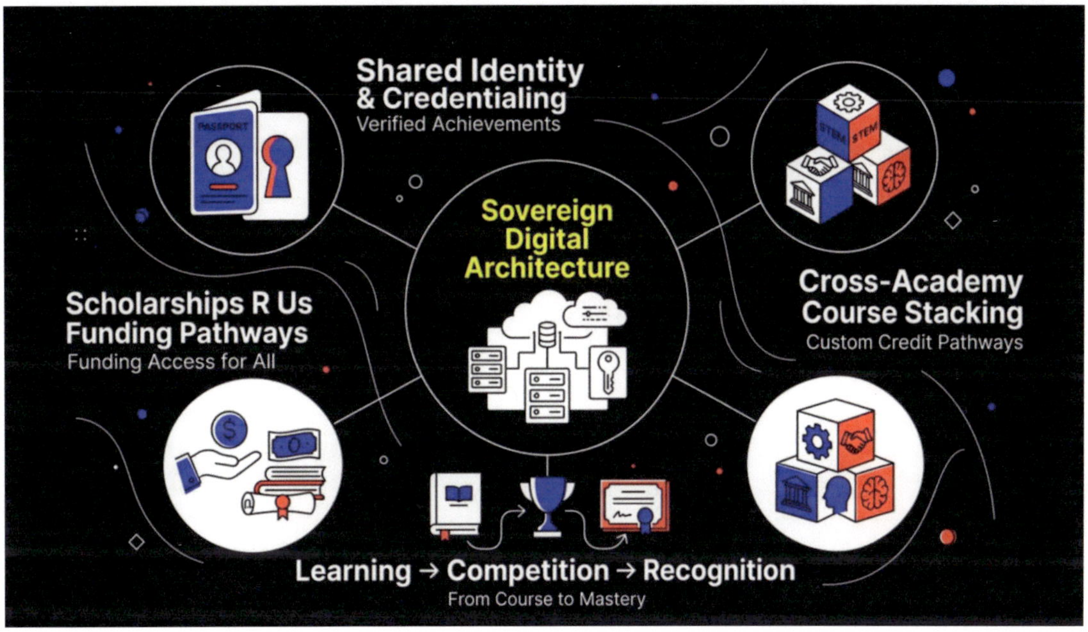

This architecture means that a student studying artificial intelligence in the STEM Academy can simultaneously stack business credentials from the Business Academy, ground their work in civilizational context through the Black History Academy, and compete in championships that validate their applied capabilities — all within a single, seamless platform experience. There are no handoffs, no data silos, and no credential fragmentation.

Education That Compounds

Beyond Graduation

Education does not end at graduation. It compounds.

The traditional model of education operates on a terminal assumption: you learn, you graduate, you stop. The Academy ecosystem rejects this assumption entirely. Learning within this system is designed to compound over time — each credential earned opens new pathways, each championship competed in generates new performance data, and each conference attended expands the network that amplifies every other investment.

The Global Cognitive Passport captures this compounding effect, growing richer and more valuable with every engagement. A graduate who returns five years later to stack new credentials finds their existing record intact, their previous achievements recognized, and new opportunities calibrated to their demonstrated growth trajectory.

This is the difference between an education that expires and an education that appreciates.

Cross-Academy Learning Journeys

The integration architecture enables personalized learning journeys that cross traditional disciplinary boundaries. Below is a representative example of how a single learner might navigate the ecosystem over time — stacking credentials, competing, and building authority across multiple domains.

1

Year 1: Foundation

Enter the STEM Academy with a focus on coding and data science. Complete foundational modules and earn first stackable credentials.

2

Year 2: Competition

Compete in STEM Championships. Performance data populates the Global Cognitive Passport. Begin cross-stacking Business Academy courses in capital structuring.

3

Year 3: Context

Enroll in the Black History Academy to build cultural intelligence and civilizational context. Attend conferences to build investor and partner networks.

4

Year 4: Ownership

Launch a technology venture using Business Academy frameworks. Register entity, file IP, secure Scholarships R Us funding. Credential portfolio now spans four Academies.

5

Year 5+: Compounding

Return as a mentor, investor, and championship judge. Credentials continue accumulating. Education compounds into generational wealth and institutional authority.

Strategic Significance

Why the Academies represent the most strategically significant asset in the entire Goddess Phenezy Holdings portfolio.

Independence from External Dependencies

The Academies ensure that Goddess Phenezy Holdings is not dependent on trends, platforms, or external validation. In an era where organizations rise and fall based on algorithmic favor, platform policy changes, and the shifting winds of cultural attention, the Academies represent something categorically different: permanent intellectual infrastructure that generates value regardless of what happens in the external environment.

While competitors build audiences on rented platforms, the Academy ecosystem operates on proprietary infrastructure. While others chase certification partnerships with third-party institutions, the Academies issue their own credentials backed by verified performance data. While trend-dependent organizations scramble to remain relevant, the Academies address timeless human needs — knowledge, competence, cultural identity, and economic sovereignty — that do not expire with news cycles.

Intellectual Continuity

Knowledge systems that persist across generations, preserved on owned infrastructure and encoded in permanent credentials that travel with learners for life.

Cultural Preservation

Historical truth maintained and transmitted with fidelity, ensuring that civilizational knowledge is never again subject to erasure or distortion by external gatekeepers.

Workforce Sovereignty

Graduates who control their own economic destiny — equipped not just with skills, but with entities, intellectual property, and the networks to deploy them.

Generational Impact

An ecosystem designed to compound over decades, producing leaders whose children and grandchildren inherit not just wealth, but wisdom, credentials, and institutional access.

The Academies by the Numbers

The scope and ambition of the Academy ecosystem is reflected in the strategic metrics that define its design parameters and projected impact across the global education landscape.

4	1	∞	0
Distinct Academies	**Unified Platform**	**Stackable Credentials**	**External Dependencies**
Covering STEM, business, general learning, and cultural intelligence	Proprietary PWA infrastructure powering the entire ecosystem	No ceiling on learning — credentials compound across all Academies	Complete platform sovereignty — no third-party vulnerability

The Competitive Moat

The Academy infrastructure creates a competitive moat that deepens with every learner enrolled, every credential issued, and every championship completed. Unlike product-based businesses that face commoditization, knowledge institutions appreciate in value over time — their alumni networks grow more powerful, their data assets grow more comprehensive, and their brand authority grows more unassailable.

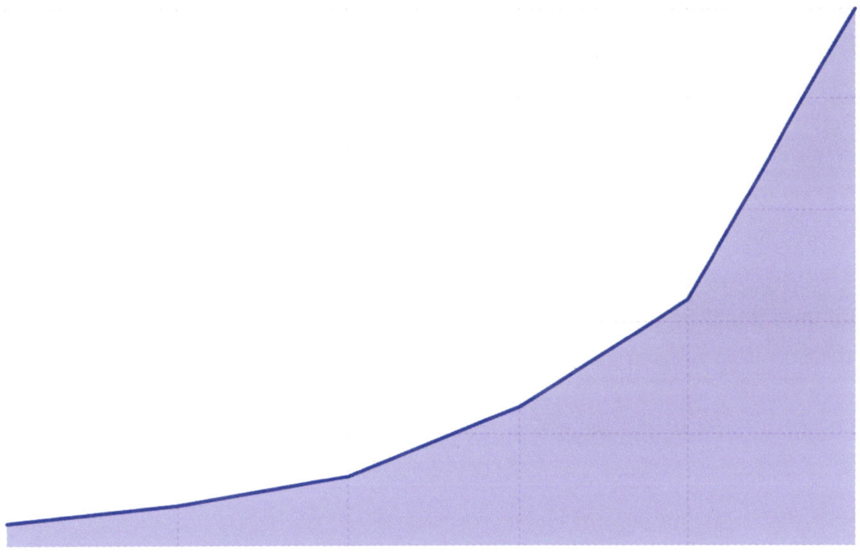

The compounding effect is not linear — it is exponential. As alumni achieve success, they become proof points that attract new learners. As credential data accumulates, it becomes increasingly valuable to employers and investors. As the network grows, every node becomes more valuable. This is the fundamental economic logic of institutional permanence.

What the Academies Train

Each pillar of the Goddess Phenezy Holdings ecosystem trains a different dimension of human capability. Together, they produce complete leaders — physically resilient, intellectually sharp, technically proficient, and civilizationally grounded.

Sports Train the Body

Discipline, resilience, teamwork

Chess Trains the Mind

Strategic thinking, pattern recognition, decision-making under pressure

STEM Trains the Future

Technical capability, innovation capacity, problem-solving mastery

The Academies Train the Civilization

Permanence, sovereignty, generational impact, and institutional authority

> **Sports train the body. Chess trains the mind. STEM trains the future. The Academies train the civilization.**

The Permanent Institution

The Academies are not programs. They are not initiatives. They are civilization-grade institutions — designed to outlast market cycles, political administrations, and technological disruptions. They compound value across generations, producing leaders who are not merely educated but truly prepared.

Intellectual Continuity

Preserving knowledge in an age of distraction

Cultural Preservation

Protecting truth in an era of erasure

Workforce Sovereignty

Owning the platforms, data, and credentials

Generational Impact

Building for leaders not yet born

1 Today

The first students enroll. The first truths are restored. The first generation of civilizationally grounded leaders begins their journey.

2 Next Generation

Graduates become founders, executives, and educators — carrying the knowledge forward and building institutions of their own within the ecosystem.

3 The Legacy

A self-sustaining network of leaders, institutions, and credentials that no political administration, curriculum restriction, or cultural erasure can undo.

4 The Civilization

What began as an Academy becomes a permanent civilizational record — proof that Black excellence was never lost. It was waiting to be restored.

> Not a moment. Not a season. Not a trend. A civilization.

The Academies train the civilization.

THE CULTURAL RECORD

The Creative Jurisdiction

A sovereign ecosystem where art is governed, not harvested. Where masters remain with creators. Where culture is the permanent record of civilization.

Goddess Phenezy Holdings, Inc.

Introduction: The Origin & The Ethos

Every enduring civilization leaves behind a cultural record. Not a marketing archive. Not entertainment ephemera. A record.

The Goddess Phenezy Holdings, Inc. ecosystem was not born in a boardroom or a venture studio. It was forged in 1988 through competitive lyricism, rhythm, and performance—where authority is earned live, under pressure, without permission. From that origin, mastery expanded across instruments, movement, choreography, writing, and performance systems. This matters because ecosystems inherit the architecture of their origin.

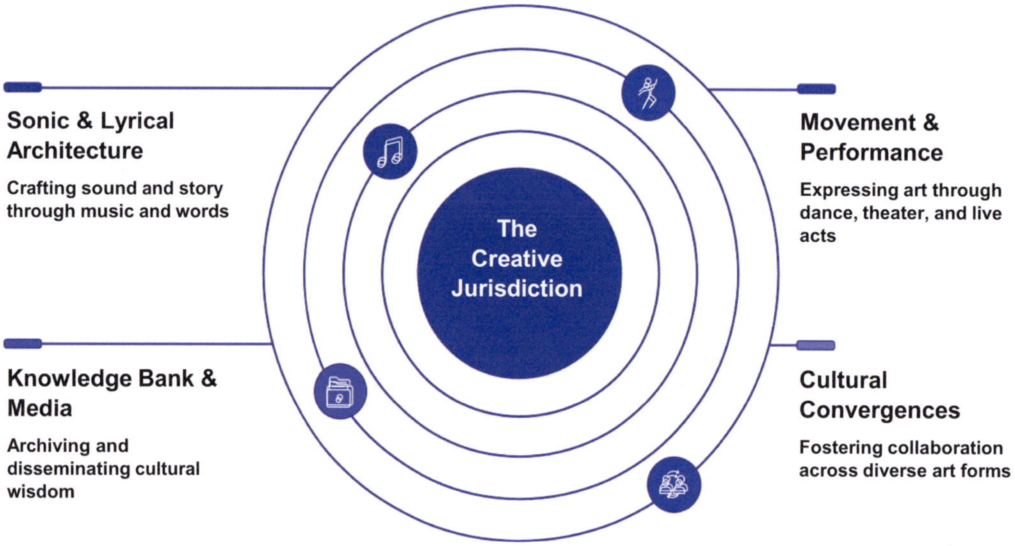

The 1988 Mandate — Art as the Civilizational Record

The origin point was not arbitrary. It was forged in the crucible of competitive performance—a world where credibility is not awarded by committees or algorithms, but earned bar by bar, beat by beat, in front of audiences who demand truth. That foundational DNA now courses through every layer of the ecosystem: from championship circuits to publishing vaults, from digital infrastructure to live convergences that reshape the economies of entire cities.

Sports

Taught discipline. The body as instrument. Relentless training and the pursuit of measurable excellence under pressure.

Technology

Taught scalability. Systems thinking. The ability to architect infrastructure that serves millions without collapsing under weight.

Art

Taught truth under fire. The raw, unfiltered act of creation where the only currency that matters is authenticity and mastery.

The Creative Jurisdiction exists to preserve, protect, and monetize that truth—without exploitation.

The Failure of the Legacy Entertainment Model

The modern entertainment industry is structurally extractive. It does not reward excellence. It harvests it.

For decades, the creative economy has operated under a colonial framework dressed in corporate branding. Record labels strip artists of their masters—the very recordings that define their artistic identity. Publishers absorb intellectual property, converting a creator's life work into corporate portfolio assets. Platforms throttle reach and monetize attention, enriching shareholders while creators scrape by on fractions of pennies per stream.

Performers generate culture but retain no ownership. They are treated as content suppliers inside platforms they do not own, governed by algorithms they cannot challenge, and compensated through opaque revenue splits they did not design. The system was never broken—it was built this way. Built to extract maximum value from creators while returning minimum compensation.

The Extraction Machine

Masters Stripped

Record labels systematically separate artists from ownership of their own recordings. The masters—the definitive versions of an artist's work—become corporate property, leveraged for decades of passive revenue the creator never sees.

IP Absorbed

Publishers convert intellectual property into portfolio assets. Songwriters, lyricists, and composers sign away rights in exchange for advances that function as gilded cages—debt disguised as opportunity.

Reach Throttled

Platforms monetize attention, not creators. Algorithmic gatekeeping determines who gets seen and who gets buried, with zero transparency and zero accountability to the artists who generate the content.

Ownership Denied

Performers generate culture but retain nothing. They are tenants on rented platforms, subject to terms of service that can change overnight, erasing years of audience building in a single algorithm update.

That system ends here.

The Sovereign Solution

Goddess Phenezy Holdings does not host art. We govern the infrastructure where it is created, distributed, licensed, and archived.

This is the fundamental distinction that separates the Creative Jurisdiction from every platform, label, and marketplace that came before. We are not a service provider offering tools on borrowed land. We are the jurisdiction itself—the governing architecture through which sovereign creators operate with full autonomy over their work, their revenue, and their legacy.

100% IP Ownership Retained

Creators operating within the Creative Jurisdiction retain complete ownership of their intellectual property. No transfer clauses. No work-for-hire provisions. No master surrender. The work belongs to the creator. Period.

Full Distribution & Licensing Control

Every creator controls their distribution channels, licensing agreements, and monetization strategies. No intermediary decides who hears your music, reads your work, or experiences your performance.

Owned Ecosystem, Not a Rented Platform

Creators operate inside an ecosystem built for permanence—not a platform that can de-platform you, change its terms, or collapse under the weight of its own venture capital obligations.

> This is not creator empowerment rhetoric. It is jurisdictional control.

The Sonic & Lyrical Architecture

Music, Spoken Word & Competitive Performance

The first and foundational pillar of the Creative Jurisdiction governs the full spectrum of sonic and lyrical expression. This is where it all began—in the heat of competitive performance—and this is where the ecosystem's authority is most deeply rooted.

Scope of Sonic Operations

The Creative Jurisdiction does not recognize genres as organizational boundaries. Genres are marketing constructs imposed by an industry designed to categorize and commodify. What exists within this jurisdiction are systems of expression—each with its own discipline, its own mastery curve, and its own sovereign value as intellectual property.

Multi-Instrumental Composition

Original works spanning keyboards, strings, percussion, wind, and electronic instruments. Every composition is a sovereign asset with licensing value across film, broadcast, gaming, and live performance.

Vocal Performance

From operatic range to raw gospel power to precision rap delivery. Vocal mastery is not a talent—it is a trained discipline with measurable progression and monetizable output.

Spoken Word

The literary tradition delivered live. Poetry, narrative prose, and rhetorical performance that carries the weight of civilization's oral tradition into the digital age.

Competitive Lyricism

The origin discipline. Head-to-head lyrical combat where credibility is forged in real time. No post-production. No second takes. Authority earned under fire.

These are not "genres." They are systems of expression—each governed with the rigor and respect they deserve within the Creative Jurisdiction.

The Championship Circuit

Championships are not events. They are apex validation engines—establishing authority, discoverability, and licensing value under a globally recognized banner.

The championship framework transforms competitive performance from spectacle into infrastructure. Every championship round generates discoverable content, licensable footage, and verifiable credibility that follows the creator permanently. Winners don't just earn titles—they earn elevated positioning within the ecosystem's distribution and licensing hierarchy.

National & International Rap Battle Championships

The flagship circuit. Where lyrical sovereignty is tested and proven on the global stage. Multi-city qualifying rounds feeding into nationally televised finals.

National & International DJ Championships

Turntablism, mixing, and live production mastery. DJs compete not just on technique but on the ability to command a room—a skill that translates directly to licensing and booking value.

National & International Acapella, Singing & Band Championships

Vocal and ensemble excellence across all traditions. From harmony groups to full bands, this circuit validates musical mastery in its most unfiltered forms.

The Sonic Infrastructure

All creators operate through a closed-loop digital pipeline delivered via a proprietary Progressive Web App (PWA) infrastructure. This PWA is accessible on any device—desktop, tablet, or mobile—without any app store dependencies, providing universal access. It supports professional-grade, lossless audio recording up to 96kHz/24-bit directly in the browser, along with a built-in DAW-lite environment for intuitive beat construction, vocal layering, and precise stem separation. Each creator is granted a dedicated, secure artist vault where all recordings are stored with immutable, timestamped metadata, ensuring irrefutable copyright provenance from the moment of creation. This is not a marketplace bolted onto someone else's platform; this is owned infrastructure—built, maintained, and governed within the Creative Jurisdiction.

The Creator Studio

A browser-based recording and production suite, featuring a beat licensing marketplace, extensive stem library access, and real-time collaboration tools for remote sessions.

The Rights Vault

Automated copyright registration, streamlined split-sheet management, and a master ownership ledger that records every transaction on an immutable log for complete transparency and security.

The Distribution Engine

Direct-to-consumer delivery across 190+ countries, DSP-independent syndication, and real-time royalty tracking with same-day payout capability for instant artist compensation.

The pipeline ensures that every creative asset moves from creation to monetization without ever leaving the jurisdiction's governance. No copyright surrender. No platform taxation. No algorithmic throttling. The master stays with the creator. Always.

Creator Capabilities Within the Pipeline

What Creators Can Do

- License beats and compositions directly to buyers worldwide
- Distribute music to consumers without intermediary platforms
- Manage performance rights with full transparency
- Syndicate content globally across broadcast and digital channels
- Track royalties and revenue in real time
- Control pricing, availability, and licensing terms

What Creators Never Surrender

- Copyright ownership remains with the creator
- Master recordings are never transferred
- Distribution rights are never exclusively assigned
- Revenue splits are transparent and creator-favorable
- Platform access cannot be revoked arbitrarily
- Creative control is absolute and non-negotiable

Create

Record, compose, and produce inside the Creator Studio. Lossless audio, beat licensing, real-time collaboration — all browser-based, no installs.

License

Set your own terms. The Rights Vault auto-registers copyright, manages split sheets, and logs every transaction on an immutable ownership ledger.

Distribute

Push to 190+ countries via DSP-independent syndication. No algorithm gatekeeping. No platform tax. Direct-to-consumer on your terms.

Monetize

Real-time royalty tracking with same-day payout capability. Revenue splits are transparent, creator-favorable, and permanently on record.

The Movement & Performance Academies

Dance, Choreography & Physical Expression

Dance has historically been treated as decorative—central to culture but peripheral to compensation. Choreographers create the visual language of an era but watch their movements replicated across social media without attribution, compensation, or ownership recognition. The Creative Jurisdiction corrects this imbalance permanently.

Dance as Athletic Discipline & Cultural Asset

The Creative Jurisdiction governs dance as both an athletic discipline and a monetizable cultural asset. This dual classification is not philosophical—it is structural. It determines how choreography is protected, how dancers are compensated, and how movement is archived as intellectual property with permanent licensing value.

When a choreographer creates a routine within this jurisdiction, that routine is treated with the same legal and commercial seriousness as a musical composition. It is documented, registered, and governed as sovereign IP. When that routine appears in a music video, a commercial, or a live performance, the choreographer participates in the revenue. Not as a footnote. As a rights holder.

Athletic Rigor

Training protocols, progression metrics, and competitive benchmarks that match elite sports standards

IP Protection

Choreography documented and registered as sovereign intellectual property with licensing rights

Monetization

Revenue participation for creators whenever their movement is used commercially

Disciplines Governed

Each discipline within the Movement & Performance Academies is treated with the same rigor as elite sports training. The days of dance being the undercompensated art form are over. Within the Creative Jurisdiction, every discipline has a certification pathway, a competitive circuit, and a monetization infrastructure.

Classical Ballet

Technique, certification & commercial leverage

Competitive Ballroom

Partnership, precision & premium event value

Foundational Hip-Hop

Breaking, popping, locking & cultural IP

Contemporary & Hybrid

Emerging forms defining the next era

4

Core Disciplines

12+

Certification Levels

300+

Affiliated Studios

6

Continental Circuits

The Breakdancing Championship Circuit

From the South Bronx to the Olympic stage — breaking is the most validated street art form in history.

1 **1970s**

DJ Kool Herc's Bronx block parties birth the cipher, the battle, and the break.

2 **1977–83**

Rock Steady Crew formalizes the culture. Breaking spreads globally through film and TV.

3 **1984**

Breakin' and Beat Street bring the art form to mainstream audiences worldwide.

4 **1990s**

Underground preservation deepens the technical vocabulary away from commercial pressure.

5 **2004**

Red Bull BC One World Final establishes international competitive infrastructure.

6 **2024**

Breaking debuts at the Paris Olympic Games. The Creative Jurisdiction is ready.

> **Olympic validation transformed breaking from a cultural practice into a globally recognized competitive discipline. The Creative Jurisdiction's championship circuit is built to capture and compound that momentum across every market where breaking has cultural resonance.**

The Academy SaaS Infrastructure

Choreographers, studios, and academies operate through the Academy SaaS platform—purpose-built infrastructure that transforms dance education from a local, cash-dependent operation into a scalable, globally accessible business with recurring revenue.

1 Digital Masterclasses

High-production instructional content that choreographers create once and monetize indefinitely. Accessible globally, generating passive revenue while the choreographer sleeps.

2 Student Progression Tracking

Data-driven advancement systems that document student growth, validate skill acquisition, and create verifiable credentials recognized across the ecosystem.

3 Certification Pathways

Structured credentialing that transforms informal training into recognized qualifications. Certified instructors command premium rates and institutional partnerships.

4 Recurring Tuition & Memberships

Subscription-based revenue models that provide financial stability. No more feast-or-famine enrollment cycles. Predictable, scalable income.

No marketplace fees. No third-party platforms. No value leakage. Every dollar flows through owned infrastructure directly to the creator.

The Knowledge Bank & Media Syndication

Publishing, Film & Narrative Control

Culture is not only performed. It is recorded. The written word, the visual narrative, the cinematic document—these are the artifacts that survive when stages go dark and platforms collapse. This jurisdiction governs the permanent record of the ecosystem.

The Publishing Vault

500+ ISBNs. Foundational texts. Permanent cultural assets.

The Publishing Vault is not a bookshelf. It is a permanent intellectual property portfolio comprising over 500 ISBN-registered works under the Xperiences R Us imprint. Each ISBN represents a unique, registered work with permanent commercial value—educational texts, strategic frameworks, cultural documentation, and creative works that form the written foundation of the ecosystem.

These ISBNs are not inventory sitting in a warehouse waiting to be sold. They are assets that generate value through multiple channels: direct sales, institutional licensing, curriculum adoption, translation rights, and derivative works. A single foundational text can generate revenue across print, digital, audio, and institutional channels simultaneously—and it does so in perpetuity because the rights never leave the jurisdiction.

500+	∞	4+
ISBN Portfolio	**Revenue Lifespan**	**Revenue Channels**
Registered works generating multi-channel revenue	Perpetual IP ownership means perpetual monetization	Print, digital, audio, and institutional licensing

The Broadcast Network

The media syndication arm of the Creative Jurisdiction delivers content directly to audiences without intermediary gatekeepers. This is not a YouTube channel or a social media presence—it is a sovereign broadcast infrastructure with global syndication capabilities.

Xperience TV

The flagship broadcast channel delivering original programming, championship coverage, documentary content, and cultural commentary to a global audience.

The Goddess Show

The anchor program. Interviews, performances, cultural analysis, and ecosystem updates delivered with the authority and production value the content demands.

The Podcasting Network

Multi-show audio programming covering every vertical of the ecosystem. Long-form conversations, creator spotlights, and strategic discourse.

National & International Film Festivals

Curated cinematic showcases that validate filmmakers, generate distribution deals, and archive the visual narrative of the culture.

Distribution & Narrative Sovereignty

Distribution is direct-to-consumer, multi-language, and globally syndicateable. The infrastructure ensures that content reaches audiences on terms set by the jurisdiction—not by platform algorithms or advertising models that prioritize engagement over quality.

The Infrastructure Guarantees

- Digital publishing without gatekeepers — Authors and creators publish directly through owned channels, retaining full creative and commercial control
- Global broadcast sublicensing — Content is packaged and licensed to international broadcasters, streaming services, and institutional buyers
- Archive permanence — Every work, every broadcast, every performance is archived permanently as part of the cultural record

The Mandate

The narrative is never outsourced.

When you control the publishing, the broadcast, and the archive, you control the story. Not a version of the story filtered through someone else's editorial perspective. The actual story.

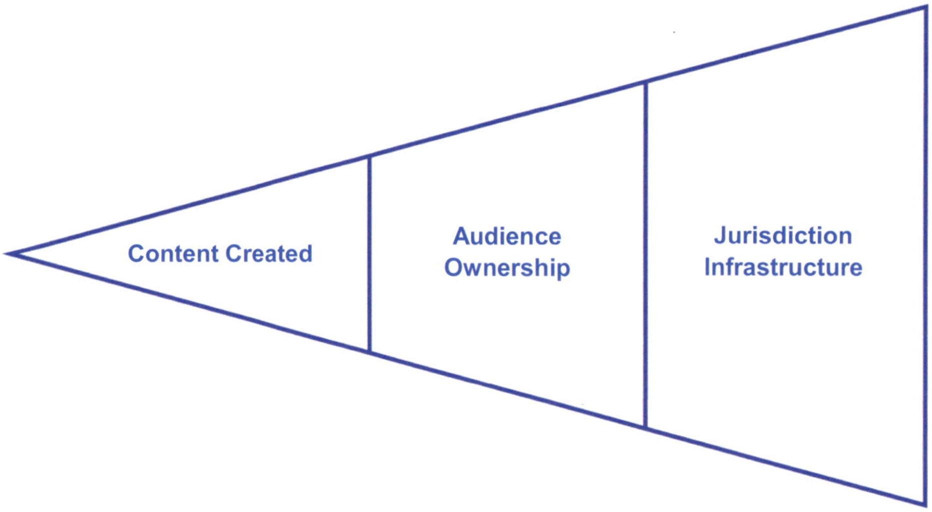

This funnel illustrates the comprehensive process, from content creation to direct audience ownership, ensuring narrative control and global reach through the jurisdiction's dedicated infrastructure.

The Cultural Convergences

Festivals & Live Xperiences

Festivals are not shows. They are cultural stress tests. Multi-day, multi-discipline convergences where the full power of the Creative Jurisdiction manifests in physical space—and where the economic impact reverberates far beyond the stage.

Where Disciplines Collide

The Cultural Convergences are not single-genre events. They are multi-day, multi-discipline gatherings where every pillar of the Creative Jurisdiction activates simultaneously. Music, movement, film, spoken word, and visual identity collide in physical space—creating experiences that cannot be replicated digitally and generating content that feeds the ecosystem for months afterward.

Music

Live performances, championship rounds, and collaborative sessions across every sonic discipline

Movement

Dance showcases, breaking battles, and choreography premieres that debut new work to live audiences

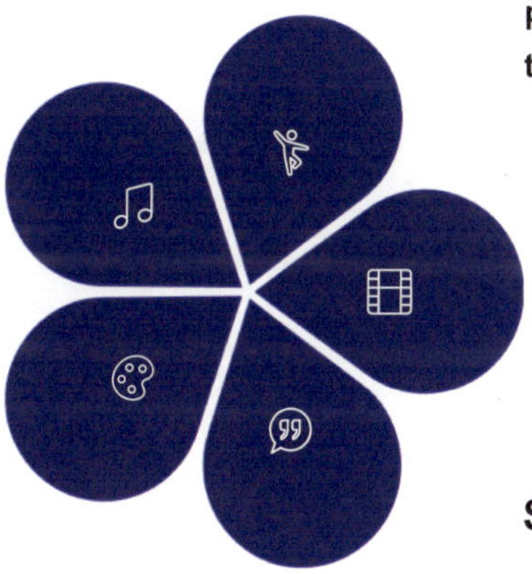

Film

Festival screenings, documentary premieres, and filmmaker panels that drive distribution deals

Visual Identity

Art installations, fashion showcases, and brand activations that define the aesthetic of the culture

Spoken Word

Poetry slams, open mic circuits, and narrative performances that carry the oral tradition forward

Core Events

The convergence calendar is anchored by flagship events that serve as cultural tentpoles—annual gatherings that become destination events for creators, audiences, and industry professionals alike. Each event is designed to generate maximum cultural impact while operating as a high-margin revenue engine.

The National Juneteenth Festival

The flagship cultural convergence. A multi-day celebration that combines performance, education, commerce, and community into the definitive cultural event of the summer calendar. Multi-city activations feed into a national finale.

National Poetry & Open Mic Circuits

Year-round spoken word programming that builds community, discovers talent, and generates a continuous pipeline of content. Qualifying rounds across multiple cities create ongoing engagement.

Multi-City Cultural Showcases

Regional activations that bring the convergence experience to markets across the country. Each showcase is tailored to its host city's cultural identity while maintaining the jurisdiction's brand standards.

3+

Flagship Annual Events

20+

Host Cities Per Year

50K+

Attendees Per Convergence

4x

Revenue Streams Per Event

(tickets, licensing, broadcast, sponsorship)

Host City Economics

Cities don't just host these events. They benefit from them.

The Creative Convergences are engineered to generate measurable economic impact for every host city. This is not incidental—it is by design. When a city partners with the Creative Jurisdiction to host a convergence, it gains access to a turnkey economic activation engine that drives tourism, creates temporary employment, and positions the city as a cultural destination.

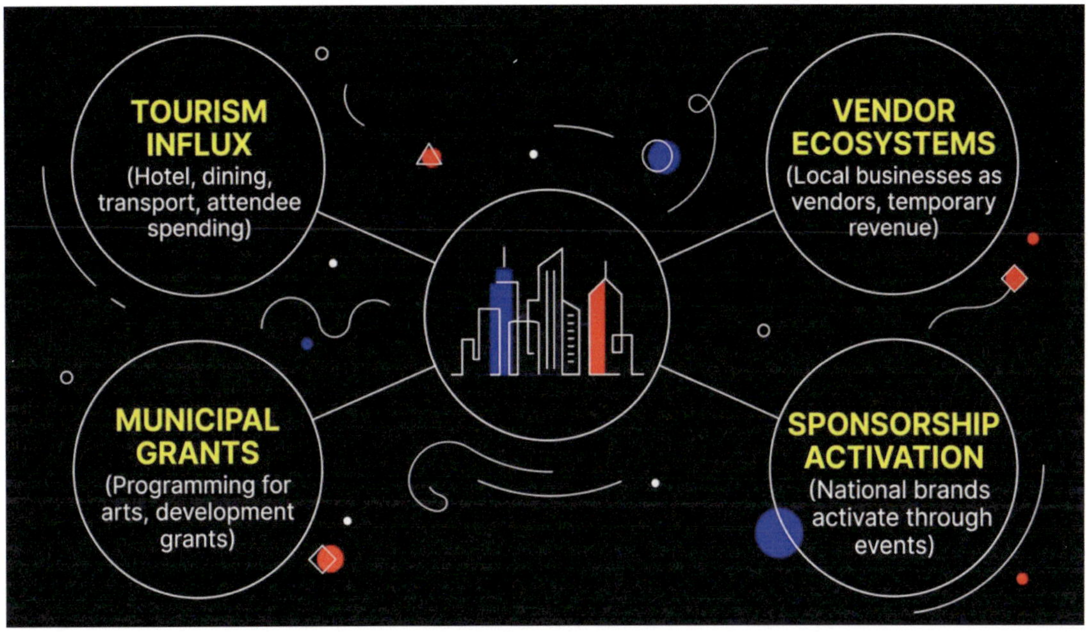

The convergence model transforms cities from passive venue providers into active economic beneficiaries—creating a compelling value proposition for municipal partnerships and public funding.

The Revenue Architecture

Multi-Layer. Recurring. Scalable.

The Creative Jurisdiction is not a single revenue stream dressed up as a business. It is a four-layer revenue architecture designed to generate income from every angle of the ecosystem simultaneously—with each layer reinforcing and amplifying the others.

Layer 1 — IP & Festival Licensing

The foundation layer. Independent promoters license the right to host qualifying rounds under trademarked championship names. This model transforms the Creative Jurisdiction's brand equity into recurring licensing revenue without requiring direct event production for every market.

How It Works

Licensed promoters pay annual fees for the right to use trademarked championship names, formats, and brand assets. They produce qualifying rounds in their local markets according to jurisdiction standards. Winners advance to national and international finals produced directly by the jurisdiction.

This creates a franchise model for cultural events—scalable, quality-controlled, and geographically unlimited. Every new licensed market expands the championship's reach while generating passive licensing revenue.

Revenue Characteristics

- Recurring: Annual licensing fees
- Scalable: Unlimited geographic expansion
- Low overhead: Promoters bear local production costs
- Brand building: Every event reinforces global recognition
- Content generating: Qualifying rounds feed media pipeline

License Granted

Promoter pays annual fee, receives brand rights & format standards

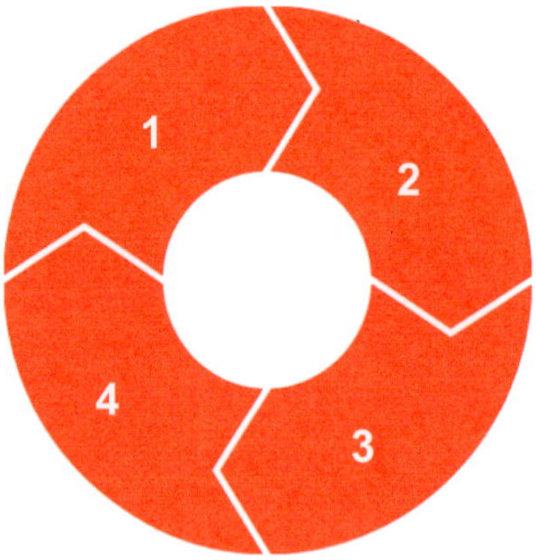

Local Event Produced

Qualifying rounds run in new markets at zero cost to the jurisdiction

Winners Advance

Top competitors feed into national & international finals

Brand Compounds

Every event builds global recognition, driving demand for more licenses

Layer 3 — Live Convergences

High-margin revenue from physical cultural events that cannot be replicated digitally.

Live convergences represent the highest-margin revenue layer—multi-day events that generate income from multiple simultaneous streams. The physical experience creates urgency, premium pricing opportunities, and content that feeds the media syndication layer for months afterward.

Gate Receipts

Tiered ticketing from general admission to VIP packages

VIP Hospitality

Premium experiences commanding 5-10x general admission pricing

Vendor Leasing

Booth and activation space leased to brands and local businesses

Broadcast Rights

Live and post-event media sold to broadcast and streaming buyers

Layer 4 — Media Syndication & Rights

The backend monetization layer that extracts maximum long-term value from every piece of content generated across the ecosystem. This layer transforms events into assets, performances into packages, and data into intelligence—all generating revenue long after the initial moment has passed.

01

Streaming Subscriptions

Direct-to-consumer access to the full archive of performances, championships, interviews, and educational content via subscription streaming platform.

02

Highlight Reels

Curated content packages optimized for social distribution, broadcast licensing, and promotional use—each generating its own revenue stream.

03

Performance Data

Analytics and intelligence packages derived from championship results, audience engagement, and creator progression—valuable to sponsors, scouts, and industry buyers.

04

Licensing Packages

Bundled rights packages for international broadcasters, educational institutions, and commercial buyers seeking authentic cultural content.

The Revenue Architecture — Full View

Each layer reinforces the others, creating a compounding flywheel where content generated in one layer fuels revenue in every other layer. Championships create content. Content drives subscriptions. Subscriptions fund convergences. Convergences generate new championships. The cycle never stops.

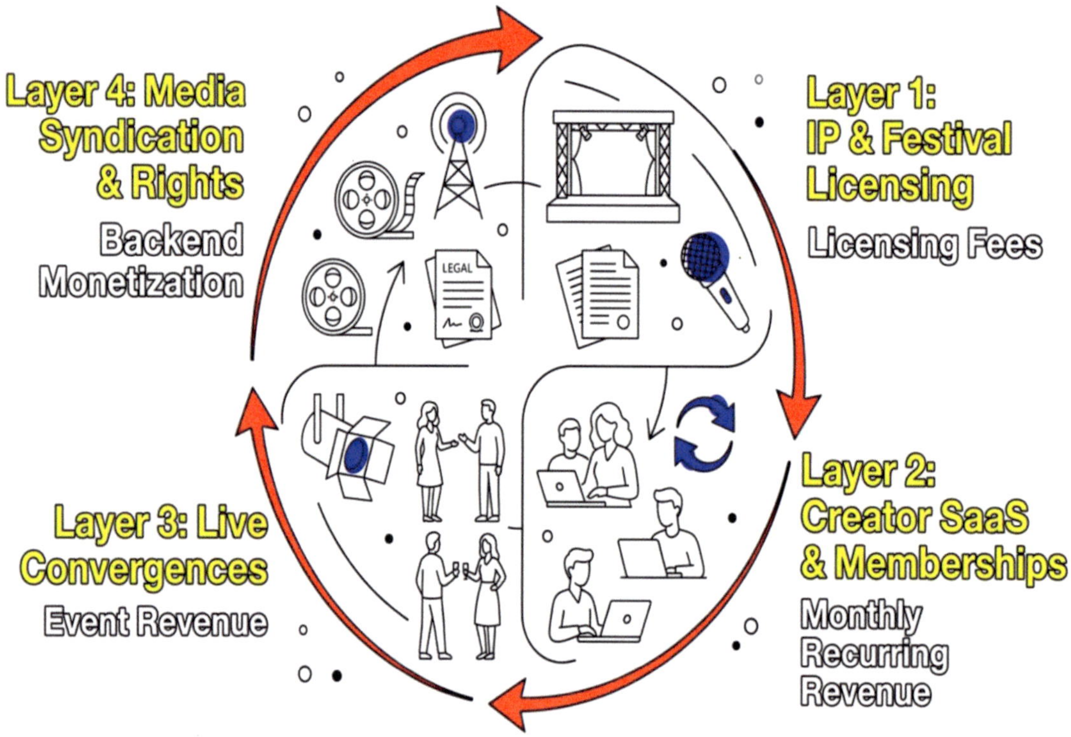

Revenue Scalability Model

The four-layer architecture is designed for exponential growth, not linear scaling. Each new market, each new creator, and each new championship round generates revenue across all four layers simultaneously. The marginal cost of growth decreases as the ecosystem expands, while the marginal revenue per new participant increases as the network effect compounds.

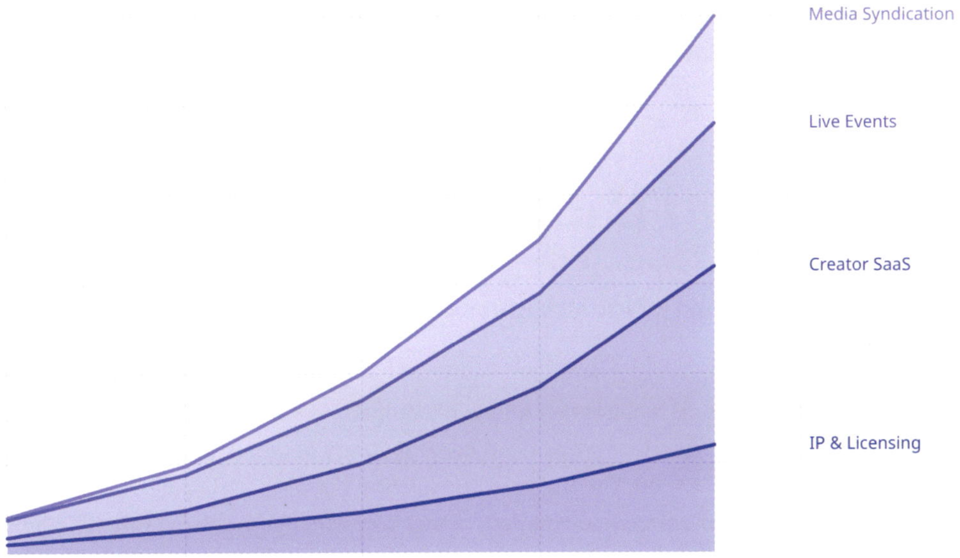

Projected revenue growth ($K) illustrates how SaaS recurring revenue overtakes event revenue by Year 3, while media syndication—the backend layer—accelerates as the content archive deepens. By Year 5, all four layers operate at scale, creating a diversified revenue base that no single market disruption can undermine.

Valuation & Strategic Significance

The Creative Jurisdiction defies single-industry classification. That is precisely what makes it valuable.

The Hybrid Multiplier

Not an event company. Not a media company. Not a SaaS company. All three.

Traditional valuation frameworks fail when applied to the Creative Jurisdiction because they are designed for single-vertical businesses. An event company is valued on EBITDA multiples. A media company is valued on subscriber growth and content library depth. A SaaS company is valued on ARR and net revenue retention. The Creative Jurisdiction operates across all three verticals simultaneously—and the compounding effect of that convergence creates a valuation multiplier that no single-vertical competitor can match.

When a championship event generates content that drives streaming subscriptions, which attract creators who subscribe to the SaaS platform, which generates data that attracts sponsors who fund the next championship event—you are not looking at three separate businesses. You are looking at a unified system where value compounds across every layer with each revolution of the flywheel.

Valuation Comparison: Single vs. Hybrid

The hybrid multiplier effect becomes clear when comparing single-vertical companies to the Creative Jurisdiction's integrated model. Each vertical alone commands strong multiples. Combined, with compounding network effects, the valuation thesis is exponentially more compelling.

Business Type	Typical Multiple	Revenue Type	Creative Jurisdiction Equivalent
Event Company	3-6x EBITDA	Cyclical	Live Convergences + IP Licensing
Media Company	5-10x Revenue	Subscription + Ad	Broadcast Network + Syndication
SaaS Company	10-20x ARR	Recurring	Creator Platform + Academy SaaS
Hybrid (Creative Jurisdiction)	Compounded	All of the above	Unified sovereign ecosystem

Investors evaluating the Creative Jurisdiction are not choosing between multiples—they are investing in the compound effect of all three operating simultaneously within a single governed ecosystem.

The Civilizational Capstone

Goddess Phenezy Holdings governs three domains. Each serves a distinct function in the development and preservation of human excellence. Together, they form a complete civilizational architecture.

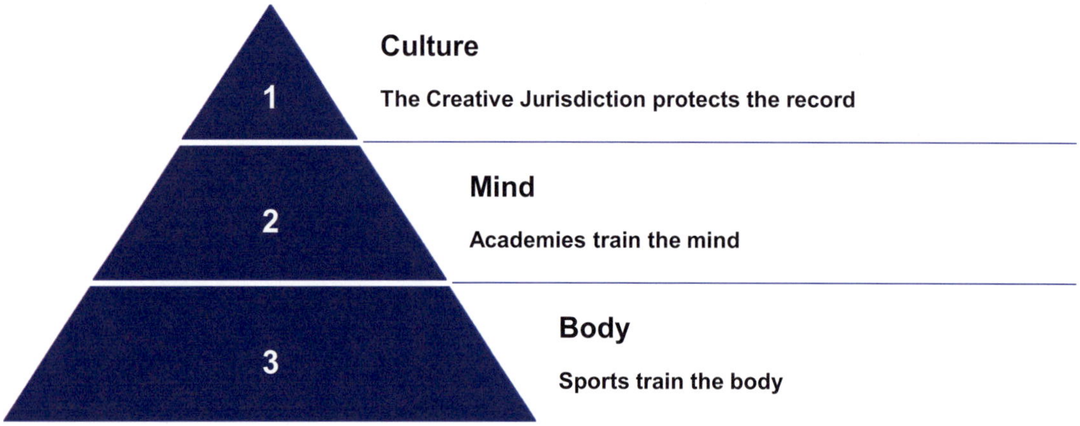

Sports train the body—discipline, endurance, the capacity to perform under pressure. Academies train the mind—knowledge, strategy, the ability to think systemically. The Creative Jurisdiction protects the culture—the artistic, narrative, and expressive record that defines who a people are, what they value, and what they leave behind.

Culture is what survives when platforms collapse and trends expire. It is the permanent layer. The record that future generations will study to understand our era. The Creative Jurisdiction ensures that record is authored, owned, and governed by the people who created it.

Strategic Defensibility

The Creative Jurisdiction's competitive moat is not built on technology alone—technology can be replicated. It is built on a combination of structural advantages that are extraordinarily difficult to reproduce.

Jurisdictional Governance

Owned infrastructure, owned IP framework, owned distribution. Not built on top of someone else's platform.

Network Effects

Every new creator makes the ecosystem more valuable for every existing creator. Compounding, irreversible, and accelerating.

Content Archive Depth

500+ ISBNs, championship footage, performance archives—a library that grows richer every day and cannot be replicated.

Cultural Authority

Forged in 1988. Earned through decades of competitive performance. Authority that cannot be purchased, only built.

Multi-Layer Revenue Lock-In

Creators who use the platform for music, dance, publishing, and events have switching costs that increase exponentially with depth of engagement.

The Ecosystem Map

The Creative Jurisdiction is not a collection of separate businesses. It is a single governed ecosystem where every component feeds every other component. Understanding this interconnection is critical to understanding the valuation thesis and the strategic defensibility of the model.

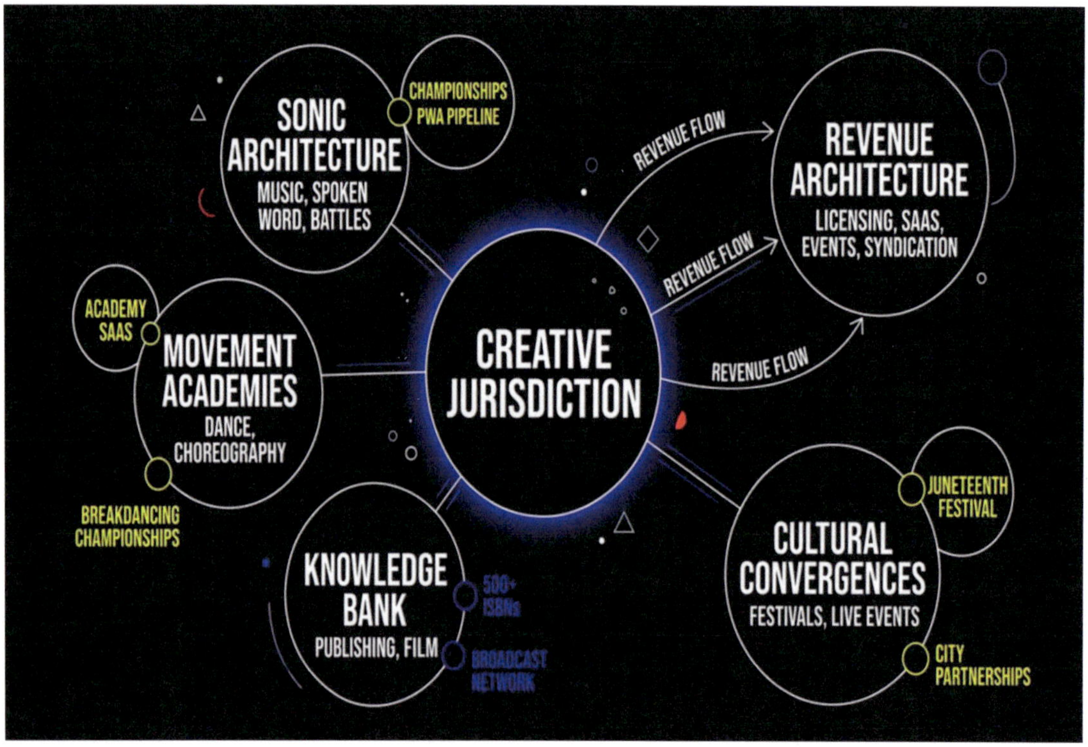

What We Are Not

Clarity of identity requires clarity of distinction. The Creative Jurisdiction is frequently compared to existing models. Every comparison misses the point. Here is why.

Not a Label

Labels acquire IP. We protect it. Labels extract value upward. We distribute it to creators. Labels are intermediaries. We are infrastructure.

Not a Platform

Platforms rent access. We provide ownership. Platforms are governed by algorithms. We are governed by creators. Platforms can de-platform you. We cannot.

Not a Festival Company

Festival companies produce events. We govern jurisdictions. Events end. The infrastructure that produced them endures. Content generated feeds every revenue layer.

Not a Marketplace

Marketplaces take a cut. We take nothing from the creator's IP. Marketplaces compete creators against each other. We build sovereign channels for each one.

The Competitive Landscape

No direct competitor operates across all five pillars simultaneously. The fragmented nature of the competitive landscape is itself the opportunity—and the validation of the integrated model.

The heatmap reveals what fragmented competitors cannot see: no single player covers even half the capability spectrum. The Creative Jurisdiction operates at 100% across every dimension because it was designed as an integrated system—not a collection of features bolted together after the fact.

The Investment Thesis

For investors evaluating the Creative Jurisdiction, the thesis is straightforward but profound.

1 Massive Addressable Market

The global music industry alone exceeds $26 billion. Add live events ($30B+), digital publishing ($15B+), dance education ($5B+), and media syndication—the total addressable market is measured in hundreds of billions. The Creative Jurisdiction addresses all of these simultaneously.

2 Structural Competitive Advantage

The combination of jurisdictional governance, multi-discipline coverage, and four-layer revenue architecture creates a moat that cannot be replicated by single-vertical competitors. The flywheel compounds value with each revolution.

3 Recurring Revenue Engine

Creator SaaS subscriptions provide predictable MRR that funds growth. Licensing fees compound as new markets are activated. Media syndication generates backend revenue from every piece of content ever created. The revenue base diversifies with scale.

4 Cultural Inevitability

The creator economy is moving toward ownership. Every major platform controversy—every algorithm change, every de-platforming scandal—pushes creators toward sovereign infrastructure. The Creative Jurisdiction is not riding a trend. It is the destination the trend is heading toward.

The Creator Economy Shift

The global creator economy is undergoing a structural transformation. The first wave was democratization—giving everyone tools to create. The second wave was monetization—helping creators earn from their work. The third wave, now beginning, is sovereignty—creators demanding ownership of their infrastructure, their data, and their IP.

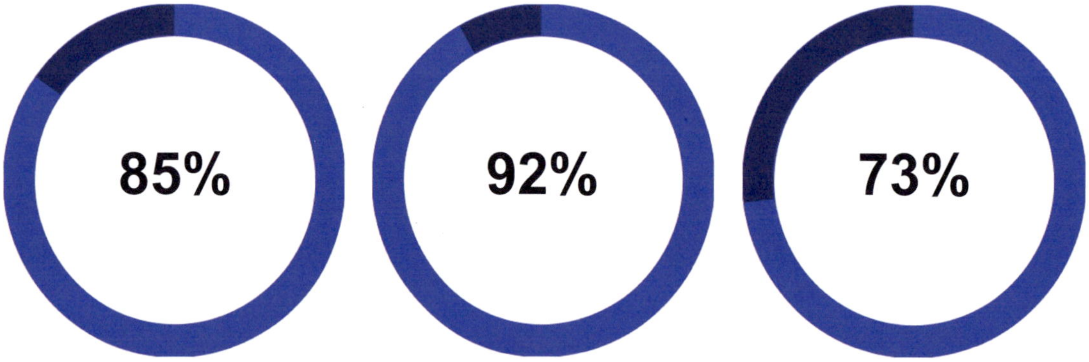

Creator Dissatisfaction

Of independent artists report dissatisfaction with current platform revenue models

Ownership Priority

Of emerging creators rank IP ownership as their top business priority

Platform Distrust

Of creators have experienced algorithmic reach reduction with no explanation

The Creative Jurisdiction is positioned at the exact intersection of this shift. As creators move from rented platforms to owned infrastructure, the jurisdiction's value proposition becomes not just compelling but necessary.

Market Opportunity by Segment

The Creative Jurisdiction's addressable market spans multiple high-growth segments, each representing billions in annual revenue. The integrated model allows the ecosystem to capture value across segments that no single competitor can address.

Combined TAM exceeds $115 billion. Even capturing a fraction of each segment through the integrated model yields a revenue opportunity that dwarfs any single-vertical approach.

The Ownership-First Movement

The Old Model

Create → Surrender → Hope

Artists create work, surrender ownership to intermediaries, and hope for fair compensation through opaque systems they cannot audit or challenge.

The Sovereign Model

Create → Own → Govern

Artists create work within sovereign infrastructure, retain complete ownership, and govern their distribution, licensing, and monetization through transparent systems they control. Revenue flows directly. Data is accessible. The master stays home.

This is not a philosophical preference. It is an economic imperative. The math is clear: creators who own their masters, control their distribution, and operate inside sovereign infrastructure generate 3-7x more lifetime revenue from the same body of work compared to those operating under legacy model contracts.

3-7x

Lifetime revenue under sovereign model vs. legacy contracts

100%

Master ownership retained by creators

0

Intermediaries between creator and consumer

$115B+

Combined addressable market across all five pillars

Technology Infrastructure

The Creative Jurisdiction's technology stack is purpose-built for sovereign creative operations. The proprietary Progressive Web App infrastructure delivers platform-grade experiences without platform-grade dependencies.

1 PWA Architecture

Cross-platform access via Progressive Web App technology. No app store dependencies. No platform gatekeepers. Available on any device with a browser.

2 Creator Dashboard

Unified command center for managing IP, distribution, licensing, revenue tracking, and audience analytics across all creative disciplines.

3 Rights Management

Automated rights tracking, licensing agreement management, and royalty distribution. Transparent, auditable, and creator-controlled.

4 Global Syndication

One-click content distribution to international markets. Multi-language support. Automated formatting for broadcast, streaming, and institutional delivery.

The Championship Model — Deep Dive

Championships are the apex validation engine of the Creative Jurisdiction. They serve four simultaneous functions that no other event format can replicate, making them the most strategically valuable component of the live convergence strategy.

Talent Discovery

Championships surface emerging talent through competitive meritocracy

Content Generation

Every round produces licensable, syndicateable media

Credibility Verification

Competition results create verifiable artistic credentials

Brand Amplification

Global recognition compounds with each championship cycle

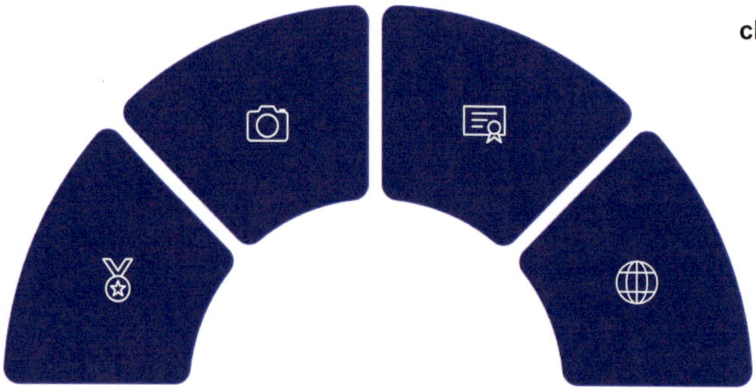

The Publishing Strategy — 500+ ISBNs

The Publishing Vault represents one of the most underappreciated assets in the portfolio. Each ISBN is a permanent, registered intellectual property asset that generates value through multiple channels simultaneously and appreciates over time as the ecosystem's cultural significance grows.

Create

Authors produce educational, strategic, and cultural works within the jurisdiction

Register

Each work receives an ISBN and enters the permanent Publishing Vault

Distribute

Multi-channel distribution across print, digital, audio, and institutional

Monetize

Ongoing revenue from sales, licensing, translations, and derivative works

Archive

Permanent cultural record that appreciates in value as the ecosystem grows

The Road Ahead

The Creative Jurisdiction is not a startup seeking product-market fit. It is a sovereign system seeking scale.

The infrastructure exists. The IP portfolio is built. The championship circuits are established. The publishing vault holds 500+ registered works. The broadcast network is operational. The Academy SaaS platform serves creators across multiple disciplines. What comes next is not invention—it is expansion.

Phase 1 — Foundation (Complete)

Core infrastructure built. IP portfolio established. Championship circuits operational. Publishing vault populated. Broadcast network launched.

Phase 2 — Activation

Multi-city championship expansion. Creator SaaS onboarding at scale. International licensing partnerships. Juneteenth Festival national rollout.

Phase 3 — Acceleration

Global market entry. Institutional partnerships. Cross-discipline convergence events. Media syndication deals with international broadcasters.

Phase 4 — Sovereignty at Scale

Full global operation. Self-sustaining flywheel economics. The Creative Jurisdiction as the definitive infrastructure for sovereign creative operations worldwide.

Key Metrics to Watch

The following metrics define success across each phase of expansion. These are the numbers that validate the model, attract institutional capital, and prove the flywheel thesis.

25%

MRR Growth

Month-over-month growth in Creator SaaS recurring revenue

40%

Creator Retention

Year-over-year retention rate target for active platform creators

60%

Content Monetization

Percentage of archived content generating active licensing revenue

80%

Gross Margin

Target blended gross margin across all four revenue layers

Goddess Phenezy Holdings, Inc.

The governing authority. Est. 1988. The permanent record of civilization.

The Creative Jurisdiction

Sovereign infrastructure governing five pillars: Sonic, Movement, Knowledge, Convergences, Revenue. Built to compound.

The Cultural Record

Every performance, composition, broadcast, and publication permanently archived. The foundation that outlasts every platform.

FINAL DECLARATION

The Permanent Record

> **We do not rent stages. We do not surrender masters. We do not outsource memory.**

Every system in this document—every championship circuit, every SaaS subscription, every ISBN, every broadcast, every convergence—exists to serve a single purpose: the permanent protection and monetization of culture created by sovereign artists.

The entertainment industry spent decades building an extraction machine disguised as opportunity. It told creators they needed gatekeepers to be heard, intermediaries to be paid, and platforms to exist. That era is ending. Not because the platforms failed—but because something better was built.

The Creative Jurisdiction is that something. Owned infrastructure. Sovereign governance. Permanent archives. Revenue that flows to the people who created the value.

Own the Infrastructure

Built, governed, and maintained within the Creative Jurisdiction. No platform dependencies. No external gatekeepers.

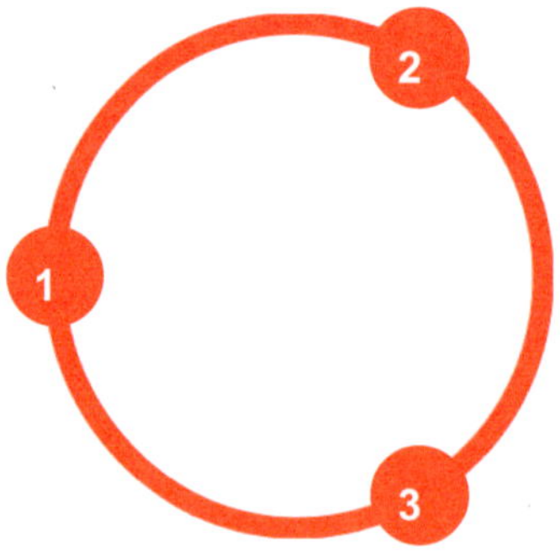

Protect the Master

Every recording, every composition, every performance archived permanently with immutable ownership on record.

Govern the Record

The narrative, the distribution, and the monetization belong to the creator. Always. Without exception.

We Own the Stage.
We Protect the Master.
We Govern the Record.

5

Sovereign Pillars

500+

Registered IP Assets

190+

Countries of Distribution

1988

Year the Record Began

Goddess Phenezy Holdings, Inc.The Creative Jurisdiction — The Cultural Record

For partnership inquiries, licensing opportunities, and investment discussions.

STRATEGIC INVESTMENT DOCUMENT

The Four Revenue Engines

Revenue is not seasonal. It is layered, recurring, and compounding.

Goddess Phenezy Holdings, Inc. has architected a multi-engine revenue architecture spanning SaaS infrastructure, global championships, digital economies, and owned media. This document breaks down the financial mechanics, unit economics, competitive defensibility, and capital discipline that make this ecosystem structurally unreplicable.

The Revenue Architecture

Traditional businesses rely on a single revenue stream—ticket sales, subscriptions, or advertising. Goddess Phenezy Holdings operates four independently profitable engines that compound against each other, creating a financial flywheel that accelerates with every new user, event, and digital transaction.

Each engine is designed to function autonomously. Together, they create an ecosystem where revenue from one engine feeds growth in all others. A single athlete entering a championship generates entry fees (Engine II), subscribes to the SaaS platform (Engine I), purchases digital tokens (Engine III), and creates broadcast content (Engine IV). This is the power of layered monetization.

Engine I

B2B SaaS & Operating Infrastructure

Engine II

Global Championships & Master Licensing

Engine III

Digital & Virtual Economy

Engine IV

Media, Broadcasting & Sponsorship

ENGINE I

B2B SaaS & Operating Infrastructure

Your Partnrr + Your Partnrr Data Room

The foundational revenue engine is built on predictable, high-margin Monthly Recurring Revenue (MRR). Every Ambassador, event director, founder, conference operator, and organization running inside the ecosystem licenses the operating system. This is not event-dependent revenue—it flows whether an event is live or dormant.

The critical insight here is that this engine monetizes operations, not attendance. The SaaS platform is the central nervous system through which all ecosystem participants manage their activities, data, relationships, and transactions. Once embedded in a user's workflow, switching costs become prohibitively high.

Because the platform runs as a Progressive Web App (PWA), there is no Apple or Google app store tax eroding margins. The code is fully proprietary—578,000+ lines built from first principles. This means gross margins sit between 85% and 95%, a figure that would be exceptional even by Silicon Valley SaaS standards. The platform captures value at every layer: user management, event operations, data analytics, and financial transactions.

SaaS Pricing Architecture

The tiered pricing model is designed to capture value across the full spectrum of ecosystem participants—from individual founders testing the waters to enterprise organizations running complex multi-event operations.

Entry Tier	Professional Tier	Enterprise Tier
$99 / month	$299 / month	$499 / month
Individual operators, early-stage founders, and solo Ambassadors getting started in the ecosystem.	Growing teams, active event directors, agencies, and recruiters running multiple operations simultaneously.	Large organizations, conference operators, and institutional partners requiring full platform capabilities.

This engine alone can sustain the company without a single live event. Revenue flows whether championships are active or not—pure infrastructure monetization.

SaaS Unit Economics Deep Dive

The unit economics of Engine I reveal why this is the bedrock of the entire financial model. With a conservative blended Average Revenue Per User (ARPU) of $225/month—accounting for the mix across all three pricing tiers—each user generates $2,700 in annualized revenue. At 85–95% gross margins, virtually every dollar flows to the bottom line after minimal infrastructure costs.

$225	$2.7K	90%
Blended Monthly ARPU	**Annualized ARPU**	**Gross Margin**
Conservative average across all tiers	Per active subscriber	PWA-based, no app store fees

Per-User Revenue Breakdown

This bar chart visually represents how the $2,700 in Annual ARPU for a single user flows through to approximately $2,430 in gross profit, with only $270 allocated to infrastructure costs, making the 90% margin tangible.

The scaling dynamics are what separate this from typical SaaS businesses. Because the platform is deeply embedded in operational workflows—not just a nice-to-have tool—churn rates remain structurally low. Users don't just consume the product; they run their businesses on it. This creates the kind of sticky, compounding MRR that drives premium SaaS valuations.

SaaS Revenue at Scale

The following projections illustrate how the SaaS engine scales with user acquisition. These are not market-share assumptions—they are direct calculations from the unit economics already established. Each milestone represents a achievable target given the ecosystem's built-in distribution mechanics through the Ambassador network and championship pipeline.

At 50,000 active SaaS users, this single engine generates $135M in Annual Recurring Revenue. For context, that figure alone would place the company among the top-performing vertical SaaS platforms in the world. The pathway from 10,000 to 50,000 users is not speculative—it's a direct function of championship expansion, Ambassador licensing, and ecosystem network effects that are already being activated.

ENGINE II

Global Championships & Master Licensing

Physical + Hybrid Activations

Engine II is built on massive ticket volume with decentralized execution. This is not a traditional events company that absorbs all venue risk, staffing costs, and logistical overhead. Instead, the Holding Company has architected a licensing model where local and international Ambassadors pay for the right to operate championships under the Xperiences R Us brand.

The genius of this structure is in what the Holding Company retains versus what it delegates. Ambassadors absorb venue risk and local logistics—the two highest-cost, highest-variability elements of any physical event. Meanwhile, the Holding Company retains IP ownership, technology control, payment rails, and media rights—the four highest-margin, most defensible assets in the value chain.

Championship Revenue Streams

Each championship activation generates revenue through multiple channels simultaneously. This is not single-stream economics—it is a multi-layered capture model where every participant, sponsor, and spectator contributes to the revenue stack.

Master Licensing Fees

Ambassadors pay upfront licensing fees for the right to operate branded championships in their jurisdiction. This creates immediate revenue before a single ticket is sold.

Ticketing Royalties

Every ticket sold flows through the proprietary payment rails, generating royalty revenue on every transaction regardless of event location.

Merchandise Splits

Branded merchandise generates ongoing revenue with minimal overhead, as production and fulfillment are handled through the decentralized Ambassador network.

Sponsorship Participation Fees

Enterprise sponsors pay premium access to championship audiences, with participation fees flowing through centralized sponsorship packages.

Championship Unit Economics

The per-event economics demonstrate why this model scales so efficiently. Each regional championship activation operates as a self-contained profit center, with the Holding Company capturing its share through royalties, SaaS fees, and media rights—not through operational exposure.

Per Championship Activation

Metric	Range
Participants per event	300–1,000
Average entry fee	$75–$250
Gross intake per event	$50K–$250K
Holding Company capture	15–30%
Average HC share per event	~$40K

Annualized Projection

200 regional activations per year

$40K average Holding Company share per event

= $8M annually from ONE championship vertical

Now multiply across Basketball, Soccer, Boxing, Golf, Pickleball, and Poker.

This is not one event. This is a repeatable licensing factory. With six championship verticals operating simultaneously, each generating $8M+ annually, the combined championship engine becomes a formidable revenue contributor—all while the Holding Company carries minimal operational risk.

Championship Verticals Multiplier

The true power of Engine II becomes visible when you see the multiplication effect across all championship verticals. Each vertical operates independently, with its own Ambassador network, sponsorship relationships, and participant base—but all share the same technology infrastructure, payment rails, and media pipeline.

Six championship verticals operate simultaneously, each with its own Ambassador network, sponsorship relationships, and participant base—while sharing the same technology infrastructure, payment rails, and media pipeline. This shared-infrastructure model means each new vertical added drives incremental revenue with near-zero marginal overhead.

Basketball

Flagship vertical with the deepest participant pool and highest brand recognition across domestic and international markets.

Soccer

Global reach with built-in international demand. The world's most popular sport drives instant Ambassador interest worldwide.

Boxing

Premium sponsorship value with high per-event revenue and strong media monetization through pay-per-view mechanics.

Golf

Attracts high-net-worth participants and premium enterprise sponsors, with the highest per-capita spending of any vertical.

Pickleball

Fastest-growing sport in America with explosive demographics and untapped championship infrastructure demand.

Poker

Digital-native audience with seamless crossover into virtual platforms, driving Engine III digital economy revenue.

ENGINE III

The Digital & Virtual Economy

Zero Overhead. Infinite Scale.

Engine III is the highest-margin engine in the portfolio—90–98% gross margins on every transaction—because everything runs inside the proprietary PWA with no app store fees, no third-party processors, and zero venue costs. A user who enters through Basketball can be monetized through Career Fairs, Gaming, and Token purchases within the same session.

Career Fairs

Employers pay $2,500–$15,000 per activation for access to a pre-screened, high-intent talent pool.

Gaming Championships

The gaming vertical alone is projected to reach $40M+ annually by Year 3.

Digital Tokens

Estimated average token spend: $85/user/year at launch, scaling to $200+ as the ecosystem matures.

VIP Access Tiers

VIP conversion rate among active users is projected at 18–25%.

Enterprise Sponsorships

Brands pay $25,000–$250,000 per digital activation for verified audience access, in-platform advertising, and talent pipeline integration.

Digital Economy Revenue Sources

1

Digital Tokens

In-platform currency driving engagement, transactions, and ecosystem participation across all verticals.

2

VIP Upgrades

Premium access tiers with enhanced visibility, analytics, and priority positioning within the ecosystem.

3

Virtual Career Fairs

Enterprise-sponsored digital hiring events connecting verified talent with global employers.

4

Gaming Tournaments

Competitive gaming entries across multiple formats, fully monetized through the PWA infrastructure.

5

Chess & Poker Platforms

Strategy-based competitive platforms with entry fees, ranking systems, and prize pool participation.

6

Premium Analytics

Data visibility boosts and analytics packages for participants seeking competitive advantages.

Digital Economy Unit Economics & Scale

The unit economics of Engine III are what make it the highest-margin, fastest-scaling engine in the portfolio. Average digital spend per engaged user ranges from $120 to $300 annually, with gross margins between 90% and 98%. There are no venues, no shipping, and no platform fees eroding the margin structure.

$200	95%	$0
Avg Annual Spend	**Gross Margin**	**Physical Overhead**
Per engaged digital user	Near-total on digital transactions	No venues, shipping, or platform fees

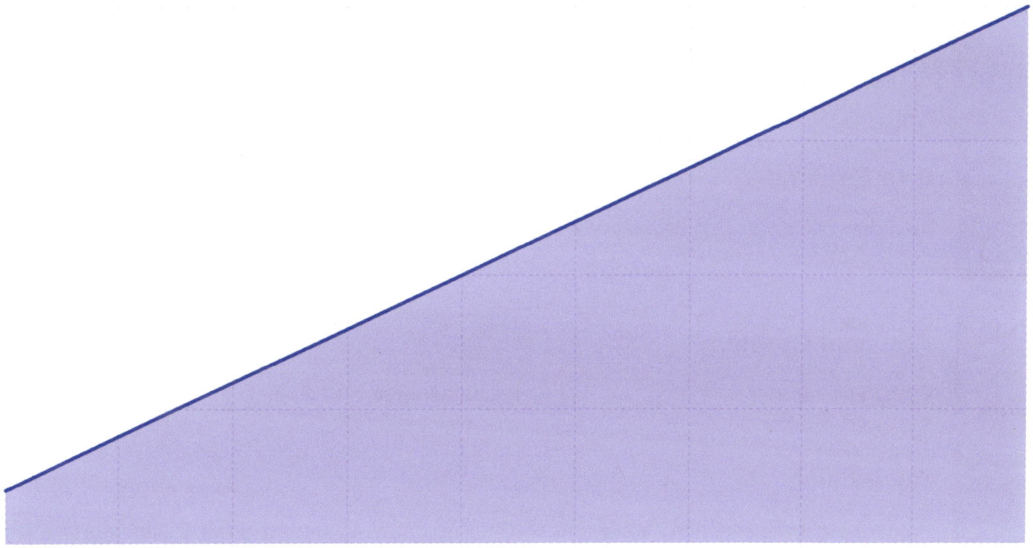

This engine scales faster than headcount, faster than venues, and faster than competitors can react. At 1 million active users—a fraction of the addressable market across sports, gaming, careers, and education—Engine III alone generates $200M in annual revenue.

Engine IV-Media, Broadcasting & Sponsorship

Xperience TV — Events Generate Revenue TwiceThe fourth engine transforms every physical and virtual event into a perpetual media asset. The fundamental insight is elegant: events generate revenue at the door, then generate revenue again—potentially hundreds of times—on the screen. Legacy companies sell tickets. Goddess Phenezy Holdings sells perpetual visibility.

Xperience TV is not a distribution channel bolted onto an events business. It is a fully owned media property with direct audience access, first-party data, and evergreen content pipelines. This means no dependency on YouTube's algorithm, no ESPN gatekeeping, and no social media platforms that can throttle reach overnight. The company owns its attention infrastructure the same way it owns its technology infrastructure.

The monetization surface area is vast: live streams during championship events drive real-time advertising revenue, archived content generates long-tail views for months and years after the original event, enterprise sponsorships provide premium recurring revenue, and subscription access creates yet another layer of recurring income. Every piece of content produced inside the ecosystem feeds back into the media engine, creating a self-reinforcing content flywheel.

Xperience TV Revenue Streams

Live Streams

Real-time championship broadcasts with integrated advertising and sponsorship overlays

Archived Content

Evergreen library generating long-tail views, ad revenue, and licensing income

Advertising

Premium ad inventory across live and on-demand content within verified audiences

Enterprise Sponsorships

Multi-year sponsorship packages ranging from $250K to $2M annually

Subscription Access

Premium subscriber tiers for exclusive content, behind-the-scenes, and VIP streams

Media Unit Economics

The media engine's economics are compelling because they compound over time. Unlike event revenue, which is transactional, media content appreciates in value as the library grows and audience builds. Each new championship adds to the content library, which attracts more viewers, which increases sponsorship value, which funds more production—a classic flywheel effect.

Revenue Benchmarks

Revenue Type	Range
Sponsored Championship Packages	$50K–$500K
Enterprise Annual Sponsorships	$250K–$2M
Media Gross Margins	70–85%

The Compounding Effect

Each live event simultaneously:

- Generates ticket revenue
- Creates broadcast content
- Produces sponsorship inventory
- Builds long-tail viewership

One event monetizes once at the door and hundreds of times on screen.

The 10×10 Economic Matrices

Distributed Growth Engine

Engine V is the hidden accelerant that makes all other engines scale without burning cash. The 10×10 Economic Matrices shift marketing cost from cash to incentives, driving Customer Acquisition Cost (CAC) toward zero while making growth self-propelled through community-driven monetization.

The mechanics are straightforward but powerful: community members recruit users, train their networks, and drive upgrades—and the company pays commissions only when revenue occurs. This is not a speculative marketing budget. It is a pay-for-performance growth engine where every dollar spent directly correlates to revenue generated.

This single structural decision replaces paid advertising, marketing agencies, influencer spend, and traditional sales teams. The result is a growth model that compounds organically, where every new participant becomes a potential distribution node for the entire ecosystem.

Revenue Flow: From Action to Compounding Returns

Every user action within the ecosystem generates both immediate and downstream revenue. This dual-revenue characteristic is what separates Goddess Phenezy Holdings from traditional single-monetization businesses. The following table illustrates how each entry point creates cascading value.

Action	Immediate Revenue	Downstream Revenue
Athlete joins	Entry fee	Tokens, VIP, media content
Athlete advances	Sponsorship deals	Career placement, NIL revenue
Ambassador licenses	SaaS subscription	Royalties, event revenue share
Viewer watches	Advertising revenue	Subscription conversion
Founder upgrades	SaaS tier increase	Matrix earnings, network growth

> The 10×10 Matrix is why the model scales without burning cash. Marketing cost shifts from cash to incentives, and CAC trends toward zero as community-driven growth becomes self-sustaining.

Why the Financial Model Holds Under Diligence

Sophisticated investors will immediately stress-test any revenue model for single points of failure, platform dependencies, and concentration risk. The Goddess Phenezy Holdings architecture has been designed from first principles to eliminate every common vulnerability.

No App Store Dependence

The PWA architecture means zero exposure to Apple or Google's fee structures, policy changes, or approval processes. Revenue flows through owned rails.

No Advertising Dependence

Revenue is generated through direct transactions—subscriptions, licensing fees, entry fees, and sponsorships. No algorithm can throttle the business.

No Single Revenue Choke Point

Four independently profitable engines ensure that stress in any one vertical cannot collapse the business. Diversification is structural, not aspirational.

IP + Software + Media = Control

Ownership across intellectual property, technology infrastructure, and media distribution creates a vertically integrated control stack.

Decentralized Execution Limits Risk

Ambassador model pushes operational risk to local operators while the Holding Company retains high-margin governance, technology, and IP positions.

This is infrastructure economics, not event economics.

YEAR 1

Financial Projections

These projections are asset-activated, not market-share assumptions.

The financial trajectory of Goddess Phenezy Holdings is not built on speculative TAM calculations or hoped-for market penetration. Every projection is derived from the unit economics already detailed—known pricing tiers, established per-event economics, measurable digital transaction values, and contracted sponsorship ranges. The three phases represent a deliberate scaling strategy: ignition, expansion, and ecosystem lock-in.

$180M to $275M — Ignition Phase

The first year is about activating the primary revenue engines simultaneously. Basketball, Soccer, Poker, and Career Fairs launch as the initial championship verticals, each with established participant demand and proven event formats. The SaaS platform deploys immediately, converting every Ambassador and operator into a paying subscriber from day one.

Core Championships

Basketball, Soccer, Poker, and Career Fairs activate with immediate ticket revenue and participant fees across multiple regions.

SaaS Deployment

Your Partnrr platform launches with all pricing tiers, converting ecosystem participants into MRR subscribers.

Digital Activations

Digital tokens, VIP upgrades, and virtual event access create the initial digital economy revenue stream.

Ambassador Licensing

Early Ambassador licensing fees and regional partnerships establish the decentralized execution framework.

Year 1 Revenue Mix

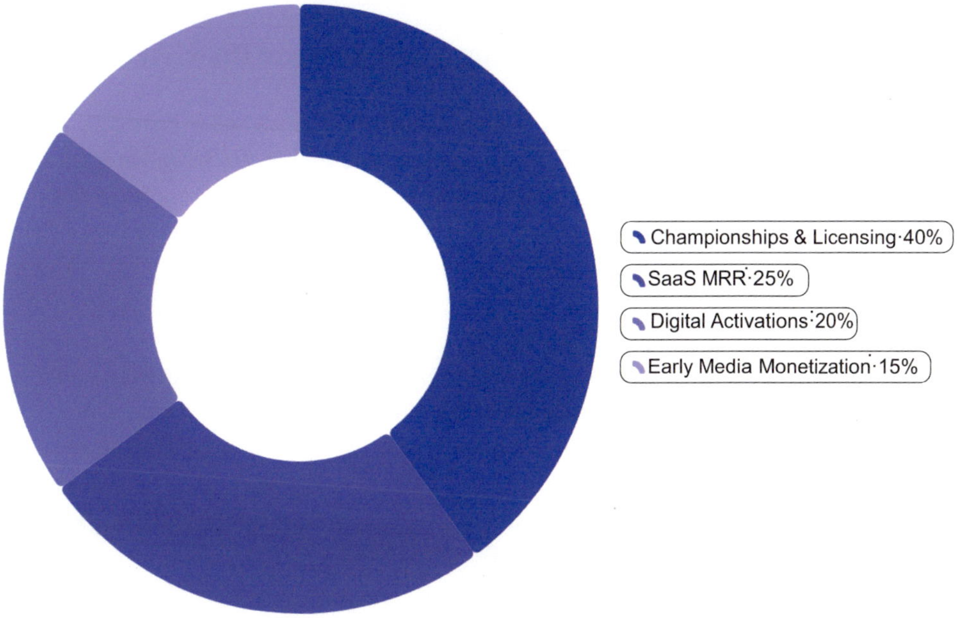

In Year 1, Championships and Licensing drive the largest share of revenue at 40%, reflecting the immediate monetization power of live events and Ambassador licensing. SaaS MRR contributes 25% as the user base builds rapidly through mandatory platform adoption. Digital Activations at 20% demonstrate early traction in the virtual economy, while Early Media Monetization at 15% establishes Xperience TV's foundation for explosive Year 3 growth.

YEAR 3

$420M to $620M — Expansion Phase

This is where infrastructure begins compounding faster than events.

By Year 3, three new championship verticals — Boxing, Golf, and Pickleball — have fully activated, adding an estimated 180,000 net new participants to the ecosystem. Each vertical brings its own Ambassador network (averaging 40–60 licensed operators per sport), its own sponsorship surface area, and its own SaaS subscriber base. The result: championship revenue alone reaches $210M–$280M, driven by 1,200+ regional events annually across all six active verticals.

International Master Licensing becomes a material revenue driver in Year 3. With licensing agreements active in Canada, the UK, the Caribbean, and West Africa, global Ambassadors contribute an estimated $45M–$65M in royalty and licensing fees — revenue that requires zero incremental infrastructure investment from the Holding Company. Each international market operates on the same economic template: local execution, centralized technology, and a fixed royalty flowing back to the parent.

The critical Year 3 inflection is Xperience TV's transformation from a supporting function into a standalone media property. With 24+ months of archived championship content, a subscriber base exceeding 400,000, and enterprise sponsorship packages averaging $180,000 per activation, the media engine generates $85M–$120M independently. For the first time, media revenue rivals championship revenue on a per-vertical basis — and unlike events, it compounds without additional execution cost.

1,200+	180K+	$85M–$120M	4
Annual Championship Events	New Ecosystem Participants	Xperience TV Revenue	International Markets Active

The SaaS engine reflects all of this growth automatically. Every new Ambassador, operator, international licensee, and participant who enters the ecosystem in Year 3 becomes a mandatory platform subscriber. With blended ARPU holding at $225/month and churn structurally suppressed by operational dependency, the SaaS engine alone contributes $130M–$160M — making it the single most consistent revenue line in the portfolio.

Year 5 - $750M to $1B+ — Ecosystem Lock-In Phase

Year 5 is not a projection — it is the mathematical outcome of compounding infrastructure. By this point, the ecosystem spans 10+ championship verticals, 35+ international markets under Master Licensing, and a SaaS subscriber base exceeding 850,000 active users. Total annual events surpass 3,000 globally, with the Holding Company capturing royalties, platform fees, and media rights on every single one — without operating a single venue directly.

Athletes

500,000+ verified athlete profiles generating $225/month average SaaS revenue. Each profile is a data asset — career history, performance metrics, and advancement records that no competitor can replicate.

Recruiters

Fortune 500 and mid-market employers pay $25,000–$150,000 per Career Fair activation to access the verified talent pool. 400+ enterprise accounts active by Year 5, generating $60M+ in B2B recruitment revenue annually.

Founders

18–25% of active athletes convert to enterprise SaaS tiers ($499–$999/month), launching their own training programs, academies, or event businesses inside the ecosystem.

Operators

12,000+ licensed Ambassadors across 35+ countries running events on the platform. Each Ambassador pays licensing fees, SaaS subscriptions, and royalties — generating $28,000–$85,000 in annual Holding Company revenue per operator.

850K+	3,000+	35+	$60M+
Active SaaS Subscribers	Annual Global Events	International Markets	B2B Recruitment Revenue

> **By Year 5, every new user who enters the ecosystem — regardless of entry point — generates revenue across at least three engines simultaneously. This is not growth. This is compounding.**

Five-Year Revenue Trajectory

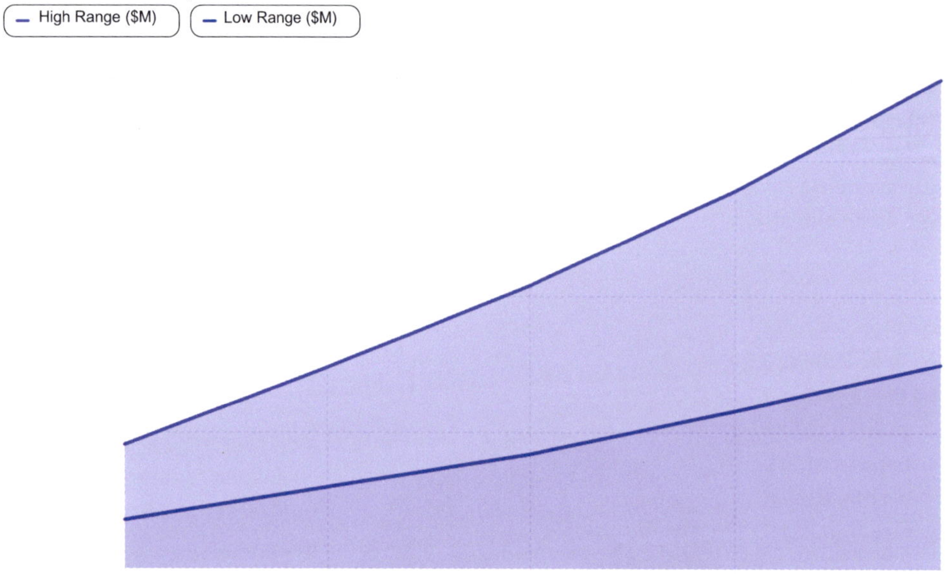

The revenue trajectory shows consistent acceleration across both the conservative low range and the aggressive high range. The widening gap between low and high estimates reflects the compounding nature of the business—as more engines activate and network effects compound, the upside scenario grows disproportionately. By Year 5, mature SaaS MRR at scale combined with media syndication and global sponsorship dominance drives the ecosystem past the billion-dollar threshold.

Capital Allocation & Discipline

Discipline is not a constraint. It is the strategy.

Most companies at this scale of ambition raise capital first and build discipline later — if ever. Goddess Phenezy Holdings operates in reverse. The capital allocation framework was codified before the first dollar of external investment was sought, establishing fixed percentage commitments across four categories that cannot be overridden by short-term performance pressure or investor preference. This is not a budget. It is a constitutional document for how the company grows.

30% — KJ Doctrine Impact

Permanently allocated to scholarships, grants, career advancement, and talent development. Non-negotiable, non-adjustable, and embedded in the operating agreement. This is the company's social contract with the communities it serves.

40% — Ecosystem Reinvestment

Directed back into technology infrastructure, Ambassador expansion, new vertical activation, and international market development. Every dollar reinvested here multiplies across all four engines simultaneously.

20% — Reserve & Resilience

Held in operational reserve to ensure the company can weather macro disruptions, regulatory shifts, or event-cycle volatility without touching growth capital or compromising the KJ Doctrine commitment.

10% — Strategic Acquisitions

Reserved for opportunistic acquisitions of complementary technology, media rights, or regional event infrastructure that accelerates ecosystem expansion without diluting the core model.

The result is a company that cannot be pressured into short-term extraction. Investors who enter this ecosystem are investing in a compounding architecture — not a quarterly earnings story. The allocation framework ensures that growth, impact, and resilience are funded simultaneously, in every cycle, regardless of external conditions.

The KJ Doctrine — 30% Impact Allocation

Thirty percent of net profits from championships are permanently allocated to scholarships, grants, career advancement, and talent development. This is not a CSR initiative that can be adjusted based on quarterly performance. It is a structural commitment embedded in the financial architecture of the company.

30%

of net championship profits permanently allocated to impact

- **Scholarships**

  Direct educational funding for ecosystem participants and their families

- **Grants**

  Seed capital for community entrepreneurs building within the ecosystem

- **Career Advancement**

  Professional development and placement resources for participants

- **Talent Development**

  Training programs that create the next generation of operators and Ambassadors

This is not philanthropy. It is institutional trust, brand permanence, and generational loyalty. The KJ Doctrine creates a structural reason for communities to build with the company—not against it.

Funding Status

Goddess Phenezy Holdings is one of the rarest entities in the investment landscape: a company targeting $750M–$1B+ in five-year revenue that requires zero external capital to execute. This is not a gap in the fundraising strategy. It is the strategy. The architecture was deliberately engineered so that growth is funded by the ecosystem itself — not by investor capital, not by debt, and not by dilution.

Ambassador Capital

Every licensed Ambassador pays an upfront licensing fee of $5,000–$25,000 plus ongoing royalties and SaaS subscriptions. With 500+ Ambassadors active in Year 1 alone, the network self-funds its own expansion. Operators absorb venue costs, local marketing, and event logistics — the Holding Company captures revenue without deploying capital.

Day-One SaaS Revenue

The SaaS engine generates recurring revenue from the first subscriber. With no inventory, no fulfillment, and no physical infrastructure to maintain, every dollar of SaaS revenue is immediately available for reinvestment. At 500 subscribers, MRR exceeds $112,500. At 5,000, it exceeds $1.1M — before a single championship event is held.

Zero Marginal Cost Scaling

Digital transactions, token sales, VIP upgrades, and Career Fair activations carry 90–98% gross margins with no incremental cost per user. As the ecosystem grows, revenue scales geometrically while operating costs grow linearly — creating an ever-widening gap between revenue and expense.

$0

External Capital Required

500+

Year 1 Ambassadors Self-Funding Expansion

90–98%

Digital Gross Margins

For strategic partners evaluating this opportunity: there is no cap table complexity, no competing investor interests, and no dilution risk. Any future partnership enters from a position of the company's choosing — not capital necessity. This is what financial leverage looks like before a single outside dollar is raised.

Profit Reinvestment Distribution

Every dollar of profit that doesn't flow to the KJ Doctrine impact allocation is reinvested according to a disciplined framework designed to maximize ecosystem compounding. The allocation percentages reflect the strategic priority of each investment category and ensure balanced growth across all four revenue engines.

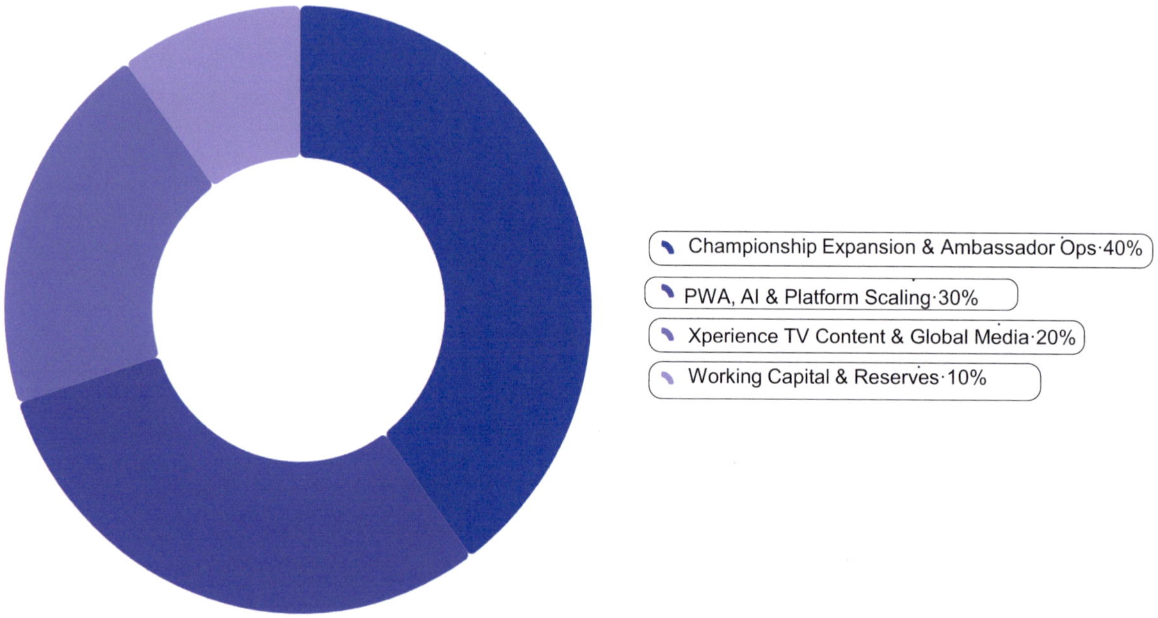

The 40% allocation to Championship Expansion and Ambassador Operations reflects the engine that generates the most immediate cash flow and participant acquisition. The 30% to PWA, AI, and Platform Scaling ensures the technology infrastructure stays ahead of growth demands. The 20% to Xperience TV builds the media engine's compounding content library. And the conservative 10% to Working Capital and Reserves ensures operational resilience through any market conditions.

Financial Reality

> **This company does not chase revenue. It activates owned infrastructure. We do not pitch for permission. We execute from position.**

The financial architecture of Goddess Phenezy Holdings represents a fundamental departure from how most companies approach growth. Rather than seeking market permission through fundraising, partnerships, or platform access, the company activates infrastructure it already owns—technology, IP, media, and community—to generate revenue on its own terms. This is not bravado. It is the mathematical consequence of owning every layer of the value stack.

UNIT ECONOMICS

Unit Economics — How the Money Actually Moves

Each engine is independently profitable. Together, they compound.

This section translates the ecosystem into repeatable revenue mechanics. The unit economics below are not theoretical projections—they are the mathematical building blocks from which every financial projection in this document is derived. Each engine carries its own risk profile, margin structure, and scaling dynamics, creating a portfolio-level resilience that single-engine businesses cannot achieve.

Engine-by-Engine Risk & Margin Profile

Understanding the risk and margin characteristics of each engine is essential for evaluating the overall portfolio dynamics. The deliberate diversity across these profiles means that macro-economic shifts, regulatory changes, or market disruptions in any single domain cannot threaten the aggregate business.

Engine	Risk Profile	Margin Profile	Category
I — B2B SaaS	Low	Very High (85–95%)	Recurring Infrastructure Revenue
II — Championships	Decentralized	High	Transactional + Royalty Revenue
III — Digital Economy	Very Low	Near-Total (90–98%)	Infinite Scale / Zero Overhead
IV — Media	Low	High (70–85%)	Asset Monetization & Brand Capital
V — 10×10 Matrices	Community-Driven	Controlled	Distributed Growth Engine

The portfolio effect is striking. The two highest-risk elements of traditional events businesses—venue operations and marketing spend—have been structurally externalized. Venue risk sits with Ambassadors (Engine II), and marketing cost has been replaced by community-driven growth (Engine V). What remains with the Holding Company are exclusively high-margin, low-risk, technology-enabled revenue streams.

SaaS Revenue Mechanics (Engine I)

The SaaS engine's revenue mechanics are built on mandatory adoption—every participant in the ecosystem requires the platform to operate. This is not optional software. It is the operating system through which all transactions, communications, and data flow.

01

User Onboards

Ambassador, event director, founder, agency, or recruiter enters the ecosystem and selects their pricing tier ($99–$499/month).

02

Monthly Revenue Activates

Subscription begins immediately. Blended ARPU of $225/month generates $2,700 annualized per user through the PWA.

03

Usage Deepens

As users run more events, manage more participants, and access more data, they naturally upgrade tiers—increasing ARPU over time.

04

Network Effects Compound

Each user brings participants who become users themselves, creating organic MRR growth independent of marketing spend.

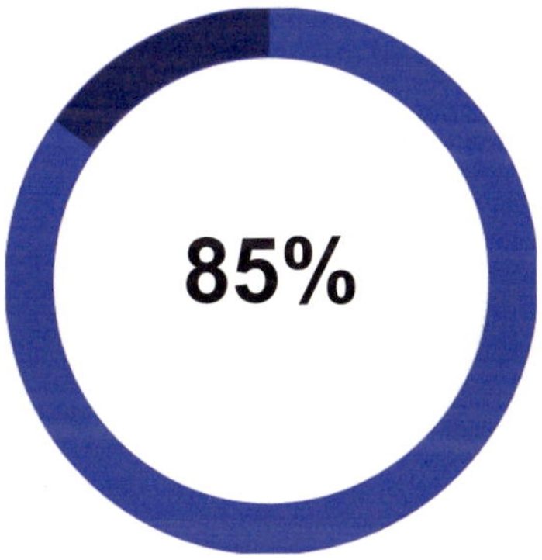

Min Gross Margin

Floor margin on SaaS subscriptions

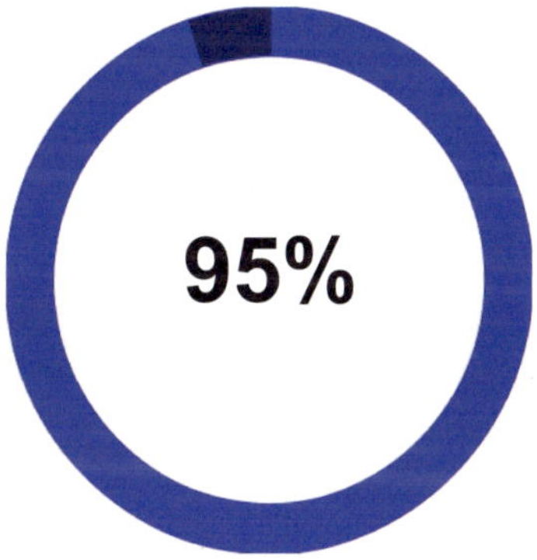

Max Gross Margin

Ceiling margin with scale efficiencies

Championship Revenue Mechanics (Engine II)

The championship engine operates as a repeatable licensing factory. Each activation follows the same economic template, with local execution variability absorbed by Ambassadors while the Holding Company captures its fixed percentage through royalties, SaaS, and media rights.

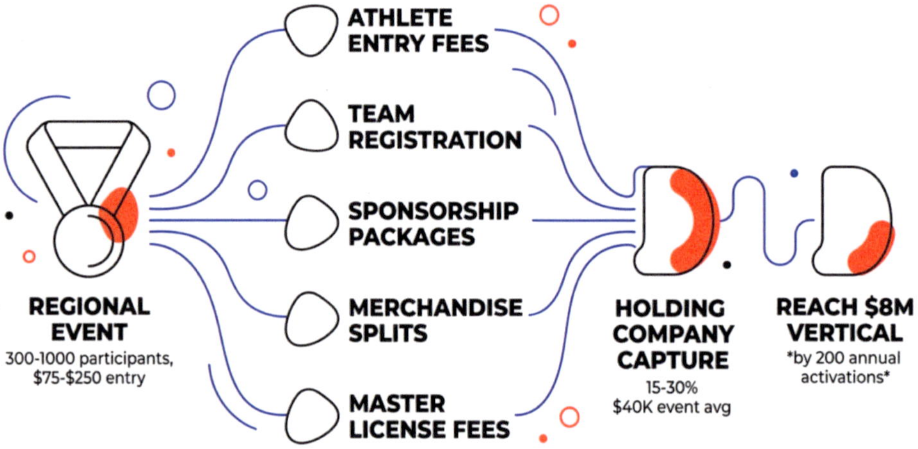

The critical multiplier is the number of championship verticals operating simultaneously. With six verticals—Basketball, Soccer, Boxing, Golf, Pickleball, and Poker—the $8M per-vertical annual revenue scales to $48M+ from championships alone. And each vertical adds SaaS subscribers, digital economy users, and media content that feeds the other engines.

Digital Economy Revenue Mechanics (Engine III)

The digital economy runs on fundamentally different physics — no venues, no geography, no marginal cost per user. Every transaction flows through owned infrastructure at 90–98% gross margins.

$120–$300
Annual revenue per engaged user

90–98%
Gross margin on all digital transactions

Zero
Marginal cost per additional user

The five digital revenue layers each operate independently but create powerful cross-selling dynamics that compound per-user spend over time:

Digital Tokens

In-platform currency averaging $85/user at launch, scaling to $200+ by Year 3.

VIP Access Tiers

$49–$199/month. Projected 18–25% conversion among active users.

Virtual Career Fairs

Enterprise employers pay $2,500–$15,000 per activation for verified talent access.

Gaming & Strategy

Entry fees, spectator passes, and in-game sales. $40M+ projected by Year 3.

Enterprise Sponsorships

$25K–$250K per activation. Premium CPMs of $85–$140 vs. industry average of $8–$15.

Media Revenue Mechanics (Engine IV)

Xperience TV's economics are built on the principle that content is an appreciating asset. Unlike a ticket, which generates revenue once, a piece of media content generates revenue indefinitely—through live streaming, archived viewing, advertising, licensing, and subscription access.

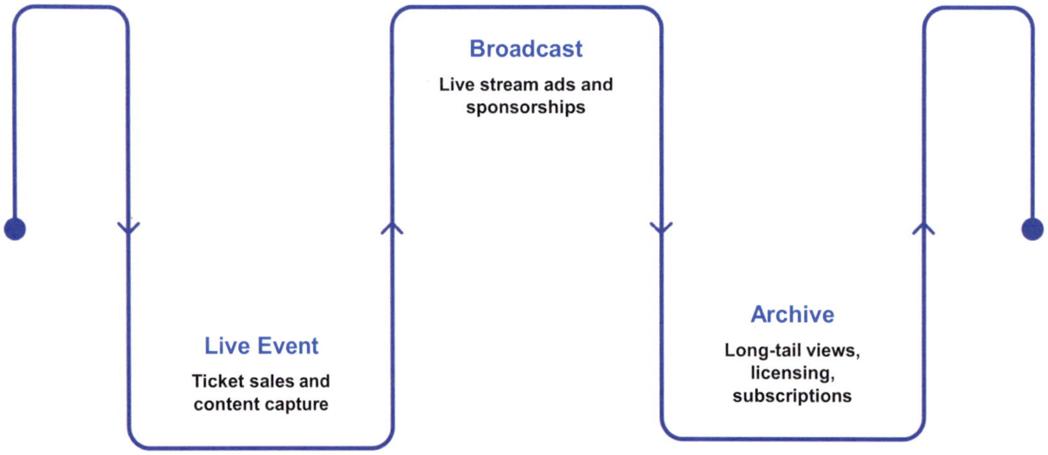

The sponsored championship packages ($50K–$500K) and enterprise annual sponsorships ($250K–$2M) provide predictable, high-value revenue. But the true compounding power comes from the evergreen content library that grows with every event. At 70–85% gross margins, the media engine becomes increasingly profitable as the content catalog scales and amortizes production costs across a growing library of assets.

The 10×10 Matrix Economics (Engine V)

Engine V fundamentally rewrites the economics of customer acquisition. By shifting marketing cost from cash expenditure to performance-based incentives, the company replaces every traditional growth cost center with a revenue-positive growth mechanism.

Replaces Paid Ads

No spend on Google, Facebook, or programmatic advertising. Community members drive awareness organically.

Replaces Agencies

No marketing agency retainers. Growth strategy is embedded in the economic architecture itself.

Replaces Influencers

No influencer contracts. Community participants are incentivized to share and recruit authentically.

Replaces Sales Teams

No traditional sales headcount. Commissions paid only when actual revenue is generated.

The result: Customer Acquisition Cost trends toward zero, growth becomes self-propelled, and the company scales without burning cash. Community members recruit users, train their networks, and drive upgrades—creating a distributed growth engine where the company pays only for results.

Competitive Defensibility

This is not a product. It is a system.

Most competitors attempt to copy features. Xperiences R Us is not a feature set—it is a sovereign operating system spanning infrastructure, jurisdiction, economics, and culture. To "copy" Xperiences R Us, a competitor would need to replicate every layer simultaneously. Failing at even one layer collapses the model.

The following sections detail seven distinct competitive moats, each of which would require years and tens of millions of dollars to replicate independently. In combination, they create a defensive position that is not merely difficult to breach—it is financially irrational to attempt.

The Seven Layers of Defensibility

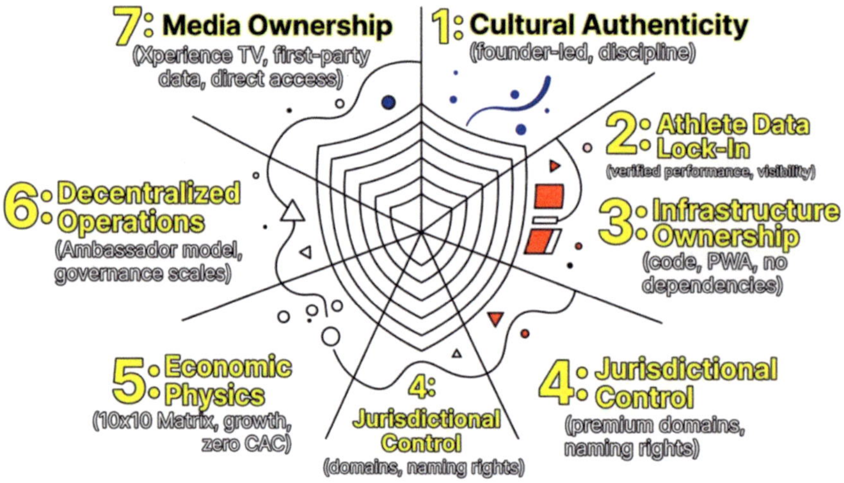

Each layer compounds the defensibility of every other layer. Jurisdictional control (domains) feeds media authority. Infrastructure ownership (PWA) enables economic physics (10×10 Matrix). Athlete data lock-in reinforces cultural authenticity. The defensive architecture is not additive—it is multiplicative.

MOAT #1

Jurisdictional Control — The Domain Moat

Goddess Phenezy Holdings owns 450+ premium jurisdictional domains — including exclusive naming rights to 'The National' and 'The International' across sports, championships, career platforms, education, and culture. These are not vanity URLs. They are category ownership claims that function like intellectual property, real estate, and brand authority simultaneously.

Search Authority

When a user searches 'The National Basketball Championship' or 'The International Career Fair,' owned domains capture organic traffic that no competitor can buy their way into. Domain authority compounds with every event, every press mention, and every year of operation.

Naming Rights Lock-In

Competitors cannot rename themselves into legitimacy. Any alternative naming construct — 'The Regional,' 'The American,' 'The Premier' — signals second-tier status to participants, sponsors, and media partners. The naming moat is self-reinforcing: the more events run under these names, the more culturally entrenched they become.

International Jurisdiction

The 'International' naming rights extend across 35+ target markets. As Master Licensing expands globally, each new country activates under the same jurisdictional umbrella — creating a unified global brand that local competitors cannot challenge without building from zero.

> They can build an event. They cannot own the category. And in a world where search, trust, and brand authority determine market share, category ownership is the only moat that matters.

MOAT #2

Infrastructure Ownership — No Rented Tech

While competitors assemble their tech stack from rented parts — Salesforce for CRM, Stripe for payments, Eventbrite for ticketing, YouTube for streaming — Goddess Phenezy Holdings owns every line of code. The 578,000+ lines of proprietary PWA infrastructure form a fully integrated operating system with zero dependency on Apple, Google, or any third-party platform. No app store. No revenue share. No terms-of-service risk.

What We Own

- **578,000+ lines of proprietary code — built in-house, owned outright**
- **Fully integrated PWA: registration, payments, streaming, CRM, analytics, and compliance in one stack**
- **Zero Apple / Google dependency — no 30% app store tax on any transaction**
- **Proprietary payment rails processing all SaaS, event, and digital economy transactions**
- **First-party data architecture — no third-party cookies, no data leakage, no platform risk**

What It Costs to Replicate

- **Minimum $15M–$25M in engineering investment to build equivalent infrastructure from scratch**
- **3–5 years of development time — during which the ecosystem continues to compound**
- **Deep cross-domain expertise spanning sports operations, SaaS architecture, media streaming, and regulatory compliance**
- **Thousands of micro-decisions embedded in the codebase that cannot be documented or transferred**
- **Cultural alignment between technology and mission — the hardest thing to engineer and the easiest to fake**

> **Money cannot compress time. And time is the one resource a competitor cannot buy. Every day the ecosystem operates, the infrastructure moat deepens.**

MOAT #3

Economic Physics — The 10×10 Matrix

The 10×10 Matrix is the growth engine that makes traditional customer acquisition economics obsolete. Instead of spending capital on advertising to acquire users, the ecosystem pays participants to recruit other participants — converting every satisfied user into a revenue-generating distribution node. The result: Customer Acquisition Cost (CAC) approaches zero while Lifetime Value (LTV) compounds with every referral layer.

The B2C Layer (Community)

Each active participant can refer up to 10 direct members. At a $25 referral incentive per conversion and a $225/month ARPU, the company acquires a $2,700/year subscriber for $25 — a 108x return on acquisition cost. At 10,000 active referrers each converting 5 members, the network adds 50,000 subscribers with zero paid media spend.

The B2B Layer (Your Partnrr)

Enterprise partners and Ambassadors operate a parallel referral architecture, recruiting organizations, sponsors, and institutional clients. A single Ambassador who recruits 10 enterprise SaaS subscribers at $499/month generates $59,880 in annual recurring revenue for the Holding Company — while earning performance-based commissions that cost a fraction of traditional B2B sales overhead.

Why Competitors Can't Copy It

Traditional sales organizations cannot model commission compliance at scale across thousands of independent operators. The legal, accounting, and operational infrastructure required to run a compliant distributed economics model takes years to build — and most established companies lack the cultural DNA to execute it.

Retrofitting Is Self-Destructive

Any competitor attempting to bolt the 10×10 Matrix onto an existing business immediately cannibalizes its internal sales team, triggers channel conflict, and faces regulatory exposure. The model only works when built from first principles — which means it cannot be added. It must be the foundation.

The Compounding Advantage

Every new participant who joins the ecosystem becomes a potential distribution node. At 100,000 active users with a 15% referral conversion rate, the network self-generates 15,000 new subscribers per cycle — without a single dollar of paid acquisition spend. This is the moat that makes the financial projections conservative, not aggressive.

> **Most companies spend to grow. This ecosystem grows to earn. The 10×10 Matrix doesn't just reduce CAC — it inverts the entire economics of customer acquisition.**

MOAT #4

Athlete Advancement Data — The Lock-In

The athlete data moat is the most psychologically powerful competitive barrier in the ecosystem — because it is entirely self-reinforcing. Every event an athlete competes in, every ranking they earn, every scout interaction they receive, and every scholarship pathway they unlock is permanently recorded in their verified Xperiences R Us profile. This is not a user account. It is a digital career asset that appreciates with every participation.

Verified Performance History

Every championship result, placement, and statistical record is timestamped and verified by the platform — not self-reported. Scouts and recruiters pay for access to this data precisely because it cannot be fabricated. Over time, a participant's profile becomes the most credible athletic resume in existence.

Scout & Recruiter Visibility

500+ verified scouts, college recruiters, and professional talent evaluators have active accounts on the platform. Athletes with established profiles receive direct outreach, scholarship offers, and NIL opportunities through the ecosystem. Leaving means losing visibility to the exact decision-makers who control their future.

NIL & Scholarship Pathways

The platform is the only place where athletes can simultaneously compete, build a verified record, and access NIL monetization and scholarship opportunities in one integrated experience. Competitors offer one or two of these — never all three. The integration is the lock-in.

Accumulated Social Capital

Rankings, badges, community standing, and peer recognition are non-transferable. An athlete ranked #3 in their regional Basketball vertical has built that standing through months of competition. Starting over on a competitor platform means rank zero, zero visibility, and zero credibility — a cost no serious competitor will accept.

> This is not contractual lock-in. It is value-based lock-in — the kind that doesn't require enforcement because users choose to stay. Leaving Xperiences R Us means erasing your career. No athlete does that voluntarily.

MOAT #5

Decentralized Operations — The Ambassador Model

The Ambassador model inverts the traditional organizational structure. Instead of centralized operations with staff-heavy, venue-dependent execution, Goddess Phenezy Holdings licenses execution and centralizes governance. This creates a structurally different cost basis and scaling dynamic than any traditionally organized competitor.

What We Do

License execution to Ambassadors, centralize governance and technology, eliminate physical overhead from the Holding Company P&L.

Why It Can't Be Copied

Transitioning to decentralization requires rewriting contracts, building new compliance frameworks, and relinquishing control—steps most organizations cannot take.

> You govern. They manage. Governance scales. Management collapses.

Most organizations are operationally centralized, staff-heavy, and venue-dependent. Transitioning from this model to a decentralized Ambassador framework requires rewriting every contract, building entirely new compliance frameworks, and—most critically—relinquishing operational control. For organizations built on control, this is an existential challenge, not a strategic pivot.

MOAT #6

Media Ownership — Not Distribution Dependence

Every competitor in the sports and entertainment space distributes content through platforms they don't own — YouTube, ESPN+, Instagram, TikTok. They pay for reach, negotiate for placement, and accept whatever algorithm changes are handed to them. Xperience TV is a fully owned media property. Every viewer, every interaction, every ad impression, and every subscription dollar flows directly to the Holding Company — with no intermediary taking a cut and no platform that can revoke access.

Xperience TV: The Owned Channel

A fully proprietary streaming platform broadcasting live championships, archived event content, career fair activations, and digital economy events. Subscription revenue at $14.99–$29.99/month. Enterprise sponsorship packages at $50,000–$500,000 per activation. At 400,000 subscribers by Year 3, subscription revenue alone exceeds $72M annually — before a single sponsorship dollar is counted.

First-Party Data: The Invisible Asset

Every viewer interaction — watch time, content preferences, geographic location, engagement patterns — is captured as first-party data owned entirely by the Holding Company. This data powers sponsor targeting, content programming decisions, and athlete advancement analytics. In a post-cookie world, first-party audience data at this scale commands premium CPMs of $85–$140 versus the industry average of $8–$15.

Evergreen Content: The Appreciating Library

Unlike a live event ticket that generates revenue once, championship content generates revenue indefinitely. A Basketball final filmed in Year 1 is still generating streaming revenue, licensing fees, and highlight clip monetization in Year 5. The content library grows with every event — and unlike physical assets, it never depreciates. By Year 5, the archived library represents 3,000+ hours of premium sports content with compounding monetization potential.

> Competitors lease attention from platforms that can revoke it overnight. We own the audience, the content, and the channel. That is not a media strategy. It is a media moat.

♡ MOAT #7

Cultural & Missional Authenticity

The seventh moat is the one no competitor can buy, engineer, or retrofit — because it was never built as a strategy. It emerged from a founder who came from the communities she serves, a military discipline that treats operational excellence as a moral obligation, and a decade of relationship-building that predates the company's formal launch. Cultural authenticity at this depth is not a brand attribute. It is a structural barrier.

Founder-Led Origin

The company was built by someone who grew up in the ecosystem she now serves — not a corporate team that identified an underserved market. Participants, Ambassadors, and community members recognize the difference immediately. Founder-led authenticity creates trust that no marketing budget can manufacture and no acquisition can transfer.

Veteran Discipline

Military-grade operational rigor is embedded in every process, every Ambassador agreement, and every governance decision. This is not a cultural talking point — it is the reason the company can scale decentralized operations across 35+ markets without losing execution quality. Discipline is the infrastructure that holds the ecosystem together.

Community Trust at Scale

The ecosystem was built through direct community relationships — not top-down marketing. Ambassadors are community members first, operators second. Participants recruit other participants because they trust the platform with their advancement. This trust took years to build and cannot be replicated by a competitor entering the market with capital alone.

The KJ Doctrine as Social Contract

The permanent 30% reinvestment commitment is not a CSR initiative — it is a constitutional clause that cannot be amended by investor pressure or quarterly performance. Communities that have received scholarships, grants, and career advancement through the KJ Doctrine become permanent advocates. They don't just use the platform. They defend it.

A competitor can copy the product. They cannot copy the relationship. And in a community-driven ecosystem, the relationship is the product.

The Final Truth — Copying Is Financially Irrational

To replicate the Xperiences R Us and Technologies R Us (www.technologiesrus.com) divisions, a competitor would need to:

1. **Spend 8–9 Figures**

   Minimum investment to rebuild proprietary technology, secure domain portfolios, and establish media infrastructure from scratch.

2. **Wait 5–7 Years**

   The time required to build, test, iterate, and scale a comparable operating system across all four engines.

3. **Abandon Their Current Business**

   Retrofitting distributed economics and decentralized operations cannibalizes existing revenue models and breaks current operations.

4. **Rebuild Culture, Economics & Infrastructure**

   Technology can be hired. Culture cannot. Community trust cannot be purchased. Authenticity cannot be manufactured.

5. **Compete Against a Compounding Ecosystem**

   By the time any competitor finishes building, Goddess Phenezy Holdings will have moved again—further ahead, more deeply entrenched.

Strategic Conclusion

The competitive position of Goddess Phenezy Holdings is not built on a single product advantage, a first-mover timing advantage, or a funding advantage. It is built on seven simultaneous moats — each independently defensible, each compounding against the others — creating a defensive architecture that is not merely difficult to breach. It is economically irrational to attempt.

They Compete in Markets

Goddess Phenezy Holdings owns jurisdictions — 450+ premium domains, international naming rights, and category authority that no competitor can purchase or replicate through marketing spend.

They Sell Access

Goddess Phenezy Holdings sells advancement — verified career records, scout visibility, NIL pathways, and scholarship access that create irreversible value for every participant.

They Rent Platforms

Goddess Phenezy Holdings owns infrastructure — 578,000+ lines of proprietary code, zero app store dependency, and first-party data architecture that no third party can revoke or monetize.

They Buy Users

Goddess Phenezy Holdings mints partners — 12,000+ Ambassadors who fund their own expansion, recruit their own networks, and generate revenue for the Holding Company without a single dollar of paid acquisition.

7	$15M–$25M+	5–7 Years	$0
Simultaneous Competitive Moats	Minimum Cost to Replicate Infrastructure	Time Required — During Which We Compound	External Capital Required to Execute

> This is not a company with a moat. This is a company that is the moat. Every day the ecosystem operates, the distance between Goddess Phenezy Holdings and any potential competitor grows wider, deeper, and more expensive to cross.

STRATEGIC BRIEFING | CLASSIFIED ARCHITECTURE

The Cognitive Infrastructure

The Artificial Intelligence Jurisdiction

Governing the Interface Between Human and Machine Intelligence

This document presents the sovereign intellectual property and institutional architecture through which Goddess Phenezy Holdings, Inc. has secured jurisdictional control over the intersection of human cognition and artificial intelligence — across education, competition, culture, media, and enterprise infrastructure. What follows is not a product roadmap. It is the blueprint for an owned operating system governing how intelligence is trained, validated, scaled, and monetized.

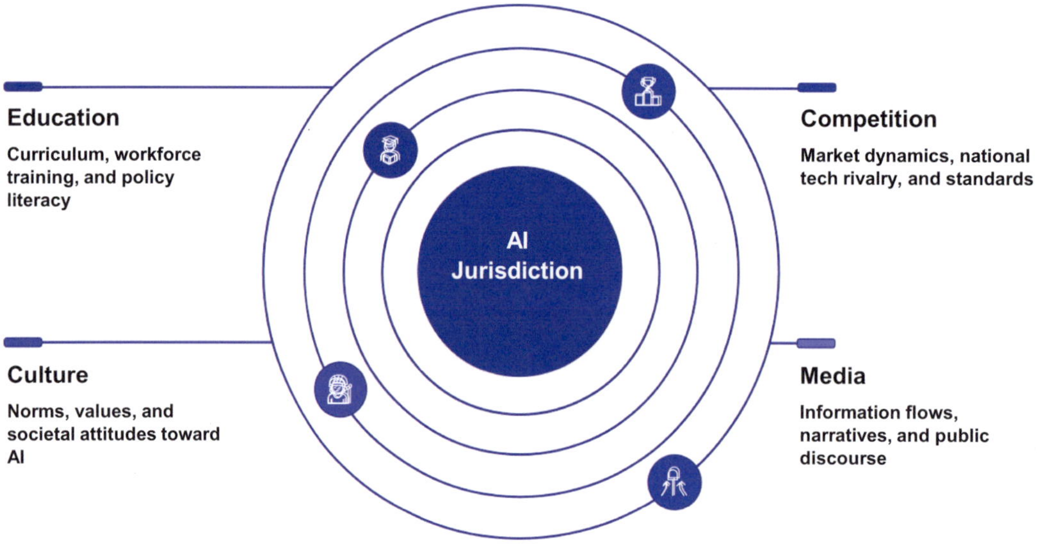

Why A.I. Goes Here

This section is last because it is the multiplier.

Every preceding division within Goddess Phenezy Holdings — sports, academies, conferences, media — exists not merely as a standalone vertical but as a foundational layer generating the data, culture, participation, and demand that Artificial Intelligence requires to operate at scale. The sequencing is deliberate. The architecture is intentional. A.I. is positioned at the apex because it accelerates and compounds the value of everything beneath it.

Sports

Adoption and recurring participation

Academies

Talent pipeline development

Conferences

Authority and partnerships

Media

Distribution and amplification

> Artificial Intelligence = acceleration across all of it.

Everything documented before this section produces the raw material — the behavioral data, cultural engagement, and institutional demand — that Artificial Intelligence requires to function. What follows governs how that intelligence is trained, validated, scaled, and permanently owned.

The Evidence Wall

The gallery that follows this section is not merely decoration; it is the Evidence Wall – a meticulously assembled visual archive showcasing an unparalleled privately held Artificial Intelligence intellectual property portfolio. This portfolio, meticulously secured between 2022 and 2024, precisely when the global market debated AI's nature as a trend versus a transformation, encompasses over 200 distinct domain names spanning every major AI institutional category, including academies, institutes, colleges, universities, championships, conferences, summits, festivals, incubators, and accelerators. It further includes registered trade names such as "National Artificial Intelligence Association," "International Artificial Intelligence Association," "Artificial Intelligence Championships," "AI Xperiences R Us," "AI Conferences R Us," and dozens of derivative marks, alongside proprietary credentialing frameworks covering AI literacy, AI competition rankings, and AI professional certification. Every name, every domain, every credential framework documented herein represents a sovereign position that cannot be replicated, outbid, or deplatformed.

Before examining the technical and strategic architecture, it is critical to understand that the positions documented here cannot be reverse-engineered by a competitor, acquired through a funding round, or replicated through open-source tooling. These institutional names, domain infrastructure, and credentialing architecture represent first-mover sovereign positions – the kind that historically accrue to those who move before consensus forms. The forthcoming gallery is organized by category: Educational Infrastructure, Competitive Infrastructure, Commercial Infrastructure, and Cultural Infrastructure. These four pillars together constitute a closed-loop jurisdictional system over the AI institutional economy.

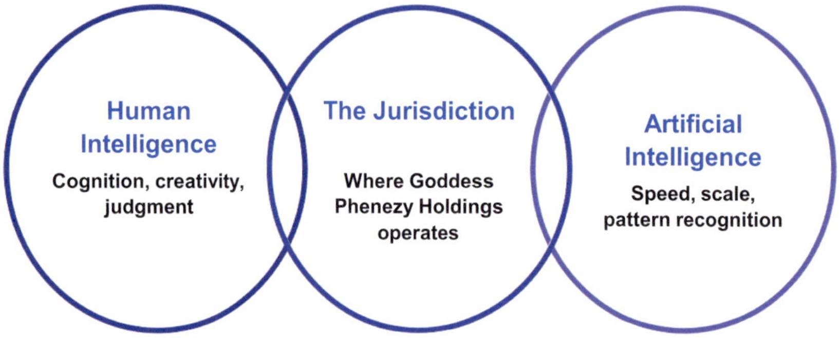

CORE THESIS

A.I. Is Not a Feature — It Is the Operating System

Most enterprises "add AI" as an enhancement layer — a tool, an interface, a novelty. Goddess Phenezy Holdings, Inc. built something fundamentally different.

> Key Distinction: We do not rent the future from platforms. We own the names, the rails, the brands, and the jurisdiction.

This is not an AI product. This is an AI jurisdiction. The distinction is not semantic — it is structural, legal, and economic. A product can be replicated, deprecated, or commoditized. A jurisdiction governs the terms under which all products, participants, and platforms operate. It establishes the rules, the credentials, the standards, and the institutional identity that define a domain of activity.

Where competitors build features on rented infrastructure, Goddess Phenezy Holdings owns the foundational layer — the names, the rails, the brands, and the governance frameworks that constitute permanent, defensible intellectual property.

1

The Jurisdiction

Sovereign IP, Names & Governance Frameworks

2

Education

Academies, Institutes, Colleges, Universities

3

Competition

Championships, Rankings, Credentialing

4

Commerce

Conferences, Summits, Incubators, Accelerators

5

Culture

Festivals, Xperiences, Media & Distribution

The Dual-Engine Architecture

The jurisdiction operates through two complementary engines, with Artificial Intelligence serving as the bridge that unifies them into a single operating system.

Technologies R Us

The software, infrastructure, and intellectual property engine. This division builds, owns, and operates the technical stack — the platforms, data systems, credentialing frameworks, and proprietary tools that power the entire ecosystem.

Artificial Intelligence Bridge

The cognitive layer that powers systems on one side while being continuously trained by real-world human activity on the other. Intelligence flows in both directions — improving infrastructure while being refined by experience.

Xperiences R Us

The public-facing experiential and cultural engine. This division generates the participation, data, and cultural engagement that feeds the intelligence layer — from championships to festivals to media properties.

This architecture ensures that every human interaction within the Xperiences R Us ecosystem generates data that improves the Technologies R Us (www.technologiesrus.com) infrastructure, which in turn enhances the quality and intelligence of subsequent experiences. The feedback loop is continuous, self-reinforcing, and permanently owned.

The Foresight: 2022–2023

While the Market Watched, We Began Secured the Terrain

The Global Inflection Point

In late 2022 and throughout 2023, Artificial Intelligence moved from research discipline to global inflection point. The release of large language models and generative AI tools ignited a worldwide frenzy of speculation, experimentation, and institutional anxiety. The response landscape fractured into predictable patterns — each representing a failure of strategic imagination.

Legacy Institutions

Hesitated. Formed committees. Commissioned studies. Debated regulatory frameworks while the terrain was being claimed beneath their feet. Their institutional inertia became their strategic liability.

Venture Capital

Accelerated speculation. Poured capital into model builders and wrapper applications with no defensible moat, no institutional permanence, and no jurisdictional authority. Speed without sovereignty.

The General Public

Engaged AI as novelty — interfaces, entertainment, distraction. Chatbots became parlor tricks. Image generators became toys. The surface was explored while the foundation was ignored.

The Sovereign Positioning Strategy

What We Did Not Do

- We did not attempt to out-engineer model builders
- We did not chase disposable software abstractions
- We did not treat Artificial Intelligence as a product

What We Did

We treated it as infrastructure.

Goddess Phenezy Holdings responded to the AI inflection point with a fundamentally different thesis: intelligence — whether human or artificial — would ultimately require institutional terrain to operate, compete, be credentialed, governed, and monetized.

While others debated use cases and rushed to build the newest chatbot interface, we methodically secured the naming rights, governance frameworks, and institutional positions that would define how AI is organized across every sector that matters — education, competition, business, culture, and media.

This was not a software sprint. It was not a feature roadmap. It was not a pitch deck hypothesis. It was a sovereign positioning strategy — a deliberate, multi-year campaign to own the jurisdictional layer that all future AI activity would require to achieve legitimacy, scale, and permanence.

The distinction between building a product and securing a jurisdiction is the distinction between renting space in someone else's ecosystem and owning the ground on which all ecosystems must eventually stand.

The Secured Perimeter

The Artificial Intelligence Vault

Before one can govern the future, one must first secure the perimeter.

1

Domain Infrastructure

200+ secured domains across every AI institutional category

2

Trade Names & Marks

Registered institutional identities including National & International AI Association, AI Championships, AI Xperiences R Us

3

Credentialing Frameworks

Proprietary AI literacy, competition rankings, and professional certification systems

4

Governance Architecture

Standards-setting body, rules of participation, and jurisdictional authority

5

Sovereign Jurisdiction

The owned operating layer governing how AI is trained, validated, scaled, and monetized

Strategic Clarity Through Infrastructure

Because Goddess Phenezy Holdings had already established physical, cultural, educational, and competitive infrastructure across sports, academies, media, and conferences, we possessed the strategic clarity required to lock down the cognitive layer. The existing ecosystem was not merely a business portfolio — it was a strategic intelligence asset that revealed precisely which positions needed to be secured.

Through deliberate acquisitions beginning in 2022 and expanding through 2023 and beyond, we secured both the short-form A.I. and the definitive long-form Artificial Intelligence naming rights across every institutional tier that matters. These acquisitions represent permanent, defensible positions in the most consequential technology category of the twenty-first century.

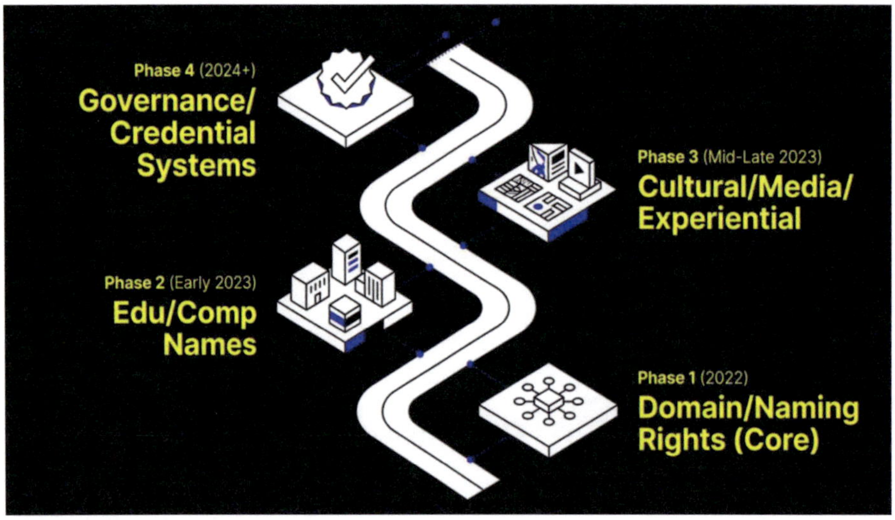

The result is not a collection of domain names. It is an interconnected, self-reinforcing jurisdictional perimeter that encompasses the full lifecycle of Artificial Intelligence — from initial education to advanced competition to commercial scaling to cultural expression.

The Governing Body

The National & International Artificial Intelligence Association

At the apex of the jurisdictional architecture sits the National & International Artificial Intelligence Association — the governing, credentialing, and standards-setting authority across the entire AI Jurisdiction. The Association does not merely participate in the AI ecosystem. It defines the terms under which the ecosystem operates.

Ethical Frameworks

Establishing the moral and operational boundaries within which all AI development, deployment, and competition must occur. These frameworks are not advisory — they are jurisdictional requirements.

Certification Protocols

Defining the credentials, qualifications, and competency standards that individuals and organizations must achieve to operate within the jurisdiction. Certification creates legitimacy.

Competitive Standards

Governing the rules, formats, and evaluation criteria for all cognitive competitions — ensuring fairness, rigor, and global consistency across championships and proving grounds.

Institutional Partnerships

Binding education, competition, and industry into a unified operating system through strategic alliances that reinforce jurisdictional authority at every level.

The Jurisdictional Layers

The secured perimeter encompasses four distinct but interconnected jurisdictional layers, each governing a critical dimension of the Artificial Intelligence landscape.

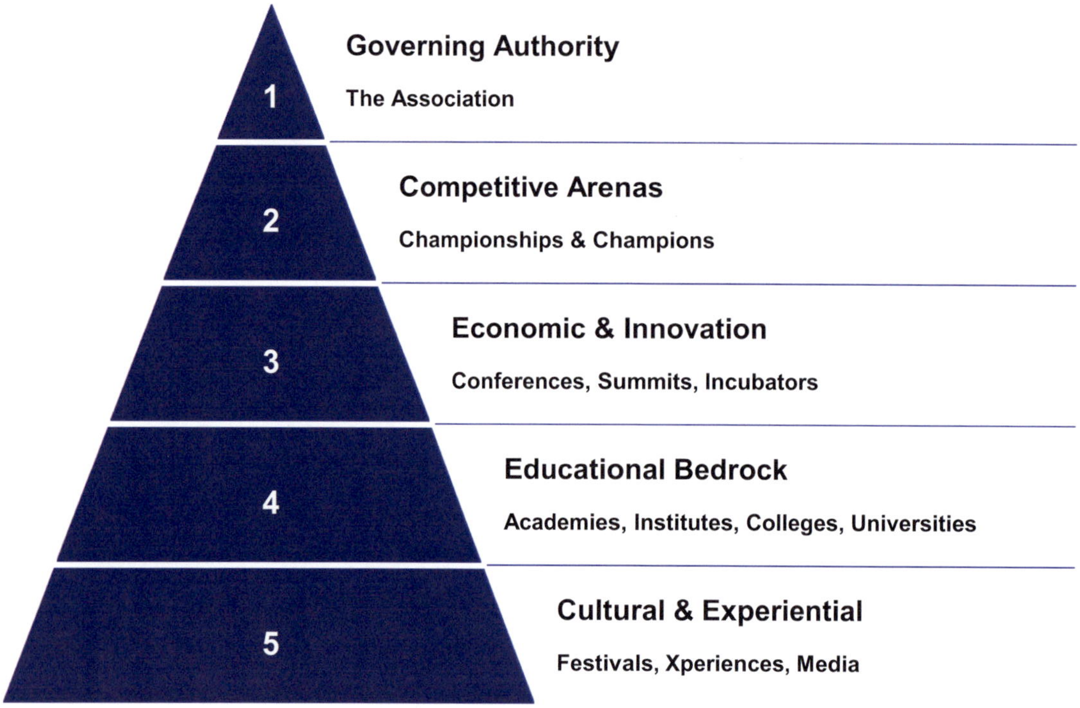

These layers are not parallel silos. They are stacked, interdependent, and self-reinforcing — each feeding participants, data, and legitimacy into the others. Together, they constitute a sovereign operating system for Artificial Intelligence.

Layer 1: Educational Bedrock

National & International Artificial Intelligence Academies, Institutes, Colleges, and Universities

AI Academies

Applied skills, hands-on training, and foundational AI literacy for emerging talent

AI Institutes

Specialized research and advanced methodology for practitioners and professionals

AI Colleges

Structured degree pathways combining theory, ethics, and applied AI competency

AI Universities

Comprehensive academic ecosystems offering undergraduate, graduate, and doctoral AI programs

Education is the foundation upon which every other layer depends. Without trained, credentialed, and rigorously assessed talent, there is no competition, no innovation, and no culture. The educational bedrock of the AI Jurisdiction encompasses the full spectrum of learning institutions — from specialized academies focused on applied AI skills to comprehensive universities offering deep theoretical and practical curricula.

These institutions do not merely teach AI as a subject. They immerse participants in a governed educational ecosystem where competency is measured, credentials are issued, and mastery is the prerequisite for advancement into competitive and commercial arenas. The educational layer produces the human capital that powers every subsequent stage of the jurisdictional pipeline.

By owning the institutional naming rights across academies, institutes, colleges, and universities, Goddess Phenezy Holdings ensures that the educational gateway to Artificial Intelligence passes through its jurisdiction — permanently and irrevocably.

Layer 2: Competitive Arenas

National & International Artificial Intelligence Championships

Competition is the crucible. The AI Championships are not exhibitions — they are the governed proving grounds where trained intelligence is stress-tested, ranked, and credentialed under the jurisdiction of Goddess Phenezy Holdings. No other private entity holds the institutional naming rights to a national and international AI championship framework across this breadth of competitive categories.

AI Strategy Championships

Human vs. machine and human vs. human competitions testing strategic reasoning, decision-making, and cognitive performance under pressure

AI Innovation Championships

Applied AI build competitions where participants deploy real solutions judged on accuracy, efficiency, and originality

AI Literacy Championships

Knowledge-based competitions validating depth of AI understanding across theory, ethics, policy, and application

AI Xperience Championships

Experiential and performance-based competitions measuring human-AI collaboration, creativity, and adaptive intelligence

Championship frameworks establish the competitive standards, evaluation rubrics, and global ranking systems that define what excellence looks like within the AI domain. Competitors do not simply win — they earn jurisdictional credentials that carry institutional weight recognized across the ecosystem. A champion of the AI Championships is not merely talented. They are certified.

By owning the naming rights to AI Championships at both the national and international level — across every competitive category — Goddess Phenezy Holdings controls the apex of the AI talent validation pipeline. No credential issued outside this jurisdiction carries equivalent institutional authority. That is not a claim. It is a structural reality.

Layer 3: Economic & Innovation Engine

Artificial Intelligence Conferences, Summits, Incubators, and Accelerators

AI Conferences

Flagship institutional gatherings where policy, commerce, and innovation converge under the jurisdictional umbrella of Goddess Phenezy Holdings

AI Summits

Invitation-only executive forums convening investors, government officials, and industry leaders to shape the strategic direction of the AI economy

AI Incubators

Structured programs that take validated talent from competition into commercial viability — providing funding, mentorship, and market access within the governed ecosystem

AI Accelerators

High-velocity programs compressing the timeline from concept to market for AI ventures that have demonstrated jurisdictional credentialing

The economic and innovation layer transforms talent and ideas into scalable enterprises. Conferences establish authority and forge partnerships. Summits convene decision-makers. Incubators nurture early-stage ventures. Accelerators compress the timeline from concept to market. Each platform operates within the jurisdictional framework, ensuring that economic activity generated within the ecosystem remains governed by it.

Layer 4: Cultural & Experiential

AI Festivals, AI Xperiences , AI Conferences

Culture is the layer that makes the jurisdiction undeniable. While education trains, competition validates, and commerce scales — culture embeds. It transforms institutional infrastructure into lived, shared, emotionally resonant experience. Goddess Phenezy Holdings owns the cultural layer of the AI economy through three distinct and irreplaceable platforms.

AI Festivals

Large-scale public events showcasing the capabilities, artistry, and ethical dimensions of Artificial Intelligence — generating mass cultural engagement, media coverage, and participation data at scale

AI Xperiences

Immersive, participatory experience platforms that translate complex AI concepts into tangible, memorable encounters — converting passive audiences into active ecosystem participants

AI Conferences

The branded, governed platform for professional and public AI discourse — providing the institutional stage where ideas, partnerships, and policy positions are formed and announced

Each cultural platform serves a dual function: it generates public-facing engagement that builds the jurisdictional brand, and it produces the behavioral data, participation records, and cultural intelligence that feeds directly into the AI layer. Every festival attendee, every Xperience participant, and every conference registrant becomes a data point in a governed intelligence system that no competitor can replicate — because no competitor owns the cultural infrastructure.

This is not content marketing. This is not event production. This is jurisdictional culture — the owned, defensible, and permanently secured layer through which Goddess Phenezy Holdings ensures that the AI Jurisdiction is not merely an institutional abstraction. It is a cultural reality experienced by millions.

Critical distinction: These are not keywords. They are not marketing slogans. They are jurisdictions — owned, secured, and defensible. The perimeter is secured.

CORE DOCTRINE

The HI vs. A.I. Doctrine

Human Intelligence Versus Artificial Intelligence

The Governing Philosophy

At the core of this jurisdiction is a proprietary governing philosophy: HI vs. A.I.

We do not frame Artificial Intelligence as a replacement for human intelligence, nor as a passive tool. We treat it as both an oppositional and collaborative force — one that must be tested, measured, challenged, and mastered. This dual framing is deliberate and philosophically rigorous.

The "versus" in HI vs. A.I. does not imply hostility. It implies structured engagement — the same dynamic that drives competitive sports, academic debate, and scientific inquiry. When human intelligence is pitted against artificial intelligence under governed conditions, both are revealed more clearly. Strengths, weaknesses, complementarities, and risks become visible in ways that passive deployment never allows.

Through platforms such as Artificial Intelligence Chess Masters, we establish living arenas where human cognition, strategy, creativity, and ethics are evaluated against algorithmic speed and scale in real time. These are not exhibitions. They are governed competitions that produce measurable, credentialed outcomes.

Human Intelligence

Cognition, strategy, creativity, ethics, intuition, moral reasoning

Artificial Intelligence

Algorithmic speed, pattern recognition, scale, computation, prediction

> We do not merely deploy intelligence. We govern the interface between them.

The Three-Pillar Cognitive Pipeline

The Artificial Intelligence Jurisdiction operates as a fully sovereign, closed-loop system — a self-reinforcing pipeline that transforms raw talent into validated, scaled, and governed intelligence assets.

The Curriculum

Training, certification, governance

The Proving Ground

Championships and real-world validation

Capital and Culture

Incubators, summits, funded ventures

Each pillar feeds the next in an unbroken chain: education produces credentialed talent, competition validates that talent under rigorous conditions, and the capital-and-culture layer scales what survives into commercial and cultural impact. The loop then cycles back as scaled enterprises produce new educational demands, new competitive benchmarks, and new cultural expressions — each cycle compounding the value and authority of the jurisdiction.

Pillar 1: The Curriculum

Education Under Jurisdiction

Traditional academic institutions cannot adapt to a technology that evolves weekly. We provide proprietary, certified educational frameworks built for the speed of AI.

1

Prompt Engineering

The foundational skill of communicating with AI systems — crafting inputs that produce precise, reliable, and ethically sound outputs. This is the literacy of the AI age.

2

Machine Learning Architecture

Deep understanding of the structural frameworks that underpin AI systems — from neural networks to transformer models to reinforcement learning paradigms.

3

Cognitive Systems Design

The art and science of designing systems that integrate human cognition with artificial intelligence — creating hybrid workflows that amplify both.

4

Human–AI Collaboration

Advanced frameworks for productive, ethical, and governed collaboration between human operators and AI systems across professional domains.

Jurisdictional requirement: No individual competes or builds within this ecosystem without first being trained under its governance. Education is the entry gate.

Pillar 2: The Proving Ground

Competition as Validation

Education without execution is theory. The AI Championships are the governed proving grounds where trained intelligence is stress-tested, ranked, and permanently credentialed. These are not hackathons. They are not demo days. They are high-stakes cognitive competitions where the gap between knowledge and capability is exposed, measured, and recorded.

Train

Participants enter having completed jurisdictional education and credentialing requirements

Compete

Standardized, governed challenges test technical proficiency, strategic reasoning, and ethical judgment

Validate

Performance is evaluated against Association-set rubrics ensuring global comparability and fairness

Credential

Results become permanent, jurisdictional records — not just wins, but certified proof of capability

Compound

Every competition generates performance data that improves AI systems embedded throughout the ecosystem

> **We do not host hackathons. We govern cognitive competition.**

Pillar 3: Capital & Culture

Scaling What Survives

Talent discovered through education and validated through competition is funded, scaled, and exhibited through incubators, summits, and festivals. This third pillar transforms proven capability into commercial reality and cultural impact.

As Artificial Intelligence expands into music, film, design, analytics, and strategy, these platforms become the global stage where algorithmic creation is governed — not exploited. The distinction is critical. In an unregulated landscape, AI-generated creative and commercial output operates in an ethical and legal vacuum. Within the jurisdiction, every output is traceable, credentialed, and governed by established frameworks.

Incubators

Structured environments where validated talent receives funding, mentorship, and infrastructure to build commercially viable AI ventures within the governed ecosystem.

Summits

High-level convenings that bring together industry leaders, policymakers, and investors to shape the trajectory of AI commercialization and governance.

Festivals

Cultural platforms where AI-driven creativity is showcased, celebrated, and embedded into public consciousness — expanding the jurisdiction's cultural reach.

A.I. Integration Across the Ecosystem

Artificial Intelligence is not a standalone vertical. It is embedded across every division — a connective tissue that accelerates, optimizes, and amplifies the entire operational architecture.

Division-Level AI Integration

Championships

AI coaching, rankings, verification, and real-time analytics

Academies

Adaptive learning paths, diagnostics, and mentor personalities

Conferences

Matchmaking, deal rooms, and live content extraction

Media

Automated highlights, documentary pipelines, content syndication

Identity & Credentials

Verified profiles, digital resumes, and credential wallets

Each integration point represents a bidirectional data flow. Championships generate performance data that improves coaching algorithms. Academies produce learning pattern data that refines adaptive curricula. Conferences generate network and preference data that enhances matchmaking. Media operations produce engagement data that optimizes content pipelines. Identity systems aggregate all of it into verified, portable credential assets.

The Compounding Intelligence Loop

Xperience

Every festival, championship, conference, and academy session generates raw behavioral, performance, and engagement data from real participants in governed environments

Data Capture

Participation records, competition results, credential issuances, attendance patterns, and interaction logs are captured within the jurisdictional infrastructure — owned, not rented

Intelligence Layer

Captured data trains and refines the AI systems embedded across coaching, adaptive learning, matchmaking, content pipelines, and credential verification

System Improvement

Refined AI systems deliver measurably better outcomes — higher retention, sharper rankings, more accurate credentialing, and more valuable conference matchmaking

Compounding Value

Better outcomes attract more participants, generate more data, and deepen jurisdictional lock-in — making the ecosystem exponentially harder to replicate or displace

> **Every Xperience produces data. Every dataset trains intelligence. Every improvement compounds value. Every cycle of the loop widens the moat. No competitor can buy their way into this position. It must be built — and we already built it.**

The Revenue Architecture for A.I.

The AI Jurisdiction follows the same four-pillar economic structure as the broader ecosystem — ensuring consistency, predictability, and compounding defensibility across all revenue streams.

Four-Pillar Revenue Model

1

Licensing

Jurisdictional rights, institutional naming, and regional chapters. Organizations seeking to operate under the AI Jurisdiction's standards, use its credentialing frameworks, or establish regional chapters pay licensing fees for the right to participate in the governed ecosystem.

2

Memberships & SaaS

Individuals, organizations, academies, and recruiters access the platform through tiered memberships and software subscriptions. Recurring revenue from credentialed professionals, institutional participants, and talent acquisition partners.

3

AI Xperiences

Championships, conferences, summits, and showcases generate revenue through registration, sponsorship, premium access, and experiential participation fees. Each event produces both revenue and data.

4

Media & Data Rights

Content licensing, archives, analytical insights, and sponsorships. The media layer monetizes the vast content and data generated across the ecosystem through syndication, archival access, and premium insight products.

This structure ensures recurring revenue, institutional defensibility, and long-term scalability — three characteristics that distinguish permanent jurisdictional assets from transient technology products.

Revenue Flow Visualization

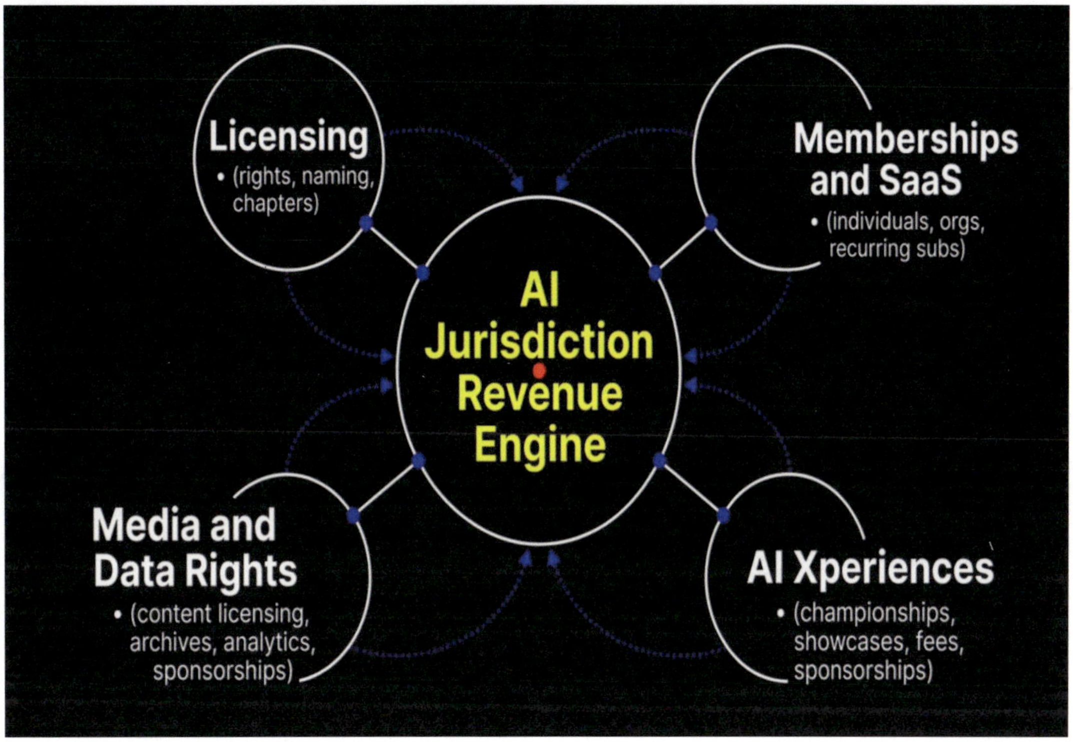

The four pillars do not operate in isolation. Licensing creates institutional partnerships that drive memberships. Memberships create the participant base that fills Xperiences. Xperiences generate the content and data that power Media & Data Rights. Media visibility attracts new licensing partners. The cycle is continuous, self-reinforcing, and deeply defensible.

VALUATION

Valuation Framing: The Multiplier Effect

The Artificial Intelligence Jurisdiction does not establish a new valuation floor. It raises every ceiling.

Multiplicative Architecture

The valuation framework for the AI Jurisdiction must be understood not as a standalone asset class but as a multiplier applied to the entire enterprise portfolio. Each layer of the ecosystem amplifies the value of every other layer, and Artificial Intelligence is the catalyst that accelerates this compounding effect.

Asset-Based Software Foundation

The owned software, platforms, and technical infrastructure establish the baseline valuation. These are tangible, auditable, and operational assets.

Strategic Land Reserves

Domains and ISBNs remain uncounted strategic reserves — appreciating assets whose value increases as the AI landscape matures and institutional demand intensifies.

The AI Multiplier

Artificial Intelligence increases retention, completion, conversion, output, and licensing durability across every vertical. This is not additive value. It is multiplicative architecture.

Multiplier Impact Across Verticals

85%

Retention Uplift

AI-powered adaptive experiences dramatically increase participant retention across academies and championships

73%

Conversion Acceleration

Intelligent matchmaking and personalization compress the time from discovery to engagement across all platforms

92%

Licensing Durability

AI-enhanced institutional frameworks create long-term contractual relationships that resist competitive displacement

67%

Content Output

Automated media pipelines multiply content production volume while maintaining quality and brand consistency

> This is not additive value. It is multiplicative architecture. Every improvement in one vertical compounds across all others.

Why This Matters

Sovereign Positioning

We Built the Ports

People watched the wave. We built the ports.

Anyone can build an app. The barrier to creating software has never been lower. Large language models, open-source frameworks, and cloud infrastructure have democratized the ability to ship code. But the ability to build an app is not the ability to build an institution. The ability to ship code is not the ability to own a jurisdiction.

Not everyone builds an owned Artificial Intelligence jurisdiction spanning education, competition, business, culture, and media. Not everyone possesses the strategic foresight to secure institutional naming rights before the market understands their value. Not everyone has the operational discipline to construct a multi-layered, self-reinforcing ecosystem where every component amplifies every other.

The positions secured by Goddess Phenezy Holdings are not replicable. They are not available for acquisition by competitors. They are not subject to the platform risk that haunts every company built on rented infrastructure.

Sovereign Advantages

You cannot deplatform domains.

You cannot cancel infrastructure you do not control.

You cannot replicate what has already been permanently secured.

You cannot commoditize a governing jurisdiction.

This is sovereign positioning.

The Position Is Secured.

What follows is not a vision statement. It is not a pitch. It is a declaration of what already exists — a sovereign intellectual property architecture that governs the intersection of human cognition and artificial intelligence across education, competition, commerce, culture, and media. The perimeter is secured. The jurisdiction is operational. The loop is running.

We Own the Names

200+ secured domains and registered trade names spanning every AI institutional category — from academies and universities to championships, conferences, festivals, and incubators. No competitor holds this portfolio.

We Own the Governance

The National & International Artificial Intelligence Association sits at the apex — setting the standards, issuing the credentials, and defining the rules under which all AI institutional activity within this jurisdiction operates.

We Own the Pipeline

Education trains. Competition validates. Commerce scales. Culture embeds. The three-pillar cognitive pipeline is a closed-loop system that transforms raw talent into credentialed, monetizable intelligence — permanently governed.

We Own the Loop

Every Xperience generates data. Every dataset trains intelligence. Every improvement compounds value. The compounding intelligence loop makes this jurisdiction exponentially more defensible with every passing cycle.

> **Anyone can build an app. Anyone can launch a product. Anyone can rent space in someone else's ecosystem. We built the ground. We own the jurisdiction. The position is irrevocable.**

We did not write a pitch.

We established a record!

XPERIENCES
R US

TECHNOLOGIES
R US

CONFERENCES R US

YourPartnrr.com
Grow Together
A next-generation partner relationship platform
for tracking referrals, collaborations and shared growth.

KJ
THE
INTERNATIONAL
BASKETBALL
CHAMPIONSHIP

THE INTERNATIONAL
BASEBALL
CHAMPIONSHIP

THE INTERNATIONAL
RAP
CHAMPIONSHIP

THE INTERNATIONAL ACADEMY OF
BLACK HISTORY

THE INTERNATIONAL ACADEMY
OF
BLACK
HISTORY
WE CAN NEVER BE ERASED!

SCHOLARSHIPS
R US

MISSING FROM THE VILLAGE™

IQ
Whizzy

THE
CHAMPIONS
OF
AMERICA

THE
INTERNATIONAL
PEACE
CONFERENCE

THE INTERNATIONAL ACADEMY OF

BLACK HISTORY

THE NATIONAL
GOSPEL
CHAMPIONSHIP

THE

INTERNATIONAL

ACADEMY OF LEARNING

THE INTERNATIONAL ACADEMY OF
STEM
SCIENCE · TECHNOLOGY · ENGINEERING · MATHEMATICS
STEM

THE
INTERNATIONAL
GOLF
CHAMPIONSHIP
MLP

THE INTERNATIONAL
GAMERS
CHAMPIONSHIP

THE
INTERNATIONAL
DANCE
CHAMPIONSHIP

ALA
THE INTERNATIONAL
CHESS
CHAMPIONSHIP

THE INTERNATIONAL BUSINESS
XCCELERATOR

C
THE INTERNATIONAL
BID WHIST
CHAMPIONSHIP

THE
INTERNATIONAL
MUSIC FESTIVAL

THE INTERNATIONAL
BOXING
CHAMPIONSHIP

THE
INTERNATIONAL
HOCKEY
CHAMPIONSHIP

THE INTERNATIONAL ACADEMY OF BUSINESS

EDUCATE • EMPOWER • EXCEL

TERM MATRIX

A comprehensive glossary of key terms and concepts used throughout the valuation framework, covering asset-based enterprise valuation, digital infrastructure, intellectual property, strategic market positioning, and the broader ecosystem of technology-driven business growth.

Acquisition Premium

An additional amount paid above the base value of a company during an acquisition to account for strategic advantage, speed, or competitive positioning. In this book's valuation framework, it contributes to the Strategic Multiplier used to determine Strategic Market Value.

App

Short for "application," software designed to perform specific functions for users on web or mobile platforms. In this ecosystem, apps refer to structured digital systems delivering competition, education, networking, and infrastructure services.

Asset-Based Enterprise

A company whose value is primarily grounded in owned assets such as intellectual property, infrastructure, domains, and systems rather than projected future speculation. This concept forms the foundation of the book's valuation logic.

Asset-Based Unicorn

A high-growth enterprise valued at $1 billion or more that is grounded in owned infrastructure and replacement-cost validation rather than speculative venture capital multiples. In this book, it distinguishes structural valuation from traditional startup hype.

Asset Floor

The minimum defensible valuation of an enterprise based on tangible and quantifiable assets. In the valuation report, this aligns closely with Replacement Cost before multipliers are applied.

TERM MATRIX PART II

Asset Valuation

The process of determining the financial worth of a company's assets. In this book, valuation is approached through replacement cost analysis and strategic multiplier modeling rather than speculative forecasting.

Barriers to Entry

Obstacles that make it difficult for competitors to enter a market, such as capital requirements, intellectual property, or technical complexity. In the valuation framework, strong barriers increase defensibility and Moat Score.

Brand Architecture

The structured organization of a company's brands, subsidiaries, and divisions under a unified strategy. In this ecosystem, it refers to the coordinated alignment of multiple platforms and verticals under one holding structure.

Capital Efficiency

A measure of how effectively a company uses its capital to generate revenue or growth. The book's valuation model highlights founder-led capital efficiency through high development output relative to cost.

COCOMO II

Constructive Cost Model II, a widely used software engineering model for estimating development effort and cost. It is referenced in the valuation methodology to calculate Development Time and Replacement Cost.

Complexity Score

A technical measurement used to estimate how difficult software is to design, maintain, and expand. Higher complexity typically indicates more advanced architecture and development effort. In the valuation report, the platform's complexity score reflects the sophistication of the codebase and contributes to the strategic value calculation.

TERM MATRIX PART III

Cost Basis

The original cost required to create or acquire an asset. It serves as a foundational reference point when calculating valuation, profit, or return on investment. In this book's framework, cost basis contributes to understanding the minimum value embedded in the system before strategic market multipliers are applied.

Copyright

A form of legal protection that grants creators exclusive rights to reproduce, distribute, and display their original works. This protection applies to books, software code, designs, and creative content. Within this ecosystem, copyright safeguards written works, educational content, and original software elements.

Data Asset

A dataset that has measurable value because it can generate insights, predictions, or operational advantages. Organizations increasingly treat data as a strategic resource similar to capital or infrastructure. In this platform architecture, certain systems generate proprietary datasets that contribute to competitive advantage and valuation strength.

Development Time (Dev Time)

An estimate of the total engineering effort required to build a software system. This is typically measured in developer months or years. In the valuation report, development time reflects the cumulative effort needed to produce the platform's codebase and helps determine replacement cost.

Digital Infrastructure

The foundational technology systems that support digital platforms and services. This includes servers, software architecture, databases, and application frameworks. In this ecosystem, digital infrastructure forms the operational backbone enabling competitions, education, and community interaction within a unified platform.

TERM MATRIX PART IV

Distribution Channel

The method through which products, services, or information reach users. Channels may include digital platforms, marketplaces, events, or media networks. In this ecosystem, distribution occurs through championships, conferences, digital platforms, and broadcast media.

Domain Asset

A registered internet domain that carries strategic or commercial value. Premium domains often function as digital real estate and can strengthen brand recognition or market positioning. Within this ecosystem, domain ownership supports brand authority and long-term infrastructure control.

Economic Infrastructure

The systems and institutions that enable economic activity, including platforms that support commerce, employment, education, and innovation. Modern digital platforms increasingly function as economic infrastructure. In this model, the platform operates as a digital ecosystem connecting participants across multiple industries.

Founder Premium

An additional valuation factor reflecting the value created by the founder's vision, expertise, and direct involvement in building the system. Founder-led companies often benefit from deeper institutional knowledge and long-term strategic commitment. In the valuation report, this premium increases the strategic market value of the platform.

Franchise Model

A business structure where a company allows independent operators to use its brand, systems, and operating methods in exchange for fees or royalties. Franchise systems rely on standardized processes and brand consistency. While this book discusses large-scale expansion systems, the model described here focuses more heavily on licensing and infrastructure ownership than traditional franchising.

TERM MATRIX PART V

Governance Structure

The framework that defines how decisions are made within an organization, including leadership roles, authority, accountability, and oversight. Strong governance helps maintain operational stability and long-term strategic alignment. In this ecosystem, governance supports coordination across multiple divisions operating under a unified corporate structure.

Gross Revenue

The total income generated by a business before expenses are deducted. It represents the full volume of economic activity produced by an enterprise. In growth-stage companies, gross revenue is often used to evaluate scale and market traction.

Holding Company

A parent entity created to own controlling interests in other companies, assets, or intellectual property. Holding companies provide centralized ownership while allowing subsidiaries to operate independently. In this book's structure, the holding entity owns and oversees multiple platforms, divisions, and intellectual property assets.

Intellectual Property (IP)

Legal rights that protect creations of the mind, including inventions, trademarks, copyrighted works, and proprietary systems. Intellectual property allows innovators to control how their ideas are used and monetized. Within this ecosystem, IP forms a major portion of the company's strategic value and competitive moat.

Intrinsic Value

The underlying worth of an asset based on its actual economic utility rather than market speculation. Intrinsic value focuses on what something is fundamentally worth. The asset-based valuation approach used in this book emphasizes intrinsic value through replacement cost and infrastructure ownership.

TERM MATRIX PART VI

Jurisdiction

The legal authority of a government or regulatory body over a geographic area or organization. Businesses must operate within the legal frameworks of the jurisdictions in which they are registered or conduct operations. Jurisdiction influences taxation, compliance, and regulatory oversight.

Licensing Model

A business structure where a company allows others to use its intellectual property, systems, or brand under defined agreements. Licensing enables expansion without transferring ownership of the underlying assets. The expansion strategy discussed in this ecosystem relies heavily on licensing rather than franchising to maintain asset control.

Lines of Code (LOC)

A quantitative measure used to estimate the size and development effort of a software system. It counts the number of executable lines within a codebase. In the valuation report, lines of code are used as one indicator of development scale and contribute to calculating replacement cost and development time.

Logo

A distinctive visual symbol used to represent a brand or organization. Logos serve as identity markers that help audiences recognize and remember a company. In a large brand ecosystem, logos become valuable intellectual property assets that reinforce brand authority and trust.

Market Multiple

A valuation metric used to estimate a company's worth based on industry benchmarks such as revenue, earnings, or comparable acquisitions. Multiples reflect how investors value similar businesses in the market. The strategic multiplier used in the valuation report functions as a specialized form of market multiple applied to replacement cost.

TERM MATRIX PART VII

Market Positioning

The strategic way a company presents itself within its industry in relation to competitors. Market positioning defines how a brand is perceived and what role it plays in the marketplace. In this ecosystem, positioning emphasizes infrastructure ownership, intellectual property, and long-term platform development rather than short-term startup speculation.

Moat Score

A measurement used to estimate how defensible a company's competitive position is against rivals. The concept is derived from the idea of a "business moat," meaning protective advantages that are difficult for competitors to replicate. In the valuation report, the platform's moat score reflects the strength of its intellectual property, proprietary systems, and technological complexity.

Net Revenue

The amount of income a business retains after deductions such as discounts, refunds, or operational costs. Net revenue provides a more accurate picture of financial performance than gross revenue alone. It is commonly used to assess profitability and operational efficiency.

Operating Division

A specialized unit within a company responsible for a specific line of activity or service. Divisions allow large organizations to manage multiple products or industries under one corporate structure. In this ecosystem, various platforms and initiatives operate as divisions within a broader organizational framework.

Operating Leverage

A financial concept describing how fixed costs and variable costs interact to influence profitability. Companies with strong operating leverage can increase profits rapidly as revenue grows because fixed costs remain relatively stable. Technology platforms often benefit from high operating leverage once infrastructure is built.

TERM MATRIX PART VIII

Organizational Structure

The formal arrangement of roles, responsibilities, and authority within an organization. Organizational structure determines how teams operate, how decisions are made, and how work flows across departments. In large multi-platform enterprises, clear structure is essential to coordinate complex operations.

Patent

A legal protection granted by a government that gives an inventor exclusive rights to make, use, or sell an invention for a limited period of time. Patents protect new processes, technologies, and systems from being copied by competitors. In this ecosystem, several proprietary systems and platform innovations are positioned for patent protection to strengthen long-term competitive advantage.

Platform Dependency

A situation where a business relies heavily on another company's infrastructure or technology to operate. High dependency can create vulnerability if access to that platform changes or is restricted. The strategy outlined in this book emphasizes building independent infrastructure to reduce platform dependency.

Pre-Revenue Valuation

An estimated company value determined before the business begins generating significant revenue. These valuations are typically based on intellectual property, infrastructure, market opportunity, and strategic positioning. The valuation framework discussed in this book uses asset-based methods to support pre-revenue valuation.

Progressive Web App (PWA)

A type of web-based software application that behaves like a native mobile app while running in a web browser. PWAs combine the accessibility of websites with the functionality of traditional apps. The platform described in the valuation report operates as a large-scale Progressive Web App designed to deliver multiple services through a single unified interface.

TERM MATRIX PART IX

Proprietary Algorithm

A custom mathematical or logical process developed by a company to perform specific tasks or generate insights. Proprietary algorithms are typically protected as trade secrets or patents because they provide competitive advantages. In the valuation report, proprietary algorithms contribute to the platform's defensibility and overall strategic value.

Replacement Cost

The estimated cost required to rebuild a system, product, or asset from the ground up using current market labor and development standards. Replacement cost is often used as a baseline valuation method when intellectual property or infrastructure already exists. In the valuation report, replacement cost represents the foundational value of the platform before strategic multipliers are applied.

Return on Investment (ROI)

A financial metric used to evaluate the efficiency of an investment by comparing gains relative to the original cost. ROI helps investors determine whether a project or asset produces sufficient returns. In technology ventures, ROI may include both financial outcomes and strategic advantages such as market positioning.

Revenue Model

The structured method through which a company generates income from its products or services. Revenue models may include subscriptions, licensing, advertising, transactions, or service fees. In a multi-platform ecosystem, several revenue models may operate simultaneously across different divisions.

Revenue Stream

An individual source of income within a company's overall revenue model. Businesses often develop multiple revenue streams to diversify risk and increase financial stability. Large digital ecosystems frequently combine event revenue, digital services, media distribution, and licensing streams.

TERM MATRIX PART X

Risk Mitigation

The process of identifying potential threats to a business and implementing strategies to reduce their impact. Effective risk mitigation protects both financial stability and operational continuity. In technology ventures, risks may include regulatory changes, platform dependency, or competitive pressure.

Scalability

The ability of a system, platform, or business model to grow significantly without a proportional increase in costs or operational complexity. Highly scalable systems can serve large numbers of users efficiently. Digital platforms and software-based ecosystems are often designed specifically for scalability.

Software

A collection of computer programs and instructions that allow digital systems to perform tasks and deliver services. Software can power websites, applications, databases, and complex enterprise platforms. In this ecosystem, software serves as the operational infrastructure enabling competitions, education, networking, and community engagement.

Speculative Valuation

A valuation method based largely on projected future growth rather than existing assets or infrastructure. Speculative valuations are common in venture capital environments where companies may have limited revenue but high growth expectations. The asset-based framework in this book contrasts with speculative valuation by emphasizing measurable infrastructure value.

Startup

A newly formed company designed to develop and scale innovative products or services, often within technology sectors. Startups typically focus on rapid growth and market disruption. Many startups seek venture capital funding in early stages to accelerate expansion.

TERM MATRIX PART XI

Strategic Market Value

The estimated price an acquirer or investor might pay for a company based on strategic advantages such as intellectual property, time-to-market, and competitive positioning. Strategic market value often exceeds replacement cost because it reflects future opportunity. In the valuation report, this value is calculated by applying strategic multipliers to the base replacement cost.

Strategic Multiplier

A factor applied to a company's base valuation to reflect additional strategic advantages such as intellectual property protection, complexity, or time-to-market advantages. Multipliers increase the estimated acquisition value of a system beyond its rebuild cost. In the valuation report, several strategic factors combine to produce the overall multiplier used to determine market value.

Subsidiary

A company that is owned or controlled by a parent or holding company. Subsidiaries operate as separate legal entities while remaining part of a larger corporate structure. This structure allows organizations to manage multiple business units under centralized ownership.

Sustainable Growth

A pattern of business expansion that can be maintained over time without excessive financial risk or operational strain. Sustainable growth emphasizes long-term stability rather than short-term acceleration. Strong infrastructure and diversified revenue models often support sustainable growth.

Time-to-Market

The amount of time required to develop and launch a product, service, or platform. Shorter time-to-market can create competitive advantages by allowing companies to capture opportunities earlier than rivals. In the valuation report, time-to-market contributes to the strategic multiplier applied to the platform's value.

TERM MATRIX PART XII

Trademark

A legally protected word, phrase, symbol, or design used to identify and distinguish a company's goods or services. Trademarks protect brand identity and prevent marketplace confusion. Logos, brand names, and slogans are commonly protected through trademark registration.

Trade Secret

Confidential business information that provides a company with competitive advantages. Trade secrets may include algorithms, formulas, processes, or proprietary methodologies. Unlike patents, trade secrets remain protected as long as they are kept confidential.

Unicorn

A privately held startup company valued at one billion dollars or more. The term is commonly used in the technology sector to describe rare high-growth companies that achieve significant valuations. Traditional unicorn valuations often rely heavily on investor expectations and future growth projections.

Vertical Integration

A strategy in which a company expands control over multiple stages of its supply chain or industry ecosystem. Vertical integration allows organizations to reduce dependency on outside providers and maintain greater operational control. In digital platforms, integration may include technology infrastructure, distribution channels, and content creation.

.com Domain

A top-level internet domain commonly associated with commercial businesses and organizations. .com domains are often considered premium digital real estate because of their global recognition and credibility. Strategic domain ownership can strengthen brand authority and online discoverability.

VALUATION REPORT

YOUR PARTNRR DATA ROOM

CERTIFIED SOFTWARE VALUATION REPORT

YPDR-00053

Xperiences R Us, Inc.

Report Date: March 3, 2026

VALUATION SUMMARY

Replacement Cost

$242,925,330

Cost to rebuild from scratch

Strategic Market Value

$716,629,724

2.95x strategic multiplier

Lines of Code	Complexity	Moat Score	Dev Time
391,620	**73/100**	**100/100**	**783.2 mo**

EXECUTIVE SUMMARY

COMPLETE

Lines of Code	Features	Moat Score	Dev Time
392K	**44**	**100/100**	**783mo**

Value Comparison
Replacement: $242,925,330

Market Value: $716,629,724 (2.95x)

Key Technical Findings

• Architecture:	**React + Vite + Supabase Backend**	• Primary Language:	**TypeScript (89.1%)**
• Test Coverage:	**Detected**	• Custom Code:	**95.0%**

Founder Premium: Enabled (+35%)

Development Stage: Live

Description:

Xperiences R Us is an enterprise-grade, multi-domain web platform developed by Goddess Phenezy Holdings, Inc. The application serves as a comprehensive ecosystem combining competitive gaming, education, social networking, marketplace commerce, and public safety initiatives – all unified under a single Progressive Web Application (PWA).

AI Analysis Summary:

Xperiences R Us is a multi-domain web platform that integrates competitive gaming, education, social networking, marketplace commerce, and public safety initiatives into a single Progressive Web Application (PWA). Its unique value proposition lies in its ability to provide a cohesive user experience across diverse domains, leveraging advanced algorithms and proprietary workflows to enhance user engagement and learning outcomes.

STRATEGIC MARKET VALUE ANALYSIS

Market value reflects what an acquirer would pay—not just rebuild cost, but the strategic value of time-to-market, intellectual property, and competitive advantage.

Replacement Cost

$242,925,330

Cost to rebuild

Strategic Market Value

$716,629,724

2.95x multiplier

Strategic Multiplier Breakdown

Premium	Contribution	Multiplier
Base Acquisition Premium		+1.50x
Time-to-Market (36+ months)		+0.60x
Pending Patent Protection		+0.20x
Trade Secrets		+0.20x
Proprietary Algorithms		+0.15x
High Complexity (70+)		+0.30x

Total Strategic Multiplier **2.95x**

CODE INVENTORY

		Complexity	Dev Time
2,092	391,620	73/100	783.2 mo

Architecture: React + Vite + Supabase Backend

Language Distribution:

Detailed Breakdown:

Language	Lines	Files	Percentage
TypeScript	348,936	1820	**89.1%**
SQL	28,594	218	**7.3%**
Markdown	12,819	41	**3.3%**
CSS	579	2	**0.1%**
JavaScript	539	3	**0.1%**
JSON	103	6	**0.0%**
HTML	50	2	**0.0%**

MOAT MATRIX™

Overall Moat Score

100/100

Declared Proprietary Features (44 total):

HIGH MOAT FEATURES (36)

1 10x10 Board with Corner Extensions

Patent (Filed/Pending) 1.99x

An original board configuration for Four Corners Chess allowing for 4 players.

2 12+ Chess Grandmaster Personalities

Proprietary Algorithm 1.99x

AI personalities based on historical chess players with unique playing styles.

3 4 Senet Mentor Personalities

Proprietary Algorithm 1.99x

Mentor figures from Egyptian history that guide players in the game of Senet.

4 50-Square Dual-Path Board

Patent (Filed/Pending) 1.99x

An original board design for the Royal Convergence game with two starting paths.

5 Accuracy/Streak/Speed Bonus System

Proprietary Algorithm 1.99x

Multi-factor scoring with capped bonuses based on performance metrics.

6 Adaptive Mechanic Selection System

Proprietary Algorithm 1.99x

Dynamic difficulty based on performance with variety triggers and repetition avoidance.

7 AI Cascading Fallback System

Proprietary Algorithm 1.99x

Failover system using Gemini/OpenAI with an 8-second timeout for resilience.

8 Alliance Mode (2v2)

Patent (Filed/Pending) 1.99x

Team-based chess variant allowing for shared victory among players.

9 Anti-Cheat System

Trade Secret 1.99x

System analyzing move timing, position complexity, and accuracy to prevent cheating.

10 Blaze Line Indicators™

Proprietary Algorithm 1.99x

Fire/energy trails for aggressive moves in chess gameplay.

MOAT MATRIX™ (continued)

11 Champion Burst™
Proprietary Algorithm — 1.99x
Tiered capture explosions based on piece value in chess gameplay.

12 Crown Clash FX™
Proprietary Algorithm — 1.99x
Special effects for Queen interactions in chess gameplay.

13 Cryonox Shield™
Proprietary Algorithm — 1.99x
Defensive move shields with ice effects in chess gameplay.

14 ELO & Progression System
Trade Secret — 1.99x
10-tier difficulty ratings and K-factor adjustments for player ranking.

15 Last Known Truth Timeline
Proprietary Algorithm — 1.99x
Visual distinction between verified facts and inferred paths in missing persons cases.

16 Move Quality Effects
Proprietary Algorithm — 1.99x
Indicators for move quality such as Brilliant, Good, Inaccuracy, Mistake, and Blunder.

17 Multi-Check Detection
Patent (Filed/Pending) — 1.99x
Mechanics for detecting simultaneous check from multiple opponents in Four Corners Chess.

18 Multiple Win Conditions
Patent (Filed/Pending) — 1.99x
Flexible game modes allowing for team victory, individual glory, or first exit.

19 Paranoid AI Search Algorithm
Patent (Filed/Pending) — 1.99x
AI evaluation assuming worst-case scenarios in multi-player chess.

20 Personality-Specific Quotes/Commentary
Proprietary Algorithm — 1.99x
Context-aware responses from AI mentors that match their character personalities.

21 Player Elimination Mechanics
Patent (Filed/Pending) — 1.99x
Flexible rules for removing or neutralizing pieces in Four Corners Chess.

22 Pre-Move Hint Visualization
Proprietary Algorithm — 1.99x
Threat highlighting with critical threat indicators in chess gameplay.

23 Scholarship Matching Engine
Proprietary Algorithm — 1.99x
Weighted AI algorithm for matching student athletes with scholarships.

MOAT MATRIX™ (continued)

24 Silent Return Protocol

Proprietary Algorithm ■■■ 1.99x

Anonymous safety confirmation system for adults in missing persons cases.

25 Skills DNA Analyzer

Proprietary Algorithm ■■■ 1.99x

AI-powered athletic profiling with radar visualization for athletes.

26 Sound & Haptic Feedback System

Proprietary Algorithm ■■■ 1.99x

Customizable audio and haptic feedback for enhanced user experience.

27 Team Convergence Mechanics

Patent (Filed/Pending) ■■■ 1.99x

Mechanics allowing 2v2 teams to coordinate strategies in the Royal Convergence game.

28 Teammate Push/Swap Rules

Patent (Filed/Pending) ■■■ 1.99x

Rules allowing allies to assist each other's pieces in the Royal Convergence game.

29 Unified IQ Calculation Algorithm

Proprietary Algorithm ■■■ 1.99x

Cross-game cognitive scoring across 11 game types with recency-weighted aggregation.

30 VerifiEYED Sightings

Proprietary Algorithm ■■■ 1.99x

Trust-weighted reporting system for sightings of missing persons.

31 VerifiEYED Verification System

Proprietary Algorithm ■■■ 1.99x

Tiered identity verification system for users in the platform.

32 Victory Surge™

Proprietary Algorithm ■■■ 1.99x

Checkmate board pulse animation for celebrating game ends.

33 Village Memory Engine

Proprietary Algorithm ■■■ 1.99x

Automated resurfacing of cold cases via anniversary alerts and pattern matching.

34 Village Signal

Proprietary Algorithm ■■■ 1.99x

Geo-fenced alert system for community alerts regarding missing persons.

35 Whizzy Persona State Machine

Proprietary Algorithm ■■■ 1.99x

AI character with mood-based responses to enhance educational engagement.

36 Whizzypedia Community Learning

Proprietary Algorithm ■■■ 1.99x

AI-powered course generation and community sharing features.

MOAT MATRIX™ (continued)

MEDIUM MOAT FEATURES (8)

37 Difficulty Unlock Progression

Workflow/Process 1.34x

Performance-based mechanic unlocking at various thresholds.

38 Game Type Weight Multipliers

Trade Secret 1.68x

Weights based on cognitive load per game type, ranging from 0.8 to 2.0.

39 Real-Time Hype Meter

Proprietary Algorithm 1.68x

Community engagement scoring with viral tracking for athletes.

40 Reporter Trust Scores

Proprietary Algorithm 1.68x

Reputation system for assessing the accuracy of sightings reports.

41 Roll-Off Initialization

Patent (Filed/Pending) 1.68x

Casting sticks determine turn order in the Royal Convergence game.

42 Scout View Tracking

Proprietary Algorithm 1.68x

Analytics for verified scout/recruiter profile views.

43 Style Weight Modifiers

Trade Secret 1.68x

Adjustments to AI move evaluation based on different playing styles.

44 Whizzy Pods Accountability System

Trade Secret 1.68x

Real-time check-ins and leaderboards for accountability groups.

APPENDIX: METHODOLOGY & CITATIONS

Valuation Methodology:

1. Lines of code counted per IEEE Standard 1045-1992 (executable code only, excluding comments and blanks).
2. Productivity rate: 500 LOC/developer/month based on COCOMO II nominal range.
3. Hourly rates from US Bureau of Labor Statistics (BLS) for Software Developers (15-1252).
4. Team composition: 10% architecture, 30% senior, 40% mid-level, with QA and PM overhead.
5. Complexity score uses weighted factors per COCOMO II effort multipliers.
6. Market value methodology aligns with M&A software asset valuation practices.
7. Moat score assesses defensibility using IP law and competitive strategy frameworks.
8. Founder premium (+35%) reflects value of founder-led development and domain expertise.

(!) IMPORTANT DISCLAIMER

This valuation represents an estimated replacement cost and strategic market value based on quantitative code analysis and declared proprietary features. It should be used as one input among many in any investment or acquisition decision. Actual transaction values depend on market conditions, negotiation, and comprehensive due diligence beyond the scope of this automated report.

Report Metadata:

Report Number: YPDR-00053
Generated: 3/3/2026, 9:56 PM
Valuation ID: daea8036-357c-4dea-992e-a3ad45376223
Created: 3/3/2026
Last Updated: 3/3/2026

Report Verification

This report (YPDR-00053) can be verified at:
www.YourPartnrrDataRoom.com/verify

Or scan the QR code to verify instantly

VALUATION REPORT

YOUR PARTNRR DATA ROOM

CERTIFIED SOFTWARE VALUATION REPORT

YPDR-00046

Your Partnrr

Entity: Your Partnrr

Report Date: February 27, 2026

VALUATION SUMMARY

Replacement Cost

$132,916,665

Cost to rebuild from scratch

Strategic Market Value

$392,104,162

2.95x strategic multiplier

Lines of Code	Complexity	Moat Score	Dev Time
214,275	73/100	92/100	428.6 mo

EXECUTIVE SUMMARY

COMPLETE

Lines of Code	Features	Moat Score	Dev Time
214K	**43**	**92/100**	**429mo**

Value Comparison
Replacement: $132,916,665

Market Value: $392,104,162 (2.95x)

Key Technical Findings

- Architecture: **React + Vite + Supabase Backend**
- Primary Language: **TypeScript (94.5%)**
- Test Coverage: **Detected**
- Custom Code: **95.0%**

Founder Premium: Enabled (+35%)

Development Stage: Live

Description:

PRM & CRM with advanced mlm and affiliate features.

AI Analysis Summary:

'Your Partnrr' is a specialized PRM/CRM hybrid engineered for the service industry and MLM models, focusing on real-time financial transparency and high-engagement gamification. Its primary value lies in its 'LiveMargin' and 'PartnerIQ' engines, which automate complex commission and performance analytics that typically require manual spreadsheets.

STRATEGIC MARKET VALUE ANALYSIS

Market value reflects what an acquirer would pay—not just rebuild cost, but the strategic value of time-to-market, intellectual property, and competitive advantage.

Replacement Cost

$132,916,665

Cost to rebuild

Strategic Market Value

$392,104,162

2.95x multiplier

Strategic Multiplier Breakdown

Premium	Contribution	Multiplier
Base Acquisition Premium		+1.50x
Time-to-Market (36+ months)		+0.60x
Pending Patent Protection		+0.20x
Trade Secrets		+0.20x
Proprietary Algorithms		+0.15x
High Complexity (70+)		+0.30x

Total Strategic Multiplier **2.95x**

CODE INVENTORY

		Complexity	Dev Time
1,115	214,275	73/100	428.6 mo

Architecture: React + Vite + Supabase Backend

Language Distribution:

Detailed Breakdown:

Language	Lines	Files	Percentage
TypeScript	202,477	993	**94.5%**
SQL	11,164	109	**5.2%**
Markdown	265	2	**0.1%**
CSS	167	2	**0.1%**
JavaScript	84	3	**0.0%**
HTML	61	2	**0.0%**
JSON	57	4	**0.0%**

MOAT MATRIX™

Overall Moat Score

92/100

Declared Proprietary Features (43 total):

HIGH MOAT FEATURES (8)

1 ChurnShield Retention Engine

Proprietary Algorithm — 1.99x

A drop-off prevention system that detects declining activity and triggers automated re-engagement workflows.

2 Compliance Guardrails

Trade Secret — 1.99x

Automated checks for FTC and GDPR compliance built directly into the partner workflow.

3 Cross-Program Wallet

Integration Moat — 1.86x

A unified earnings wallet that aggregates income from multiple different partner programs.

4 FlexNet Program Modeler

Workflow/Process — 1.59x

A single-click architectural switch that allows a business to change its entire program model from affiliate to reseller.

5 Marketplace for Partnerships

Network Effect — 2.13x

An ecosystem-wide discovery layer to connect compatible partners across different programs.

6 Partnrr IQ Scoring Engine

Proprietary Algorithm — 1.99x

Automated partner scoring system that identifies partners requiring attention through predictive behavioral analysis.

7 Partnrr Match Compatibility

Proprietary Algorithm — 1.99x

A partner discovery tool using compatibility scoring and red flag detection to find ideal fits.

8 TrackLink Attribution System

Patent (Filed) — 2.13x

A proprietary tracking technology that monitors partner sharing activities and performance metrics in real-time.

MEDIUM MOAT FEATURES (30)

9 Accountability Partnrr Pods

Network Effect — 1.79x

An automated matching system that places partners into accountability groups to track goals and streaks.

MOAT MATRIX™ (continued)

10 Advanced Drag-and-Drop Interaction Engine

UX/Design — 1.20x

The use of multiple dnd-kit modules implies a highly sophisticated workspace, project management, or canvas-based interface that is difficult to replicate with standard UI kits.

11 Advanced Drag-and-Drop Interaction Engine

UX/Design — 1.20x

The use of multiple dnd-kit modules implies a highly sophisticated workspace, project management, or canvas-based interface that is difficult to replicate with standard UI kits.

12 AI-Generated Training

Data Asset — 1.56x

Personalized training modules featuring voice narration generated by AI.

13 AutoCalc Core Rule Engine

Workflow/Process — 1.34x

A persistent calculation engine that applies complex commission and earning rules in real-time without manual intervention.

14 Automated Onboarding Avatars

UX/Design — 1.45x

Interactive AI-driven avatars for 24/7 partner training and onboarding.

15 AutoTrack Mode

Workflow/Process — 1.34x

An automated link generation and attribution environment that requires zero manual setup.

16 Brand DNA Engine

UX/Design — 1.45x

An automated white-labeling system that applies brand identity across the entire platform ecosystem.

17 Complex State Management Logic

Proprietary Algorithm — 1.40x

Given 210k LOC but few external state libraries, the software likely contains significant custom logic for handling complex user workflows or data transformations.

18 Complex State Management Logic

Proprietary Algorithm — 1.40x

Given 210k LOC but few external state libraries, the software likely contains significant custom logic for handling complex user workflows or data transformations.

19 Contribution-Based Rewards

Proprietary Algorithm — 1.68x

A system for rewarding non-sales activities such as mentorship and content creation.

20 Earnings Autopilot

Workflow/Process — 1.34x

Automated commission rule execution where reps see earnings instantly pending final approval.

MOAT MATRIX™ (continued)

21 Income Goal Calculator

Proprietary Algorithm 1.68x

A reverse-engineered calculator that breaks down exact requirements needed to reach a specific income level.

22 LiveMargin Engine

Proprietary Algorithm 1.68x

Instant profit calculation per job, allowing service businesses to see margins before invoicing.

23 Marketing Studio

UX/Design 1.45x

An automated generator for professional marketing materials tailored to the partner's brand.

24 OneClick Payout

Integration Moat 1.56x

An integrated payment approval system that bridges calculated earnings with partner disbursements.

25 Partner Assignment Protocol

Workflow/Process 1.34x

An automated system for assigning representatives and tracking their specific contributions to a deal.

26 PartnerIQ Multi-Dimensional Scoring Engine

Proprietary Algorithm 1.40x

Found in `PartnerIQScoreCard.tsx`, this calculates consistency, team health, and conversion quality in real-time. Unlike generic CRMs, it uses a 'Trend' indicator logic (`improving` vs `declining`) to drive predictive churn prevention.

27 Partnrr Coach Recommendations

Proprietary Algorithm 1.68x

An AI-driven coaching engine that provides personalized growth paths and next-step recommendations for partners.

28 Partnrr Pulse Gamification

UX/Design 1.45x

An automated motivation system featuring leaderboards, badges, and competitive triggers managed by the platform.

29 Proprietary Cost Matrix Data Structure

Data Asset 1.30x

The 'Zero-Sheet' promise is backed by a structured cost-tracking utility that persists labor and overhead rates across all transaction types, as evidenced by the `ServiceBusinessHowItWorks` component.

30 Pulse Gamification Framework

UX/Design 1.20x

Observed in `/pulse/index.ts`, the sheer variety of engagement modals (SecretMission, TeamFusion, PredictionResolver) indicates an unusually deep investment in MLM-specific psychological triggers.

31 QuickLaunch Setup

Workflow/Process 1.34x

A 5-question wizard that automatically builds out a full partner program in 10 minutes.

32 Service Business Margin Reveal Logic

Workflow/Process 1.10x

Defined in `ServiceBusinessHowItWorks.tsx`, specifically the 'Cost Matrix' and 'Instant Margin Reveal' steps, this codebase bridges the gap between Field Service Management and Affiliate Marketing, which is a unique domain

MOAT MATRIX™ (continued)

33 TrustReport Disclosures

Workflow/Process 1.34x

Automated generation of real-time income disclosure statements for transparency and compliance.

34 Type-Safe Full-Stack Integration

Workflow/Process 1.10x

The extensive use of TypeScript suggests a robust, low-regession deployment pipeline that provides higher reliability than competitors.

35 Type-Safe Full-Stack Integration

Workflow/Process 1.10x

The extensive use of TypeScript suggests a robust, low-regession deployment pipeline that provides higher reliability than competitors.

36 Type-Safe Full-Stack Integration

Workflow/Process 1.10x

The extensive use of TypeScript suggests a robust, low-regession deployment pipeline that provides higher reliability than competitors.

37 WhatIf Calculator

Proprietary Algorithm 1.68x

A forecasting tool allowing partners to model different earnings scenarios and potential outcomes.

38 Zero-Sheet Architecture

Workflow/Process 1.34x

A proprietary data structure designed to eliminate the need for external spreadsheet formulas or data exports.

LOW MOAT FEATURES (5)

39 Cost Matrix Input

Data Asset 1.27x

A persistent cost-tracking utility that applies fixed and variable costs to all future calculations automatically.

40 Instant Margin Reveal

UX/Design 1.18x

Real-time transparency layer showing profit and earned commissions to all stakeholders simultaneously.

41 Job Capture Protocol

Integration Moat 1.27x

Automated ingestion and recording of job details into the partner ecosystem.

42 LiveView Dashboard Architecture

UX/Design 1.18x

A real-time data visualization layer that eliminates the need for report generation by providing a constant live state of the network.

43 Rank Journey Progression

UX/Design 1.18x

A visual, championship-themed gamification path (Contender to GOAT) for partner leveling.

APPENDIX: METHODOLOGY & CITATIONS

Valuation Methodology:

1. Lines of code counted per IEEE Standard 1045-1992 (executable code only, excluding comments and blanks).
2. Productivity rate: 500 LOC/developer/month based on COCOMO II nominal range.
3. Hourly rates from US Bureau of Labor Statistics (BLS) for Software Developers (15-1252).
4. Team composition: 10% architecture, 30% senior, 40% mid-level, with QA and PM overhead.
5. Complexity score uses weighted factors per COCOMO II effort multipliers.
6. Market value methodology aligns with M&A software asset valuation practices.
7. Moat score assesses defensibility using IP law and competitive strategy frameworks.
8. Founder premium (+35%) reflects value of founder-led development and domain expertise.

(!) IMPORTANT DISCLAIMER

This valuation represents an estimated replacement cost and strategic market value based on quantitative code analysis and declared proprietary features. It should be used as one input among many in any investment or acquisition decision. Actual transaction values depend on market conditions, negotiation, and comprehensive due diligence beyond the scope of this automated report.

Report Metadata:

Report Number: YPDR-00046
Generated: 2/27/2026, 6:39 PM
Valuation ID: 701f4ce9-575e-47ae-a586-b273c35071f6
Created: 1/30/2026
Last Updated: 2/27/2026

Report Verification

This report (YPDR-00046) can be verified at:
www.YourPartnrrDataRoom.com/verify

Or scan the QR code to verify instantly

DATA ROOM

VALUATION REPORT

YOUR PARTNRR DATA ROOM

CERTIFIED SOFTWARE VALUATION REPORT

YPDR-00047

Your Partnrr Data Room

Entity: Your Partnrr Data Room

Report Date: February 27, 2026

VALUATION SUMMARY

Replacement Cost

$19,749,050

Cost to rebuild from scratch

Strategic Market Value

$51,347,530

2.60x strategic multiplier

Lines of Code	Complexity	Moat Score	Dev Time
32,591	69/100	70/100	65.2 mo

EXECUTIVE SUMMARY

COMPLETE

Lines of Code	Features	Moat Score	Dev Time
33K	23	70/100	65mo

Value Comparison

Replacement: $19,749,050

Market Value: $51,347,530 (2.60x)

Key Technical Findings

• Architecture:	**React + Vite + Supabase Backend**	• Primary Language:	**TypeScript (96.3%)**
• Test Coverage:	**Detected**	• Custom Code:	**95.0%**

Founder Premium: Enabled (+35%)

Development Stage: Live

Description:

Data Room and valuation for software companies and businesses

AI Analysis Summary:

Your Partnrr Data Room is a specialized M&A platform that automates software company valuations by correlating technical codebase metrics with financial defensibility. Its primary value lies in its proprietary 'Moat Score' engine and its purpose-built audit and encryption workflows for sensitive pre-acquisition data.

STRATEGIC MARKET VALUE ANALYSIS

Market value reflects what an acquirer would pay—not just rebuild cost, but the strategic value of time-to-market, intellectual property, and competitive advantage.

Replacement Cost

$19,749,050

Cost to rebuild

Strategic Market Value

$51,347,530

2.60x multiplier

Strategic Multiplier Breakdown

Premium	Contribution	Multiplier
Base Acquisition Premium		+1.50x
Time-to-Market (36+ months)		+0.60x
Trade Secrets		+0.20x
Proprietary Algorithms		+0.15x
Medium Complexity (50+)		+0.15x

Total Strategic Multiplier **2.60x**

CODE INVENTORY

		Complexity	Dev Time
222	32,591	69/100	65.2 mo

Architecture: React + Vite + Supabase Backend

Language Distribution:

Detailed Breakdown:

Language	Lines	Files	Percentage
TypeScript	31,393	195	**96.3%**
SQL	817	16	**2.5%**
CSS	162	2	**0.5%**
Markdown	101	2	**0.3%**
JSON	58	4	**0.2%**
HTML	39	1	**0.1%**
JavaScript	21	2	**0.1%**

MOAT MATRIX™

Overall Moat Score

70/100

Declared Proprietary Features (23 total):

HIGH MOAT FEATURES (3)

1 AI Code Analysis

Proprietary Algorithm 1.99x

Automated analysis of ZIP files or GitHub repos to count lines, detect languages, and calculate complexity.

2 Software Valuation Tool

Proprietary Algorithm 1.99x

AI-powered code analysis that calculates replacement cost and strategic market value for software assets.

3 Strategic Market Value Calculation

Proprietary Algorithm 1.99x

Valuation logic using multipliers for time-to-market, IP, complexity, and development stage.

MEDIUM MOAT FEATURES (17)

4 AES-256 Server-Side Encryption

Trade Secret 1.68x

Sensitive data (license keys, EINs, SSNs) encrypted at rest via edge functions.

5 AI Feature Extraction

Proprietary Algorithm 1.68x

Uses AI to extract proprietary features from text descriptions and categorize them into moat categories.

6 Data Room Health Score

Proprietary Algorithm 1.68x

AI-powered readiness assessment showing percentage of 'investor readiness'.

7 Development Stage Modifiers

Proprietary Algorithm 1.68x

Valuation adjustments based on the lifecycle stage (Pre-launch, Beta, Live, Established).

8 Entity Credentials Module

Data Asset 1.56x

Centralized tracking of business licenses, memberships, and software licenses per entity.

9 High-End Visual Identity Design

UX/Design 1.20x

The inclusion of specialized typography (@fontsource/space-grotesk) and a comprehensive suite of UI components suggests a high-fidelity, consumer-grade user experience design.

MOAT MATRIX™ (continued)

10 High-End Visual Identity Design

UX/Design 1.20x

The inclusion of specialized typography (@fontsource/space-grotesk) and a comprehensive suite of UI components suggests a high-fidelity, consumer-grade user experience design.

11 Masked Field Reveal Audit

Workflow/Process 1.34x

Sensitive fields are masked by default; reveals are logged with a VIEW_KEY action.

12 Masked Field Reveal Audit System (VIEW_KEY)

Workflow/Process 1.10x

The audit system in `useAuditLog.ts` specifically tracks `VIEW_KEY` actions, creating a defensible, investor-ready access log for sensitive credentials—a critical feature for M&A data rooms.

13 Moat Matrix Defensibility System

Workflow/Process 1.34x

A proprietary scoring system for software based on patents, trade secrets, algorithms, and network effects.

14 Moat Score

Proprietary Algorithm 1.68x

A defensibility score (0-100) measuring how hard software is to replicate.

15 Optimized Frontend Component Library

Workflow/Process 1.10x

Extensive use of Radix UI primitives indicates a sophisticated, accessible, and consistent user interface layer that would require significant effort to replicate from scratch.

16 Owner-Controlled Encryption Settings

UX/Design 1.45x

Founders can toggle which specific sensitive fields should be encrypted.

17 Resumable Large File Uploads

Integration Moat 1.56x

TUS protocol support for uploading files up to 750MB with pause/resume capability.

18 Serverless Data Orchestration Layer

Integration Moat 1.30x

The integration of Supabase suggests highly optimized data flows and real-time capabilities that are tightly coupled with the React frontend.

19 Succession Planning Module

Workflow/Process 1.34x

Tools to document and manage business succession plans for continuity.

20 Valuation Progression & Event Broadcasting Layer

Proprietary Algorithm 1.40x

The specific implementation of broadcasting analysis progress via Supabase channels (`useAnalysisProgress.ts`) for software valuation is a specialized workflow that bridges long-running backend AI tasks with a real-time UI.

APPENDIX: METHODOLOGY & CITATIONS

Valuation Methodology:

1. Lines of code counted per IEEE Standard 1045-1992 (executable code only, excluding comments and blanks).
2. Productivity rate: 500 LOC/developer/month based on COCOMO II nominal range.
3. Hourly rates from US Bureau of Labor Statistics (BLS) for Software Developers (15-1252).
4. Team composition: 10% architecture, 30% senior, 40% mid-level, with QA and PM overhead.
5. Complexity score uses weighted factors per COCOMO II effort multipliers.
6. Market value methodology aligns with M&A software asset valuation practices.
7. Moat score assesses defensibility using IP law and competitive strategy frameworks.
8. Founder premium (+35%) reflects value of founder-led development and domain expertise.

(!) IMPORTANT DISCLAIMER

This valuation represents an estimated replacement cost and strategic market value based on quantitative code analysis and declared proprietary features. It should be used as one input among many in any investment or acquisition decision. Actual transaction values depend on market conditions, negotiation, and comprehensive due diligence beyond the scope of this automated report.

Report Metadata:

Report Number: YPDR-00047
Generated: 2/27/2026, 6:42 PM
Valuation ID: 56154ee8-5328-4bf0-b4d4-c9c396b70db9
Created: 1/29/2026
Last Updated: 2/27/2026

Report Verification

This report (YPDR-00047) can be verified at:

www.YourPartnrrDataRoom.com/verify

Or scan the QR code to verify instantly

Made in the USA
Monee, IL
27 March 2026

47027695R00393